GLOBALISATION, POVERTY AND CONFLICT

Globalisation, Poverty and Conflict

A Critical "Development" Reader

Edited by

Max Spoor
Institute of Social Studies,
The Hague, The Netherlands

KLUWER ACADEMIC PUBLISHERS

DORDRECHT / BOSTON / LONDON

A C.I.P. Catalogue record for this book is available from the Library of Congress.

ISBN 1-4020-2857-1 (HB)
ISBN 1-4020-2858-X (e-book)

Published by Kluwer Academic Publishers,
P.O. Box 17, 3300 AA Dordrecht, The Netherlands.

Sold and distributed in North, Central and South America
by Kluwer Academic Publishers,
101 Philip Drive, Norwell, MA 02061, U.S.A.

In all other countries, sold and distributed
by Kluwer Academic Publishers,
P.O. Box 322, 3300 AH Dordrecht, The Netherlands.

Printed on acid-free paper

Printed in the Netherlands.

For Francisca Quilaqueo
Saskia and Friso

CONTENTS

IN CONCLUSION

ABBREVIATIONS

BUPP	Bangalore Urban Poverty Project (India)
CCD	Christian Commission on Development (Honduras)
CCL	Contingency Credit Line
CEE	Central and Eastern Europe
CESTRAD	Centre for the Study of Transition and Development (Netherlands)
CIFOR	Centre for International Forestry Research (Indonesia)
CIS	Commonwealth of Independent States
CODESRIA	Council for the Development of Social Science Research in Africa
DISHA	Developing Initiatives for Social and Human Action (India)
DSR	debt service ratio
EBRD	European Bank for Reconstruction and Development
ECA	export guarantee agency
ECLAC	Economic Commission for Latin America and the Caribbean
EGDI	Expert Group on Development Issues
ESAF	Enhanced Structural Adjustment Facility
EU	European Union
FDI	foreign direct investment
FTAP	Fair and Transparent Arbitration Process
GATS	General Agreement on Trade in Services
GDP	gross domestic product
GEAR	Growth, Employment and Redistribution (South Africa)
GONGOs	government-organised non-governmental organisations
HDI	Human Development Index
HICs	Highly Indebted Countries
HIID	Harvard Institute for International Development
HIPCs	Highly Indebted Poor Countries
HOS	Heckscher-Ohlin-Samuelson (paradigm)
IADB	Inter-American Development Bank
ICCO	Inter-Church Organisation for Development Cooperation (Netherlands)
ICRG	International Country Risk Guide
IDASA	Institute for Democracy in South Africa
IDPs	internally displaced persons
IDRC	International Development Research Centre
IFAD	International Fund for Agricultural Development
ILBE	Institute for Law-Based Economy (Russia)
IMF	International Monetary Fund
IRB	internal rating based (approach)
ISPA	Instrument for Structural Policies for Pre-Accession
ISS	Institute of Social Studies (Netherlands)
LDCs	less developed countries
MMD	Movement for Multi-Party Democracy (Zambia)
MSY	maximum sustainable yield

NATO	North Atlantic Treaty Organisation
NDCEs	New Developing Countries of Europe
NEPAD	New Partnership for Africa's Development
NGO	non-governmental organisation
NIK	Supreme Chamber of Control (Poland)
NTB	non-tariff barriers
ODA	official development assistance
OED	Operations Evaluation Department (World Bank)
OPEC	Organization of the Petroleum Exporting Countries
OSCE	Organization for Security and Co-operation in Europe
OTBs	*Organizaciones Territoriales de Base* (Bolivia)
PAP	Poverty Alleviation Programme (Bolivia)
PD	probabilities of default
PFRON	Fund for the Rehabilitation of Disabled People (Poland)
PIN	Public Information Notice
PISA	Project International Student Assessment
PPP	purchasing power parity
PROGRESA	Education, Health and Food Programme (Mexico)
PRSC	Poverty Reduction Support Credit
PRSP	Poverty Reduction Strategy Paper
QUANGOs	quasi-non-governmental organisations
RBEC	Regional Bureau for Europe and the CIS (UNDP)
RICE	relative income conversion efficiency
RPC	Russian Privatisation Centre
RSAA	Royal Society for Asian Affairs
SAPARD	Special Accession Programme for Agriculture and Rural Development (EU)
SDRM	Sovereign Debt Restructuring Mechanism
SDRs	Special Drawing Rights (IMF)
SEE	South and Eastern Europe
SEWA	Self-Employed Women's Association (India)
SME	small- and medium-sized enterprise
SRF	Supplementary Reserve Facility
TRIPs	Trade-Related Aspects of Intellectual Property Rights
UN	United Nations
UNCED	United Nations Conference on Environment and Development
UNCHS	United Nations Centre for Human Settlements
UNCTAD	United Nations Conference on Trade and Development
UNDP	United Nations Development Programme
UNECE	United Nations Economic Commission for Europe
UNEP	United Nations Environment Programme
UNESCO	United Nations Educational, Scientific and Cultural Organisation
UNHCR	United Nations High Commissioner for Refugees
UNICEF	United Nations Children's Fund
UNRISD	United Nations Research Institute for Social Development
UNU	United Nations University
US	United States
WCED	World Conference on Environment and Development
WHO	World Health Organization
WIDER	World Institute for Development Economics Research
WTO	World Trade Organization

INTRODUCTION

GLOBALISATION, POVERTY AND CONFLICT

MAX SPOOR

1. INTRODUCTION

Within the dynamic field of "development studies", an important body of literature has emerged which deals with complex processes of economic, social and political transformation in the developing and, more recently, in many of the transition countries. This literature reflects the evolution of debates and paradigms on the societal changes and transformations that have taken place over the past half-century. The current volume presents a set of state-of-the-art contributions to that literature, offering critical thinking on the theory and practice of development policy as it is evolving within the current era of rapid globalisation. Chapters focus on poverty reduction, the build up of multi-level democratic institutions and containing and preventing conflicts, critically confronting the predominantly neoliberal mainstream ideas on "development" and proposing alternatives.

The chapters, written by a select group of scholars and development practitioners, will be of direct interest to those involved in this field of the social sciences and in "development practice". At the same time they are composed so as to remain accessible for the generally interested reader. All chapters in this book except one were originally presented at the Institute of Social Studies' Fiftieth Anniversary Conference *Globalisation, Poverty and Conflict* (7–9 October 2002). This volume contributes – in a critical manner – to understanding of the complex development problems at hand. It analyses highly controversial issues on the "development agenda" that must be taken seriously in the present unstable global economic and political environment. *Globalisation, Poverty and Conflict* is therefore intended as a critical development reader, which complements current debates in this field, but also brings once again to the forefront the importance of development in an era in which "development fatigue" seems to have become more profound then ever before. Achieving the Millennium Development Goals by 2015 is a concrete target for "development", and many governments have agreed to these objectives. Whether these goals can be realised is another issue. Nonetheless, the Millennium Development Goals have brought focused

Max Spoor (ed.), Globalisation, Poverty and Conflict, xiii–xxv.
© *2004 Kluwer Academic Publishers. Printed in the Netherlands.*

attention to "development" in the past few years and a renewed intense debate about the relationship between the process of globalisation, widespread poverty and the emergence of many violent conflicts.

The field of development studies has seen quite fundamental policy debates over the past decades. Thinking on development has changed substantially, as well as the main paradigms that provide its foundations. In the early days of post-colonialism, with the emerging independent and nationalist regimes, there was a near absolute belief in the virtues of the "developmental state", interventionist policies, import substitution, protectionist models of development and a strong emphasis on overcoming or mending "market failures". Several stages of the development debate emerged, such as the introduction of the "basic needs approach" and the classic dispute over whether growth and equity objectives can be reconciled. By the mid- and late 1970s the focus shifted towards "state" or "bureaucratic failures" and the realisation that there was substantial rent-seeking within the state, which had previously been seen as benign or good. The early 1980s saw the emergence of a neoliberal development agenda, which radically turned the clock towards a near sacrosanct belief in the virtues of the market. The foundations were laid for the era of structural adjustment and what later became known as the "Washington Consensus". The pendulum shifted from regulation and state control to deregulation, from state property to private property rights through a process of privatisation of assets, and from protectionist trade, exchange rate and capital account policies to liberalisation and openness. Adjustment took place in the midst of a profound debt crisis, and renewed lending, particularly by the international financial institutions, followed, making the debt issue a sometimes forgotten but nonetheless structural issue for the developing world. The role of the World Bank and the International Monetary Fund (IMF) became more important than originally intended at Bretton Woods, and also different, for example, in their invoking conditionalities related to structural adjustment before lending could follow.

Globalisation, with its unbridled international expansion of trade and financial flows, and reduction of national governments' degree of freedom to influence economic development, has advanced at unprecedented pace in the past two decades. Trade negotiations, which had proceeded slowly for a long period, moved towards the reduction of tariffs (in particular, those levied by developing countries, as the OECD countries retained a substantial degree of protection). Finally, this process led to the founding of the World Trade Organization (WTO) in 1995 and subsequent trade negotiation rounds. Poverty re-emerged on the development agenda during the 1990s. In the early stages of adjustment it was fashionable to see poverty as a temporary problem which would be resolved after economic growth resumed and economic recovery was realised. The "trickle down" theory, according to which any type of economic growth has a high elasticity of poverty reduction, regained popularity in the neoliberal era, despite evidence against it. Actually, the period of economic adjustment (known by the structural adjustment programmes, or SAPs) has shown growing income inequality to be a main factor underlying rising poverty. Though some scholars have tried to show there is convergence, it is now largely accepted that the gap between poor and rich (in the various manners that this can be measured) became more profound in the past decades. While globalisation provides increased opportunities, it seems that quite some countries and segments of populations have been unable to benefit and a process of "exclusive" rather than "inclusive" growth has taken off. There are positive signs, however, such as the rapid development of countries like China and India, which house most of the world's poor and show reduced poverty incidence (at least in China). Africa, however, is lagging, alongside quite a number of the "transition" countries that emerged after the fall of the Berlin Wall and the collapse of the former Soviet Union. In both cases, nationalist and ethnically based conflicts have caused much suffering and widespread poverty in politically often unstable and fragile societies where sustainable economic development is still a "bridge too far".

The democratisation of many developing and transition countries is another important phenomenon that has emerged in the past two decades. While in the 1960s and 1970s there were still countless military and authoritarian regimes, a wave of democratic resurgence has swept through many regions of the world. In parallel (and partly in response), civil society organisations have risen in importance, demanding a share of responsibility in the development process. Economic and political decentralisation and deregulation has provided them with more space and room for manoeuvre.

2. REVIEW OF THE CHAPTERS

The current volume, *Globalisation, Poverty and Conflict*, discusses these very topical, inter-related and controversial elements of development by looking to several sets of issues:
- globalisation, inequality and poverty;
- governance, civil society and poverty; and
- resource degradation, institutions and conflict.

Though the chapters are grouped under these main headings, such separation does not deny their many linkages and overlaps.

The first section of this volume deals with globalisation, growing inequality and poverty. It touches upon the Millennium Development Goals and their feasibility in terms of poverty reduction, globalisation and marginalisation (Van Ardenne, Pronk, Ritzen, Spoor and Murshed) and then looks to the international financial system in relation to the position of developing countries (Griffith-Jones and Raffer).

In the brief opening chapter, Van Ardenne discusses a number of the relationships. First, it is evident that violent conflicts can cause widespread poverty, particularly amongst internally displaced persons and refugees. A causal relation between poverty and conflict is less straightforward, but growing inequality can lead to social and political unrest, resulting in higher poverty incidence. Second, globalisation provides opportunities, but it does not automatically lead to equal distribution of economic benefits. Third, globalisation can give impetus to conflicts, as some conflicts are fuelled more by greed than by grievance, and open borders stimulate the spread of AIDS and illegal trafficking (of arms, drugs and human beings). However, the information age and exchange of ideas might ease tensions. Van Ardenne makes a strong plea for "inclusive governance" based on "inclusive policies" (through participation, democratic development and local governance), in contrast to the "exclusive development" that is now taking place.

Pronk (Chapter 2) follows by discussing the feasibility of the Millennium Development Goals in view of the politics of globalisation. He distinguishes two phases of the current wave of globalisation; namely, a first since 1945 and a second since the end of the Cold War and the fall of the Berlin Wall in 1989. While discussing various development initiatives, particularly those by United Nations bodies, Pronk underlines the significance of the Millennium Declaration. He sees this as a milestone development agreement, constituting a step away from the paradigm of "trickle down" of whatever growth is generated. A United Nations evaluation of development progress at the start of the new millennium points to mixed outcomes. Some countries or regimes are doing well, while others show a negative record in poverty reduction and human development. Pronk asks, 'Is this just collateral damage of globalisation, or the result of faulty development policies (and calculated default)?' Excluding China, poverty has increased worldwide (measured as the number of people with an income below the threshold of one US dollar per day). Pronk criticises the adjustment policies backed by the Washington Consensus as 'bound to result in a stagnation of poverty reduction'. Adjustment has led in some cases to the resumption of growth, but this was mostly not "pro-poor growth". He discusses the fate of different development paradigms during what the United Nations calls the various "development decades". When sustainability rose to the top of the development agenda, more optimism emerged, which in a sense was reflected in the Social Summit of 1995 and the Millennium Declaration of 2000. The latter addresses not only income poverty, but also access to social services and utilities (water, sewerage). In order to realise these goals, Pronk argues, there is a need for a 'dramatic change of direction', and 'this requires political leadership'.

Ritzen (Chapter 3) defends the role of "good" governance and policies (based on "best practice") that contribute to development. He adheres to the position that the differentiation between successful and failed cases of poverty reduction in developing countries can be largely explained by the quality of governance and the policies implemented. In five of the six developing regions, income poverty is likely to be halved by 2015, thus bringing the income-related Millennium Development Goal into sight. Yet this is not the case with the Millennium Development Goals related to education and health. Though agreements have recently been made to increase finance for human development, Ritzen is weary of the ineffectiveness of public expenditure in these areas. He provides examples in which private delivery of public goods and services has actually improved access by the poor. This is done by "contracting out" or "contracting in" non-governmental organisations (NGOs) or the private sector. Hence, public finance remains necessary, but the public sphere need not be the exclusive one to deliver. Ritzen also emphasises the role of direct incentives (such as school attendance-related payments for girls), which have shown success.

Spoor, in Chapter 4, deals with one of the most dramatic changes of the twentieth century, namely the process of transition since the fall of the Berlin Wall in 1989 and the subsequent collapse of the Soviet Union in 1991. The transition countries suffered many conflicts during the 1990s, the most dramatic being that in the former

Yugoslavia. With the huge contraction in their economies and redistribution of their national wealth to (mostly a small number of) private actors, until the recovery of the mid-1990s the number of poor in this region increased from 14 million to 147 million (including a yet largely unrecognised new category, the "working poor"). Poverty thus became the most pressing problem of economic recovery and development in this part of the world. New (or the continuation of old) peripheries became evident. In the territory of the former Soviet Union, the periphery is formed by the so-called CIS-7 (the poorest areas of the Commonwealth of Independent States). Along Central and Eastern Europe (CEE) another periphery is visible in the South-eastern European (SEE) states. After 'die Wende' of 1989, the CEE countries moved quickly towards a Western-type market economy and political system, using in some cases even "shock therapy" and benefiting from nearby markets and large streams of foreign direct investment (FDI). The two peripheries had a more mixed record, suffering conflicts, slow or no reforms, authoritarian regimes and less favourable initial conditions in terms of market access and possession of mineral wealth. Comparing incomes per capita shows that the CIS-7 are the worst off, while the difference between this category of countries and the new members of the EU-25 is around 1:10. This is also illustrated by the HDI (Human Development Index), which places the two peripheral groupings in the "medium" human development category. Absolute poverty in most of the "transition peripheries" is high, above 50 percent and with an income level of just over two US dollars a day – a staggering figure indeed.

Bringing the analysis to a more global level, and making the link with the two subsequent more technical chapters on the international financial system and debt crises, Murshed (Chapter 5) outlines the inter-linked processes of globalisation and marginalisation of part of the developing world. Since the 1980s the developing countries have (become) rapidly integrated into the global trade regime. Some middle-income countries have particularly benefited from this development, but also large (low-income) countries such as China and India. Yet just a few countries (including these and Mexico and Brazil) receive most of the FDI flow to the developing world. Within this unequal international investment environment, it is no surprise that global inequality, the income gap between rich and poor countries, rapidly grew from the early nineteenth century to the early 1990s. Murshed draws on Milanovic's three indices of inter-country inequality: "category 1 inequality" which treats countries, whether large or small, equally; "category 2 inequality" which adjusts for population size; and "category 3 inequality" which is based on individuals rather than the countries themselves. In all cases inequality is high, highest in category 3. Following this analysis, Murshed discusses various theories of international trade, pointing to the unequalising effects of North-South trade. Structural adjustment programmes have forced more openness on the developing countries (while the North has retained much of its trade protection), and in the emerging debt crises the South has been treated harshly by its creditors in the North. The establishment of the WTO and results of trade negotiation rounds, in particular the Trade-Related Aspects of Intellectual Property Rights (TRIPs), discriminate against the South developing technological capacity to provide a basis for their "development". Murshed concludes with an analysis of the origins of violent conflicts, which are usually fought out under ethnic and nationalist banners. He uses the "greed versus grievance" dichotomy and the failure of the social contract in explaining the emergence of conflicts.

Griffith-Jones (Chapter 6) contributes a profound analysis of global financial governance in its current state. The objectives of a new financial architecture should be to prevent recurrent banking and currency crises and to promote, and even guarantee, stable private and official financial flows towards developing countries. Although during the 1990s some changes have been introduced in the global financial system, important and even dramatic financial crises occurred in this decade, such as the Asian, Russian and Brazilian crises. Financial flows to developing and emerging economies have dropped since 1998, with the latter category even becoming a net exporter of capital to the developed world. Why have the reforms been insufficient? Griffith-Jones mentions at least four reasons. There is no agreed reform agenda. There is substantial asymmetry in reforms, which are focused primarily on changes in developing countries. There is too much emphasis on (short-term) crisis management, rather than promoting stable flows of finance. Finally, the developing countries themselves have little influence on the reforms, since they are hardly represented in the deciding bodies, and the large "bail-out" packages of the IMF are viewed with increasing suspicion. New financial regulations should be counter-cyclical, Griffith-Jones posits, although the newly proposed Basel Capital Accord seems to promote pro-cyclicality. The author also points out that risk is lower in bank loan portfolios that are more equally spread among developed and developing countries. The IMF introduced new financing facilities in the 1990s. However, any developing country trying to use these options would likely be stigmatised by the financial markets. Finally, she asks why there are so many difficulties in reforming the international financial system. Maybe it is because powerful financial actors in developed countries do not see the reforms as being in their interests. However, many (non-financial) actors take the side of the developing world. Griffith-Jones pleas for renewed negotiations for a new "first-best" international financial system, rather than continued focus on marginal adaptations to the current system.

In another contribution dealing with international finance, Raffer (Chapter 7) analyses the debt crises that have struck the developing countries over the past decades. The current position of the international financial institutions is founded on the early neoliberal doctrine as implemented in Chile under the post-1973 Pinochet military regime, leading to the deep financial crisis of 1982. After that, globalisation produced a great number of – readily forgotten – financial crises, including the US Savings & Loans scandal and the Mexico crisis of 1994–95. During the past decades, international financial institutions have become more important in the global financial architecture. Several waves of "debt-creating flows" changed the structure of the overall debt of developing countries and also increased its volume. Originally, with the loan spray of the Euromarket and the negative real interest rates in the international financial markets, debts contracted by southern countries grew sky-high, in particular from commercial banks. After the series of financial crises in the early 1980s there was a shift towards debt held by international institutions, following a massive bail-out of private banks. In the 1990s this was followed by more influence of bondholders (such as mutual and pension funds). Although various steps were taken to address the increasingly severe debt problems of developing countries, such as the Highly Indebted Poor Countries (HIPC) initiatives, the international financial institutions continued to advise developing countries to implement policies similar to those that had led to previous crises. "Debt relief" financing usually meant more lending in order to finance debt service, further deteriorating the position of the debtor countries in the long run rather than resolving the debt problem. Raffer's main argument is that the international financial institutions do not accept default, while commercial banks do. Therefore, procedures for sovereign insolvency should be established internationally. The newly proposed Sovereign Debt Restructuring Mechanism (SDRM), as scrutinised by the author, offers the debtors insufficient protection. Instead, Raffer proposes his Fair and Transparent Arbitration Process (FTAP), applying the US Code on Corporate Bankruptcy to sovereign debt overhang. This, he suggests, would provide adequate debtor country protection (which is now totally absent).

The second group of chapters deals with governance, civil society and poverty, looking to democratic institutions and the role of civil society organisations, local government and rapidly transforming forms of governance (Mkandawire, Mihyo, Helmsing, Reuben, and Wedel).

The opening chapter in this section, that by Mkandawire (Chapter 8), analyses democracy, economic policy reform and performance in terms of poverty reduction. There have been (and are) authoritarian regimes that have done (and do) quite well in poverty reduction. Economic growth, however, does not necessarily lead to poverty reduction, since growth may be anti-poor biased. Only explicit redistributional measures and growth in the sectors where many of the poor are active will benefit the poor. With such measures and "pro-poor" growth in place, much lower growth rates will be needed to realise the income poverty Millennium Development Goal. Although the contrary is often believed, the new democracies have implemented the most orthodox adjustment policies. Mkandawire makes this point by screening economic policies implemented by old and new African, Latin American and Asian democracies. However, structural adjustment has delivered neither sustainable growth nor poverty reduction. Why then were these democracies so orthodox in their economic policies? The author gives various explanations. There were ideological shifts and radical changes in leadership in the early 1980s. As interventionism was the trademark of authoritarian regimes, the new democracies identified themselves with the defining principles of adjustment. While the older democracies were entrenched in social pacts, these were absent in the new ones. Finally, the new democracies used the figurative "new broom" in implementing harsh and austere reform programmes, availing of the euphoria following the radical change towards democracy. In the era of globalisation the room for manoeuvre for nation-states has been substantially reduced. Also, many of these new democracies emerged as an outcome of the crisis of the interventionist model and inherited large deficits. Mkandawire calls these "choiceless democracies". He concludes that the core adjustment model has been left unchanged, even now that poverty has re-emerged on the development agenda. Though the international financial institutions have recognised the importance of poverty, they seek no fundamental change in their proposed economic reforms. Poverty is being targeted, rather than making the implementation of proper social policies part and parcel of the development model.

Mihyo (Chapter 9) turns attention to rural poverty and poor local governance. He paints a vivid picture of Africa's poverty, in particular, rural poverty, which is widespread and often extreme (not so much below a dollar a day, but below a dollar a week). Many rural-based ethnic minorities in Africa have been marginalised. Ruling elites have perpetuated rural illiteracy in order to keep control, and the gap between urban and rural areas – in all respects – has grown. Viewed with hindsight, even the "basic needs" approach that was popular in the 1970s was geared more towards managing poverty than to reducing it. The author compares various development policy paradigms for their impact on rural poverty. The structural adjustment programmes of the 1980s weakened states' capacity to steer the development process. The environmental concerns that were added in the 1980s and 1990s were very much elite-based with no concern for rural problems. Finally, with good governance becoming the buzzword of the 1990s, the relation between rural people and rulers remained that of "slaves and master". Market intrusion has increased vulnerability, alongside of which there has been a continuity of, rather than a break in, the "dialogue failure" with rural people, in particular the rural poor. African rural poverty is caused by many factors, the author contends, and rural development failure is evident overall. Rural areas have weak local governance, minimal or no productive and social infrastructure and low literacy. The move towards decentralisation has mostly meant a transfer of responsibility to the local level, without providing the necessary resources. Local governance and the capacity to develop and implement policies at the local (and rural) level remains weak. Finally, Mihyo concludes, 'none of the major development policy paradigms seriously addressed the issue of rural poverty'.

Pushing the argument for improved local governance and decentralisation further, Helmsing (Chapter 10) develops an "enabling policies–citizen approach" to poverty reduction, distinguishing it from mainstream social policy, community development, and empowerment approaches. In fact, the "enabling state" was a cornerstone of neoliberal reforms, a new view of the role of the state that contrasted with the previous "interventionist state". However, in the "citizen approach" the state must become the prime regulator, rather than assuming the passive role of "enabling". The author develops his approach by examining examples of local governance and decentralisation. Bolivia introduced decentralised governance and extensive popular participation in the early 1990s, leading Helmsing to conclude that 'decentralisation, if well-resourced, can make a difference'. The "citizen approach" is quite different from interventionist policies, as the state is involved in regulation of the delivery of services to the poor rather than the delivery itself, or for example enabling contacts between local producers and international buyers. It is also a different form of governance, as is shown by the example of the Porto Alegre participatory budgeting process. The approach has advantages, such as the possibility of leveraging additional resources and, through localising the tendering of certain infrastructure projects, it generates opportunities for local producers. Finally, in order to reach the poor, not as "targets", but as active "citizens" who have a right to development, organisations of the poor are crucial. Helmsing provides examples of strong indigenous organisations in Ecuador and Bolivia and analyses the role of NGOs, traditional trade unions and political parties, versus community organisations and management, such as experienced in Porto Alegre and Kerala.

Reuben (Chapter 11), in a similar vein, stresses the role of "civic engagement" in "development" and the importance of civil society in general. Whether in Africa and Latin America, civil society should be considered equally important as the market and the state. NGOs are 'just one sub-set' of civic engagement, often supported by donors or domestic private funds. Beyond these there is a wide spectrum of other civil society organisations, such as labour unions and other professional, economic and cultural associations. Organisations of the poor mostly suffer from "civic exclusion", because of asymmetric access and connections. The author takes the position that there has been a fundamental change towards 'multi-layered, highly diversified civil societies', beyond the traditional forms of class-based interest representation. Also, civil society organisations are now more non-membership-based organisations, which derive legitimacy only from transparency and accountability. Civic engagement can influence "the market" by improving access (for the poor), reducing market failures and influencing the ethics of corporate management. Likewise, civil society organisations can be complementary to the state, for example, in public service delivery. A "governance crisis" occurs in particular when trust in the state (in providing public goods and services) is lacking. Governments have different approaches towards civil society, which can range from laissez-faire to proactive engagement, while civil society organisations have strategies varying from confrontation to full support or endorsement. As examples of successful civic engagement, Reuben also mentions participatory budgeting in Brazil, financial tracking systems (monitoring allocations to ensure they really arrive at their designated destination) and civil society performance evaluations of public entities. He concludes that, in order to be effective, civic engagement requires an enabling environment that includes a regulatory framework, a proper political and institutional setting and a civic culture.

Wedel (Chapter 12), the final author in this section, goes beyond the "state-market-civil society" triangle. She focuses on "flex organisations", which 'can switch their status from state to private and vice versa, according to the particular situation', in the context of the transition in Eastern Europe. Flex organisations are like chameleons in their behaviour, changing colours when adapting to the environment can provide protection. Outsourcing has become popular in the United States and the United Kingdom, and there is a worldwide tendency to privatise the delivery of public goods and services (and "governance" in a broader sense), as part and parcel of the dominant neoliberal agenda. How does flex organising fit into this? There are disconnects between government growing by private (or privatised) work force and the need for accountability and the loss of public control on the processes of outsourcing and subcontracting. Wedel uses social network analysis to disentangle the complex relationships between formal and informal and between private and public structures and processes. In the concrete cases of Poland and Russia, there are informal social groupings of elite, called "institutional nomads" and "clans", respectively, with their 'fingers in a kitchenfull of pies'. The process of privatisation (without transparency and accountability) led to an uncontrolled "grabitisation" and the emergence of billionaire oligarchs. Wedel's research found tightly-knit interlocking networks that governed the multi-million western aid to Russia. Furthermore, special agencies and targeted funds in Poland present 'myriad opportunities for corruption'. Flex organisations are also found in the West, such as the powerful informal grouping of neo-conservatives in and around the current Bush administration. Improving transparency in flex organisations might be possible, but she concludes that these organisations are 'inherently unaccountable', as they can shift their status within the 'private-state nexus'.

The third section of this volume analyses resource degradation, institutions and conflict (with contributions by Ostrom and Janssen, Salih and Lélé). In particular, it looks at the type of multi-layered institutions that are needed in resource management, and their role in conflicts and sustainable development.

Ostrom and Janssen offer a detailed analytical piece (Chapter 13) on natural resources management and complex institutions. Social-ecological systems are seen as highly complex; simple linear policies cannot successfully intervene in such complex institutions. The policies of economic development and environmental protection are both based on surprisingly similar "top-down" technocratic "mental models" (or "belief structures"), as the authors contend. Since the era of the founding of the Bretton Woods institutions, "development" has been a synonym for "modernisation". The bottlenecks encountered in the development process were seen as caused by gaps in capital, resources and technology. Hence, large quantities of foreign assistance have been extended as the means to overcome these gaps. Thinking on "development" was based on "top-down" solutions. Mainstream economic theory, in particular neo-classical theory, ignored institutions. While growth did occur, poverty remained widespread. It was even (more recently) questioned whether development aid helped or was counterproductive to development. In the field of resource management, top-down and simplistic policy solutions have also prevailed, implemented mostly by central governments and backed by "science". These policies were often contrary to indigenous resource management practiced at micro-levels. The "belief system" behind such an approach is centralistic and technocratic, as policy analysts search for the "optimal policy", rather than an understanding of the complexities of the system in which the policy is implemented. Ostrom and Janssen analyse social-ecological systems as "complex adaptive systems" as they move through multiple equilibria (between connectivity and resilience, focused on production or innovation for sustainability). They cite the example of the Dutch waterboards, which throughout history have been (and still are) successful because of their complex institutional arrangements and the shared norms that support them within Dutch society. However, in another example, that of Dutch engineering knowledge forcefully transferred to Bali, the results were largely negative. This island has a thousand-year history of multi-level irrigation. The "top-down" intervention during the Green Revolution years had a negative impact because it ignored indigenous culture and knowledge. The authors conclude, 'It is time to declare this belief system bankrupt.'

Salih (Chapter 14) follows suit to investigate the relationship between environment and politics, in order to discuss whether conflicts arise over resource degradation and what is the role of institutions in managing, causing or even aggravating conflict. Conventional wisdom might propose the former causal relationship, but conflict itself might also generate environmental degradation. The focus of analysis should therefore be on "environmental change" rather than degradation, as – according to Salih – there are likely to be 'more intense conflicts over a healthy environment than over degraded resources'. The linkage between population growth, environmental degradation and conflict is still popular. However, the author finds it 'absurd, to say the least' to claim that conflicts would be increasingly linked to resource degradation, in particular without analysing the political economy of the cause of depletion. He cites various conflicts that can only be explained by factors such as clan politics (Ethiopia, Somalia), appropriation by a business elite of large areas of fertile land (Sudan) or the alienation of traditional rights of communities (India). Actually, environmental degradation is often caused by various factors of intervention, and is not the cause itself of the conflicts that occur. What follows is a detailed discussion of the role of institutions, taking a line similar to Ostrom, citing her statement that 'neither the state nor the market is the answer to the commons dilemma and the conflicts associated with it'. The role of institutions is complex, as they are able to both cause and manage resource conflicts. However, there are limits to institutions with regard to resource conflicts, and therefore they 'should be demystified rather than glorified'.

To conclude this section, Lélé (Chapter 15) contributes with a conceptual and operational analysis of institutions of resource management, from local-level community institutions to the state, ending with the concrete design of an institutional framework for community-based forest management. In order to promote "environmentally sound development", institutional arrangements must be concerned with efficiency, sustainability and equity. However, much of the literature on resource management focuses on only one of these concerns. Institutions can be voluntary or involuntary, productive or regulatory, (quasi)state or civil society, etc. After discussing specific characteristics of eco- and social systems, he concludes that these 'necessitate coordination *and* regulation of individual actions through rules (and institutions that make the rules)'. Institutional arrangements for resource management are needed at various levels, overlapping and strengthening each other, ranging from local-level self-organised collective-action institutions to the central state itself. Simple "co-management" is clearly not sufficient, and multi-layered institutions are the obvious answer, Lélé notes. The principles of institutional design are (1) strong linkages between authority, responsibility and incentives; (2) proper jurisdictions between different levels; and (3) financial viability and maximum transparency. Forests represent a typical mixture of common-pool resources, toll goods and purely public goods. Hence, local-level community management systems are most appropriate for forest resource management, and centralised management systems would be ineffective. He provides a detailed design for multi-layered forest management, going 'against the grain' of the simplistic co- or joint management institutional frameworks. The different institutions need to be able to deal with the harvesting of forest products, resource management, marketing, conflict resolution and forest conservation. In his design of an appropriate institutional framework for forest resource management, Lélé keeps a close eye on the three basic concerns he laid out, namely efficiency, sustainability and equity. He concludes by saying that indeed institutions matter, but institutional failure is by no means the only cause of 'environmentally unsound development'.

In a concluding chapter, Opschoor reminds the reader that "development" in the early post-World War II years was equated with "modernity". Though there has been much critique of this rather simplistic notion, it seems that with globalisation and the predominant neoliberal agenda, again a (questionable) uniformity has emerged in the concept of "development" that is invoked, namely through the implementation of that agenda and with "good governance". The author, as part of his public address to the fiftieth *Dies Natalis* of the Institute of Social Studies in The Hague, discusses briefly the variety of high-level contributions that were presented at the conference *Globalisation, Poverty and Conflict*, of which this volume is the final product. The main topic of his contribution, however, is knowledge and education in development. With the current liberalisation of international educational markets, many providers are active and competing. Nevertheless, there are disadvantages in far-reaching privatisation and liberalisation, such as a loss of diversity, exclusionary effects for lower-income students, and quality reduction because of cost minimisation. Although education might not be considered a "global public good", it is a "global public need", Opschoor ascertains. Actually, the needs in developing countries for higher education are high, and rapidly growing, but the "knowledge gap" is also increasing between rich and poor countries. Globalisation increases demands for different educational systems focused on flexible learning and permanent education. He sees a shifting of the centre of gravity of educational programme delivery (especially in the international higher education system of which the ISS is part), towards the developing (i.e. receiving) countries, and away from its home base in the OECD countries, through institutional capacity development and partnerships. Opschoor concludes that we have to move further to develop knowledge and promote 'knowledge-sharing in the field of social change and sustainable human development', emphasising the special role international ("development oriented") higher education institutes can play in this process.

In summary, this volume presents a collection of high-quality, analytical and policy-relevant contributions on current "development" issues. They are overall quite critical on the current process of globalisation, which since the 1980s – apart from rapidly expanding trade, capital markets and information streams – has seen continued widespread poverty and violent conflicts in many parts of the world. Although globalisation in its current form presents opportunities to certain countries and segments of populations, it also stimulates inequality and marginalisation. Without steering the current process of globalisation and development, the income gap will continue to grow. As Branko Milanovic said in his presentation to the conference, 'a world without middle class' is emerging. Poverty reduction is not a "trickle down" effect of economic growth, since much of the growth is not "pro-poor". Despite the increased attention by the Bretton Woods institutions to poverty (with the Poverty Reduction Strategy Papers), the "fundamentals" of the adjustment era are still largely in place, in spite of their questionable poverty reduction performance record.

Many authors in this volume contend that institutions are of crucial importance, in particular those institutions that are multi-layered and operating at different levels. Institutions should take into account that development processes and social-ecological systems are highly complex and continuously transforming and adapting. "Top-down" development (or resource management) policies are seen counterproductive, with possible very negative outcomes. Conflicts can arise in socially unstable situations, in the absence of appropriate institutional arrangements that can manage them. Conflicts can be caused by resource degradation, but most resource degradation is the consequence of other factors (such as alienation of rights and resource appropriation by elites), which in themselves form the origin for conflict. In turn, conflict can equally be the consequence of greed, leading to grabbing of resources. Democracy, civil society organisations and local governance are seen as important elements of sustainable human development. Finally, human capital development is crucial for "development". Given the increasing knowledge gap between poor and rich countries, human capital is becoming even more important for seizing the opportunities offered in the process of globalisation.

3. THE CONFERENCE AND THIS READER

The conference *Globalisation, Poverty and Conflict*, of which this volume is the outcome, was successfully organised with a team of several colleagues at the Institute of Social Studies, such as Arjun Bedi, Kristin Komives, Peter Knorringa, Paschal Mihyo, Mansoob Murshed and Hans Opschoor. It would not have been possible without the support of many people, such as Martin Blok, Paula Bownas, Berhane Ghebretnsaie, Matty Klatter, Sandra Nijhof, and several other members of the technical and administrative staff of the Institute of Social Studies. The assistance of Michelle Luijben in the language editing of the volume and the editorial support rendered by Henny Hoogervorst and Esther Verdries of Kluwer Academic Publishers have been invaluable. Finally, the conference was generously sponsored by the Dutch Ministry of Education, the Dutch Ministry of Development Cooperation, the European Union and the Dutch National Commission for Sustainable Development (NCDO). Their support is gratefully acknowledged here. Agnes van Ardenne, Dutch Minister for Development Cooperation, opened the conference, which commenced the day after Prince Claus of the Netherlands passed away. In her opening statement she reminded us of one of Prince Claus' better known theses, 'You can't develop people, they develop themselves.'

We hope that this "development reader" will contribute to an improved understanding of the very complex issues that are currently debated in "development studies". This could stimulate the emergence of innovative and alternative policies and institutions that effectively address the enormous problems of poverty and emerging or ongoing conflicts in our globalising world.

The Hague, May 2004

PART I

GLOBALISATION, INEQUALITY AND POVERTY

CHAPTER 1

FROM EXCLUSIVE TO INCLUSIVE DEVELOPMENT

AGNES VAN ARDENNE

Before opening this conference on globalisation, poverty and conflict, I would like to dedicate a few words to Prince Claus. In Prince Claus the Netherlands has lost not only a beloved member of the Royal Family but also an expert on development issues.[1] His knowledge and commitment made him a pillar of Dutch development cooperation. Unlike my predecessors, I shall have to manage without his wise advice. It will be a great loss.

The Prince was ahead of his time with his ideas on modern forms of poverty reduction, ideas which have since gained worldwide acceptance. He believed strongly in the ability of poor countries to shape their own future. 'You can't develop people, they develop themselves', he used to say. Prince Claus attached great value to the relationship between culture and development. 'Without respect for and confidence in local culture and traditions, progress is difficult to achieve', in his own words.

Let me now first of all congratulate the Institute of Social Studies on its fiftieth anniversary. We are proud to have this renowned institute here in the Netherlands. The ISS was founded as an autonomous institution for postgraduate studies. Let me quote part of its mission statement: 'ISS aims at widening civil participation in the process towards just and sustainable development in the context of global change.' That mission is as relevant now as it was in 1952, if not more so. This does not mean that the ISS has not been affected by change. On the contrary, its annual report for 2000 is appropriately called 'Meeting New Demands'. New demands are indeed being asserted from the outside world: more market orientation, globalisation, more emphasis on good governance. There are also new demands from within the Netherlands: integration with universities and new international education programmes offering more competition. I wish the ISS every success in meeting these demands, both on the global and on the domestic front.

True to its ambitious nature, the ISS chose globalisation, poverty and conflict as the theme of this conference. This covers virtually every aspect of current development policy and thus provides an excellent opportunity for me to explain what I believe

Max Spoor (ed.), Globalisation, Poverty and Conflict, 3–7.
© *2004 Kluwer Academic Publishers. Printed in the Netherlands.*

binds these three themes. I will do this and afterwards describe the policy we have set out in response.

Let me start with the relation between conflict and poverty. Conflict occurs in every society. There is no development without conflict. Conflict only becomes a problem when a society is no longer able to reconcile contradictory interests, often because its institutions and governments are weak. Conflict causes severe poverty. Of that there can be no doubt. What we see on television is only a fraction of the suffering that people endure in countries like Angola, Congo and Sierra Leone. Conflict can wipe away years of development efforts all at once.

Does poverty also cause conflict? Many volumes have been written on this subject, and the consensus appears to be that poverty as such does not lead to armed conflict on a large scale. But impoverishment and growing inequality within society do lead to social and political unrest. You only need to drive through Lagos, Nairobi or Kigali to see the inequality. And then there is the alarming growth in inequality between rich and poor countries. Political, ethnic and religious leaders can channel unrest into violence if it suits their ends.

What is the relationship between globalisation and poverty?

Globalisation offers plenty of opportunities for reducing poverty in developing countries and any structural strategy for poverty reduction must be based on open economies and open societies. To someone from the Netherlands, one of the most globalised countries in the world, this seems fairly obvious. But the Netherlands also demonstrates that globalisation does not have to mean an unbridled free market. I am a devotee of the Rhineland model, which holds that globalisation does not automatically lead to wide distribution of prosperity and social progress.

On the contrary, globalisation also has another face: the sharpening of existing contrasts; that countries and people who are unable to join in will become marginalised. International trade and capital flows might bring greater prosperity, but they are not shared equally. They hinge on having appropriate institutions. Poor countries certainly do not have the capacity to participate on an equal footing with the more prosperous participants in the world market.

Addressing the relation between globalisation and conflict, it is clear that stability and security, too, are by no means automatic side-effects of globalisation. Take for instance the global arms trade, trade in diamonds and other precious commodities, and trafficking in human beings and drugs. All have become more difficult to control as a result of globalisation. Another example is the rising rates of infection with the AIDS virus. This not only causes enormous human suffering. It also poses potential security threats to countries like Russia, China, India, Ethiopia and Nigeria.[2]

Conflict often occurs within states. Weak states already find it difficult to hold violence in check, and globalisation is undermining their autonomy and sovereignty even further. Internal conflict soon attracts external stakeholders. Weak states ultimately fall prey to elites that benefit from complex and chaotic situations. In such situations they are free to plunder natural resources and enrich themselves through illegal activities like arms smuggling and drug trafficking. It therefore becomes an economic necessity for elites to keep the conflict going. Conflict is often fuelled more by greed than by grievances. The certification system for diamonds now under discussion is a welcome step in curbing illegal trade and fostering stabilisation in African countries like Angola and Sierra Leone.

One fascinating but largely unexplored question is whether globalisation can curb conflict. Globalisation opens closed societies to external influences. As a result, it can lead to less repression of women, more involvement in international debate and less violence. As Thomas Friedman wrote in the *International Herald Tribune*, 'Countries that don't trade in goods and services also tend not to trade in ideas, pluralism or tolerance.'

These issues by no means affect only developing countries. Dutch interests are also at stake. We cannot simply ignore conflict elsewhere in the world, however far away. Refugees, terrorism, migration, poverty: they all impact our society. There can be no global security as long as part of humanity lives in a vicious circle of poverty and violence. The answer to the challenges lies in what can be defined as "inclusive policy", namely policy based on including people rather than excluding them.

Firstly, inclusion should take place at a global level. The *Human Development Report 2002* rightly pleas for global democracy, good governance within international institutions and more transparency in decision-making. Rarely have developing countries been so frustrated about the imbalance in economic and political decision-making power. This is not only about the imbalance in voting rights in, for instance, the Bretton Woods institutions. They are also frustrated about the imbalance in negotiating power and access to information and knowledge. Developing countries must therefore be given a stronger voice in global governance. They must have the opportunity to defend their interests on an equal footing. This applies to both governments and civil societies. Of the hundreds of NGOs participating at the World Trade Organization conference in Seattle, 87 per cent came from rich countries. Fortunately, things are moving in the right direction. In Doha, Monterrey and Johannesburg poor countries were represented more prominently than ever before.

At the same time we must not expect global solutions to cure all ills. It is often more useful to think in terms of regions. For Africa the challenge of globalisation has different implications than for Asia. It is therefore encouraging to see African leaders making a breakthrough with their New Partnership for Africa's Development (NEPAD). Research should also focus more on regional problems, like the study of the structure of agriculture in Africa that is now being set up by NEPAD.

But for ordinary people the global and regional level often seem strange and distant. For most people inclusiveness begins at the level of national and local government. It is crucial that people regain confidence in their government and its institutions. This applies to prosperous countries like the Netherlands, and even more so to poorer countries. Countries that approach globalisation with the right institutions and governance can get the best out of it and cushion the worst.

So countries must avoid liberalising their financial markets without regulating them. They must not privatise services unless they can monitor abuses of monopoly positions. In other words, they should not put their faith in the market unless they have the institutions in place to ensure that the poor benefit and that inequality is reduced.

Think global, act local is still the motto. Democracy and participation in government are essential if confidence is to be restored. Decentralisation of powers to elected local councils can help make government legitimate and effective in the eyes of citizens. In Rwanda, for example, strengthening local governance is an integral part of the peace-building process. Rwanda is also developing a new way for justice to be administered by and for the community, with its focus on restoring communities. No lasting peace can be agreed over the heads of the people.

Women are playing a key role in the reconstruction and peace-building process in Rwanda. It is important to support them so they really do get the opportunity to take part in decentralised government and the administration of justice. Inclusive policy in matters of peace and security means working towards peace with both men and women. However, women are often disregarded or deliberately shut out. In Somalia women had no voice in the traditional clans, so they set up their own, transcending the existing clan divisions. The women's clan has gained official recognition and is now known as the "sixth clan". Strategies for inclusive governance give women the opportunity to contribute to peace-building and governance.

The Netherlands is working hard in Sudan, Rwanda and Afghanistan to put Resolution 1325 on women, safety and security into practice. I think that we, too, could do more to involve women in peace processes. For instance, they are still grossly underrepresented in the European Union and in military operations and peace missions undertaken by the Organization for Security and Co-operation in Europe (OSCE).

Above all, inclusive policy means not just the government. I strongly favour giving the private sector – both companies and NGOs – more room and responsibilities. International cooperation is increasingly taking the form of public-private partnerships. I fully support this. For me, this is one of the undervalued outcomes of the Johannesburg Summit.

Finally, inclusive policy means safeguarding the close links between development policy and other policy areas. The only effective approach is an integrated approach. We must continually ask ourselves whether we are being consistent and coherent. Trade, agriculture, research and food security are all areas that go beyond traditional assistance.

Here I want to stress the importance of coherence on the part of the rich countries. The West has pushed for free trade for its own exports. But at the same time, it has continued to protect sectors of its own economies from the threat of competition from developing countries. World trade in sugar and cotton are but two examples. Calculations suggest that developing countries currently lose some ten billion dollars in export income because of subsidies to producers in the European Union and the United States. That means farmers in African countries like Mali, Benin, Burkina Faso and Tanzania are losing out. If we want Africa to be included in the world market, we will have to put an end to the hypocritical combination of aid and market exclusion.

We must also take care not to build barriers between foreign policy and development policy. The Minister of Foreign Affairs and I will be focusing on Africa and on the regional approach to conflict resolution and the relationship between conflict and poverty reduction. Alone neither diplomacy nor poverty reduction programmes can win the day.

Our agenda is just as ambitious as this conference. Of course the Netherlands cannot do it alone. We have to make choices. We must choose the themes, regions and countries where we can really make a difference. We also have to put our heads together with other donors to identify where and how we can be most effective. Perhaps the most important thing is to avoid making policies without regard for the poor. It is their future we are talking about, and their voices have to be heard more clearly. That is the heart of the inclusive policy presented here. "Voices of the Poor" shows how little command the poor have over their own fate, how exposed they are to forces beyond their control. For us, the challenge is to take their grievances seriously and to raise our voices in favour of globalisation with a human face, globalisation that helps reduce poverty and brings peace to the world. I am counting on the ISS, as it enters its second half century, to continue providing inspiring and critical contributions to that aim. New demands are at stake.

NOTES

1. This chapter is a slightly edited version of the opening speech to the fiftieth anniversary conference of the Institute of Social Studies, 7 October 2002. Prince Claus of the Netherlands, honorary fellow of the ISS, passed away on the evening before, which is the background of his obituary in the beginning part of this contribution. Agnes van Ardenne is currently Minister for Development Cooperation of the Netherlands.

2. These security threats are related to a further weakening of the state apparatus due to AIDS-related deaths. It has been shown that HIV infection is particularly high in armies.

CHAPTER 2

COLLATERAL DAMAGE OR CALCULATED DEFAULT? THE MILLENNIUM DEVELOPMENT GOALS AND THE POLITICS OF GLOBALISATION

JAN PRONK[1]

1. INTRODUCTION

> I call on the international community at the highest level – the Heads of State and Government convened at the Millennium Summit – to adopt the target of halving the proportion of people living in extreme poverty, and so lifting more than 1 billion people out of it, by 2015. I further urge that no effort be spared to reach this target by that date in every region, and in every country (United Nations 2000b: 12).

This appeal was made in September 2000, by the secretary-general of the United Nations, Kofi Annan, in his report to the Millennium Assembly *We the Peoples: The Role of the United Nations in the Twenty-First Century*. These were more than just empty words or a reflection on the state of the world at the turn of the century. In Annan's words, 'The arrival of the new millennium is an occasion for celebration and reflection.... There is much to be grateful for.... There are also many things to deplore, and to correct' (ibid.: 3). In his report, Annan pointed to economic progress for many as one thing to be grateful for, and ruthless conflict, grinding poverty, striking inequality and a degraded natural environment as things to deplore and correct. Poverty figured prominently in the analysis, in part because of its relation with wealth, inequality, conflict and natural resources.

The world leaders gathered at that occasion responded by adopting the Millennium Declaration pledging, among other things, to halve world poverty by 2015. They also adopted a number of Millennium Development Goals: to halve the proportion of people living on less than a dollar a day, suffering from hunger and without access to safe drinking water; to reduce by two-thirds the mortality rate among children and the maternal mortality rate; to halt the spread of HIV/AIDS, malaria and other major diseases; to ensure primary education for all boys and girls; and to eliminate gender

9

disparity. They also adopted a final goal: to develop a global partnership to achieve the
other goals, including specific commitments on trade, finance, aid, debt, technology
and essential drugs. The partnership was explicitly meant to imply a commitment to
good governance, development and poverty reduction, nationally and internationally
(United Nations 2000b).

    Two years later, at the World Summit for Sustainable Development in
Johannesburg, world leaders reconfirmed their pledge by unanimously adopting a plan
of implementation. In this document they stressed that 'eradicating poverty is the
greatest global challenge facing the world today and an indispensable requirement for
sustainable development' (United Nations 2002a). They again explicitly committed
themselves to halving the proportion of the world's poor and to concerted and concrete
measures to achieve the other millennium goals.

2. THE MILLENNIUM DEVELOPMENT GOALS

What are such agreements worth? Does it make sense to set goals and targets at the
highest political level? Are the Millennium Development Goals the right ones? Is there
any chance that they will be met? If so, how? If not, what could be the consequences?
Will the Millennium Development Goals help to bridge the gap or will they only serve
as a diversion in the global battle for riches, leaving the world's poor as "road-kill"
(Friedman 1999) along the way?

    The Millennium Development Goals are new in many respects. First, together they
cover a broad and rather complete terrain of basic human well-being. They represent
nearly all the relevant dimensions of poverty. At the Johannesburg Summit, after
lengthy negotiations, the original set of goals was extended with the pledge to halve the
proportion of people without access to basic sanitation. This is a crucial addition, since
providing access to safe drinking water is not enough. Lack of sanitation, having
neither a place nor the means to discharge human excrement without creating new
health risks, is essential in the fight against poverty. The goals are part of an integrated
whole. The main dimension which is still lacking is poverty resulting from inadequate
access to energy and natural resources.

    The second aspect of the Millennium Development Goals that is new is that they
concern the world as a whole, but they are not so global that they become vague or
unbalanced. The goal is not to halve the proportion of the world's population that is
poor by concentrating only on certain countries. Statistically that would be an option,
but not politically. The call is to reach the target in every region and in every country.

    Third, the Millennium Development Goals are output targets, result-oriented. Not
input targets, like the pledge to spend 0.7 per cent of national income on foreign aid.
They refer to welfare increases and poverty reduction, not to the means to be used for
that purpose. Nor are they process targets, such as combating the tendency to exclude
the poor by enhancing their participation and integration in society. To agree on the
need for full integration is politically important, but it would be difficult to measure
progress in achieving such an objective.

    Fourth, they are direct. The Millennium Development Goals are not growth targets,
chosen in the expectation that by meeting them less people would stay poor.

Apparently the time that world leaders thought that poverty could be reduced with the help of trickle-down mechanisms is behind us. They seem to have decided in favour of direct poverty reduction rather indirect measures like safety nets intended to compensate the poor for the negative effects of growth. Adopting the Millennium Development Goals implies that the nature and composition of economic growth should not be subordinate to growth itself.

Fifth, they are precise and quantified, not vague; not 'less' poverty; not qualitative, such as to change and reverse a trend, or further improve the lot of the poor. The Millennium Development Goals are precise: halving percentages of people in poverty in 15 years. Performance against such goals is measurable and accountable.

Sixth, they are ambitious. Individual countries may have been rather successful in this respect – China is a case in point – but never the world as a whole. So, it is an ambitious goal. But not over-ambitious. It will require intensive concerted action at all levels, by all policymakers and actors. It will require structural changes in priorities, investment allocations and resource use patterns. But it is doable. To my knowledge there are no ecological, physical, technical or other autonomous reasons why it would be inherently impossible to halve poverty rates within a reasonable period. There may be economic or political reasons, but those are always a matter of choice. The other half of the world's poor cannot be neglected. Can poverty be halved while simultaneously stimulating poverty reduction for the other 50 per cent? Or is it at the expense of better prospects for those others? That again is a political choice. The first option may be the more difficult of the two. Choosing that path would make the target even more ambitious. Not doing so would risk frustration among those whose prospects become more sombre. So, a higher ambition would be justified by considerations of equity as well as by the political necessity to enhance the cohesion in society.

Finally, the Millennium Development Goals represent a Political Target with a capital P and a capital T, not just another promise like all those made earlier but easily forgotten. Here we have set goals and made pledges at the highest possible level. The decision to adopt them was well prepared and well thought through. The goals were chosen consciously, in awareness of the needs of the poor and of alternative options. All sorts of alternatives (input targets, indirect approaches) had been tried in the course of the twentieth century. But they did not work well and did not deliver the hoped-for results. That is why, at the turn of the millennium, world leaders chose a radically different approach. They must have felt it was now or never.

Cynical analysts may reach a different conclusion and argue that in politics agreed goals and explicit promises have little significance. It is true, politics is a matter of power and interests. That is why, in the implementation of political decisions, practical reality so often differs from theoretical models. But the decision to adopt the Millennium Development Goals was not made arbitrarily, incidentally or by accident. Those who had the power to take decisions must have come to the conclusion that it was in their nation's interest to take this course and that the alternative options were inferior. Non-implementation of the goals was considered to be counterproductive. It would resemble the broken promises and unmet targets of the past and lead to even more frustration, threatening the stability and well-being of their nations.

3. GLOBAL POVERTY

Is there any chance that global poverty will really be halved by 2015? Will we fail in the future because we failed in the past, for the same reasons? That failure has not been across the board. There has been progress. According to the most recent *Human Development Report* (2003), 'the past 30 years saw dramatic improvements in the developing world. Life expectancy increased by eight years. Illiteracy was cut nearly in half, to 25 per cent. And in East Asia the number of people surviving on less than $1 a day was almost halved just in the 1990s.' But, the report continues, 'Still, human development is proceeding too slowly. For many countries the 1990s were a decade of despair' (UNDP 2003: 2). More than 50 countries are poorer now than in 1990. In over 20 countries a larger proportion of people go hungry. In quite a few countries child mortality is increasing, life expectancy is falling and school enrolments are shrinking. The authors of the report do not hesitate to speak of a 'development crisis' and of 'reversals in survival... previously rare' (ibid.). In more than 20 countries the Human Development Index declined, an alarming phenomenon because – as the authors point out – the capabilities captured by the index are not easily lost.

This means three things. First, there has been progress in some places. Development has worked. Second, in other places, there has been regress. In those countries, either development policies have not worked, the countries have not benefited from international economic growth, or they were victims of progress elsewhere, as "collateral damage". Third, global inequality has increased. In regions experiencing regress this could mean that people fall below a decent level of living, slip through safety nets, lose any capacity to catch up later and even lose their dignity as human beings. Indeed, that would be a crisis, a crisis in development and a crisis in societies. Insofar as it would be due to the inherently dualistic character of the global economy, whereby large parts of the world's population are condemned to poverty and despair, it would also be a crisis in the world community.[2]

Is this too gloomy? According to the World Bank, less than 1.2 billion people now live on one dollar or less a day, compared to 1.3 billion a decade ago (World Bank 2002a).[3] This is positive because it shows that high economic growth can lead to less poverty: the fall in the poverty figure was largely due to developments in China, where annual economic growth was 9 per cent in the 1990s, lifting 150 million people out of poverty. However, excluding China the overall figure increases, from somewhat more than 900 million to about 950 million people. That is a relative decrease, from less than 30 per cent of the population of the developing countries as a whole, to less than a quarter. Nevertheless in sub-Saharan Africa poverty increased substantially, by 30 per cent in only a decade. In South Asia it is still around half a billion people. That is disappointingly high, four decades after the beginning of coordinated national and international strategies for growth, development and poverty reduction.

In many parts of the world poverty is high and increasing, despite the fact that the 1990s saw the highest annual world average growth figures since the end of World War II and decolonisation. Clearly the vast accumulation of income and wealth have not been used for sustainable poverty reduction. During the last decade of the previous century an alarmingly high number of people suffered from worsening living conditions, notwithstanding political promises made at the beginning of that decade. In 1992 in Rio de Janeiro, at another global summit – the United Nations Conference on Environment and Development (UNCED) – world leaders came together to pledge adherence to a new development paradigm: sustainable development. Heads of state and government adopted the new principles and policy programmes contained in Agenda 21. In this they declared that in the future the world's resources would be utilised in such a way that people would be at the centre and that future generations would not be deprived of the opportunities open to present generations (United Nations 1992). Ten years later we can only conclude that the "sustainable development decade" of the nineties – the UN Fourth Development Decade – was less successful than the First Development Decade of the 1960s.

Nonetheless, conventional wisdom is that poverty is on its way out, because the number of people living on less than a dollar a day is falling. Yet, what is the significance of the dollar a day criterion, which has found its way into the Millennium Development Goals? The authors of the *Human Development Report* do not take a position on this. They refer to critics who think that this yardstick 'reveals little about income poverty and its trends', but note that others call it 'rough but reasonable'. I belong to the second group in that I believe it is of course necessary to have some global yardstick. But the dollar a day yardstick is rough beyond reason. Why not, say, one and a half or two dollars a day? Are people who no longer figure in the poverty statistics, because they now earn a dollar a day, out of poverty for good?

As questionable as estimates of the poverty line itself are estimates of the number of people living on or below it. What about the people who did not reach this level because they died, due to poverty, shortly after birth or at least far earlier than the average life expectancy level enjoyed by others? Their numbers do not even reach the income statistics. Nor did they in the past, which means that these statistics alone cannot shed light on historical comparisons of income poverty. At present, annual poverty-related deaths run into dozens of millions. Such a figure would add substantially to an annual stock figure of poor people at any particular moment of time. One sign of progress is that, since the 1960s, maternal death and infant and child mortality – strong indicators of poverty – have decreased and that life expectancy has increased. But this is by no means true everywhere and the trend seems to be turning again: AIDS kills poor people in particular. In some countries, previously rising life expectancies have been reversed – sometimes by as much as one-third – and have dropped back to pre-1980 levels. At present, children in southern Africa can expect to live shorter lives than their grandparents (United Nations 2002c: 8).

What is the significance of the concept of income poverty anyway? Poverty cannot be captured in terms of money and income alone. Poverty is more than income: the lack of opportunity to acquire lasting control of resources in order to strengthen one's capacity to acquire the basic necessities of life, such as water, energy, food, a safe place to eat, rest, sleep, wash, have sex, go to school, have basic health services and medicine in case of illness; a job enabling all of this or the income to acquire it by means of exchange, access to economic markets and social networks, knowledge to survive in this world, information and education to acquire more knowledge and to gain insights to cope with disasters, threats, violence and challenges and, when that is beyond the capacity of the individual, some protection. All this requires more than money, more than an income. It requires assets or entitlements, the value of which cannot be easily estimated in monetary terms. In other words: rights that ensure access to all these things.[4] Rights that certainly cannot be acquired for a dollar a day.

Income poverty is only an indirect indicator of human poverty.[5] Other indicators reveal that there is more stagnation in the battle against poverty than we would expect by looking at the dollar a day yardstick only. Some examples: Every day 800 million people go hungry. During the last decade the number of hungry people fell in China, but it increased in 25 other countries (UNDP 2003: 88). One out of six children of primary school age in developing countries does not attend school, and only half of those who do start primary school finish it. Close to 900 million adults cannot read or write, one out of four adults in the developing world is illiterate (ibid.: 92). In sub-Saharan Africa one in every hundred live births results in the mother's death. In many countries the already high maternal mortality figures are on the rise. HIV/AIDS, tuberculosis and malaria are killing more people than a decade ago (ibid.: 97). More than 1.2 billion people lack access to safe drinking water, while 2.4 billion people have inadequate sanitation services (United Nations 2002b: 7).

The *Human Development Report* labels one of these indicators a 'shameful failure of development' (UNDP 2003: 97). In fact all of them are. The world as a whole has never been as rich as it is today. In the past 15 years globalisation has accelerated without precedent. The opportunities offered by money, capital, technology and communication to enable more and more people to benefit from progress are without precedent. But these opportunities have not been used to correct this shameful failure of the past and bridge the gap between those who benefit from modernity and those without entitlements enabling them to break out of the vicious circle of malnutrition, disease, illiteracy and poverty. On the contrary, if present trends regarding the nature of globalisation continue – and there are no indications to the contrary – progress and modernity will not bridge that gap, but widen it further.

Unlike Reddy and Pogge (2002) in their recent *How Not to Count the Poor*, I do not criticise the usual poverty yardstick because of measurement difficulties. Statistical methods of measuring income levels and comparing them between countries and over time can be criticised, but there is no reason to suspect that they have been used to influence political decisions. My criticism of the dollar a day yardstick is not so much statistical or conceptual as political. The figures and statistics concerned have been published and quoted so often that politicians, policymakers, public opinion and world leaders have been led to believe that the trend has been in the right direction. However, there is a large difference between theory and practice, between statistics showing a decrease in the number of people below the poverty line and the realities of misery. That reality has been put out of view. During the 1990s the degree and extent of world poverty were played down and political leaders were lulled to sleep.[6] Were they fooled? Or were they fooling themselves by not asking the obvious question. What kind of life can you live on a dollar a day anywhere, in Africa, Asia, in the cities of Latin America, or even in China? Has that question not been raised because of the fear that a more ambitious goal, affecting more poor people, could never be attained without far-reaching changes in the distribution of world income and entitlements, while the one-dollar level would require only better governance in the poor countries themselves and a slight increase in development aid?[7] The blame for not meeting the dollar a day target could easily be apportioned to the poor countries themselves, while failure to reach a more civilised goal could be attributable to the richer countries and their reluctance to share with the poor.

Though in terms of macro world statistics – that is, in theory – fewer people now live below a poverty line which has been selected on political grounds, in practice poverty has increased in many regions of the world, including in many countries where there has been improvement in macro terms. This is because inequality across the world increased during the last part of the previous century, both between and within nations. The distribution of income among the people of the world, regardless of national borders, has become more unequal. Nowadays incomes are distributed more unequally among the global population than in the most unequal countries (UNDP 2003: 39).[8] Since the First Development Decade inequality has only increased. While in 1960 the top 20 per cent received 30 times the income of the poorest 20 per cent, by the turn of the millennium this inequality had widened to more than 70 to 1.[9] Baker and Nordin (1999) foresaw even a global quintile disparity of 150 to 1, which would be 'fraught with risks for rich and poor alike'.

This is not new, as various authors have highlighted the economic, social, environmental and legal dimensions of the relation between dire poverty and worsening income distribution.[10] Why am I adding my voice? Because there is such a difference between the irrefutable facts about poverty and inequality and the political answers.

"Poverty is declining, but is still a challenge." That has been the general message during the last decade, explicitly repeated in the most recent *World Development Report* from the World Bank (2003b: 2). The political meaning of such a message is that we are moving in the right direction and we should continue to do so, albeit somewhat faster. However, that direction is wrong. Since 1990 the trend has been negative. The authors of the *Human Development Report* (UNDP 2003: 40–44) conclude that incomes are falling in 54 countries, hunger is on the rise in 21 countries and child mortality is increasing 'in a way not seen in previous decades' (ibid.).

4. POLITICS: GAINING NEW INSIGHTS OR IGNORING THEM?

Early development theories were based on the assumption that economic growth was not only a necessary precondition for poverty eradication, but that it was a sufficient condition, since growth would trickle down. These found a counterpart in theories on the dualistic character of societies and economies in development. Various limits to growth soon became evident, as well as the fact that growth itself can lead to impoverishment through exploitation and the cultivation of dependence.

Consequently, in the 1970s, policies to strengthen the economic growth and self-reliance of developing countries were complemented by policies to fulfil the basic human needs of the poor in these countries. These policies were short-lived however. New insights into the significance of domestic macroeconomic policies affecting the conditions for growth – strong and open markets, free entrepreneurship and competition, stable financial and monetary relations – together with the world economic recession at the turn of the decade, gave free rein to adjustment policies. It was believed that these should precede development policies, including the reduction of poverty. Adjustment policies implied cuts in public expenditure to meet basic human needs, for instance, health and education, and also withdrawal from programmes in the areas of food, agriculture and rural development. They resulted in a disinvestment in the capacities of a society to provide a floor to poverty. After a while, safety nets were introduced to avoid a worsening of poverty due to adjustment itself, but these only functioned as stop-gap measures. And while adjustment-cum-safety nets had originally been seen as transitional, soon a new mantra was introduced as a structural precondition for development: good governance. The Washington Consensus regarding good governance made adjustment a lasting feature of development, which was bound to result in a stagnation of poverty reduction.

The reason was political. The Washington Consensus aimed at stability and high growth while accepting increases in inequality, assuming that the positive effects of the former would outweigh the negative consequences of the latter. To decide on the optimum pattern of adjustment was a specific aspect of the general political task to find the optimum combination of growth, inequality and poverty reduction. This is the most difficult task facing political leaders, in all countries and at all times. Even if increased inequality and higher poverty are meant to be only transitory, in order to enhance the capacity to grow and thus the future potential to redistribute and to redress poverty, the institutional consequences of the Washington Consensus imply a political bias towards a permanent postponement of poverty reduction.

That is the reason why so-called "pro-poor growth policies" never had a chance on any reasonable scale.[11] Not because it is difficult to determine a theoretical optimum: these optima will differ in time and between countries, because of the specific social and economic structures, their institutional capacities, their starting position and their resource endowment. But a choice can be made, depending on the political priorities attached to people's basic needs. Pro-poor growth policies did not get off the ground because of different opinions on these priorities. Those with the power to decide that now is the time to give priority to poverty reduction and redistribution will always be tempted to further postpone that decision and to further prioritise growth, under the pretext that this will increase the potential even further. And so on and so on. The political class that has the power to decide does not have an interest in changing priorities. On the contrary.

In the 1990s the formulation of the "rights approach" to development contributed to the understanding that unequal power distributions can be counteracted with the help of law and institutions, based on a growing international consensus on basic values. At UNCED heads of state and government embedded the right to development in the Rio Declaration as a principle underlying international policymaking aimed at sustainable development in the new century. By the end of the decade, the rights approach to poverty had become a guideline in authoritative international documents (see also De Gaay Fortman 2002, 2003; Opschoor 2002, 2003; Schrijver 2001, 2003).[12]

So, further analysis of poverty as a phenomenon, the processes involved and the underlying forces and mechanisms has deepened our insights. These have found their way into new and better policies to combat poverty. But not all lessons learned have been put into practice. More and stronger safety nets have been built, more sophisticated adjustment policies construed and more emphasis laid on governance, institutions and capacity building. But the most important lesson was not put into practice: start with the distribution of assets and make it more equal. Efforts were made, but these remained an exception, at local level only.[13] At the global level, there was no effort whatsoever.

5. AID AND COOPERATION

The only instrument at the global level to make asset distribution more equal was development aid. Some, like Tinbergen (1962, 1990), advocated development assistance as a means to redistribute world income. Others, like Chenery (1967, 1973) saw aid as a means to stimulate economic growth in developing countries beyond what would be possible through domestic resources only. But, as documented, for instance, by Ohlin (1966), there were also political motives. The end of World War II and the winding up of colonisation ushered in a new era of globalisation. The old powers were willing to use international technical and financial assistance to control the new situation. In 1961 the combination of such motives led to the consensus decision in the General Assembly of the United Nations to adopt a strategy for the First Development Decade, 1961–1970. This strategy made self-sustaining economic growth (per capita) the overriding objective for the Third World, following the more or less successful outcome of the struggle for liberation and independence after decolonisation. The strategy's rationale strongly resembles that of the Millennium Declaration. In the words of the then secretary-general of the United Nations, U. Thant:

> It is an extraordinary fact that at a time when affluence is beginning to be the condition... of whole countries and regions rather than of a few favoured individuals, and when scientific feats are becoming possible which beggar mankind's wildest dreams of the past, more people are suffering from hunger and want than ever before. Such a situation is intolerable and so contrary to the best interests of all nations that it should arouse determination, on the part of advanced and developing countries alike, to bring it to an end.

These words could have been written in the year 2000.

The objective of the strategy was to enable developing countries to achieve an annual growth rate of at least 5 per cent by 1970 and to sustain it thereafter. It was thought that, if the population of the developing countries continued to grow at a rate of 2.5 per cent, this could result in a doubling of personal living standards within 25 to 30 years. To this end the richer countries would make available 1 per cent of their national income in the form of development assistance (United Nations 1962).[14]

However, the aid targets were not met and no international agreement could be reached on a more ambitious growth target and more intensive cooperation in the Second Development Decade.[15] Political interest shifted towards direct rather than indirect ways to combat poverty and to sectoral rather than global approaches and targets. Many of these targets had the end of the millennium as the deadline: the removal of hunger and malnutrition (Rome 1974), "Health for All" (Alma Ata 1976), safe drinking water for everyone (Mar del Plata 1977), increasing developing countries' share in global industry to 25 per cent (Lima 1975), universal access to basic education (Jomtien 1990). All these targets were set at high political level. Nearly all are based on rights laid down in the International Covenant on Economic, Social and Cultural Rights, which came into force in 1976 (United Nations 1966). None have been accomplished.[16]

All the targets were directly related to poverty. This also applied to the target concerning manufacturing, given the importance of manufacturing for employment. Creating employment was seen as a direct and effective road towards poverty eradication. A "world employment programme" was designed,[17] which was a new and unique approach to combat poverty not through high economic growth but by aiming directly at the fulfilment of basic human needs.

What followed is well known. Both new approaches – the basic needs and the sectoral programmes – received a severe blow in the 1980s due to the world recession. Developing countries were affected by a decline in world trade and by mounting debts. They received an even stronger blow from adjustment policies imposed upon them and from a refusal by northern countries to lift trade restrictions, relieve debts and stabilise commodity prices. The overall global approach to development and poverty reduction that had started with the First Development Decade had died out. International measures to sustain development were made conditional on developing countries accepting adjustment programmes, opening their markets, liberalising their economies and reducing budgetary deficits and government expenditure. Far from being a Third Development Decade, the 1980s became the era of adjustment to new realities set by the North, or, in the words of Corea (1985), a 'lost decade' for development. Many concluded that it was a lost decade for poverty reduction too. Adjustment policies often implied more unemployment and disinvestments in education, health and other social sectors, which were increasingly left to the market.

Then the world entered a new phase in the post-1945 globalisation process. The first phase had started with the end of the world wars and the end of colonisation. The second started with the end of the Cold War. Global political relations drastically changed with the end of the Cold War in 1989. A new élan emerged for global cooperation and new development paradigms became prominent: "a new world order", "sustainable development", "peace and development". Wild-west adjustment was replaced by less harsh programmes. Together this created room for the implementation of a world Agenda 21 for sustainable development, poverty eradication and environmental protection.

However, the new optimism soon faded. The 1990s would not herald the beginning of an era with a new perspective for people in developing countries, as originally had been expected (Pronk 1997, 2000). The main reason is that the capacity of the international community to implement a new political agenda would be eroded by that same agenda. The 1990s would see more and more conflicts within countries, due to economic inequalities and cultural differences between religious, ethnic and other identity groups. In the 1990s, transnational forces would also threaten the economic, social and environmental sustainability of national societies. These forces and domestic conflicts would undermine democratic public authority, both within nation-states and globally. I argued then that at best the 1990s would be a decade of transition. Whether this would be a transition towards sustainable development in all respects would depend on whether it would be possible to design and implement an alternative to a lop-sided globalisation process, mainly of an economic character, carried by market considerations only.

In Rio itself it was already clear how ill-prepared public authorities were for such a task. It was impossible to reach a meaningful consensus on finance for sustainable development. I am certain that the negotiators knew in 1972 that, with or without the United States, the 0.7 per cent aid target would never be met. Some individual countries would be able to make achieving and keeping the development assistance target a political issue; but there were no political indications whatsoever that the others would even come close. From 1972 to 1992 all negotiations on commitments were only for show. Did that change in Rio? Far from it. After weeks of futile discussions and negotiations, it was finally decided that each individual country would be free to choose its level of development assistance, with some countries stating only that they would make their "best efforts" to increase their assistance.

In 1992 the 0.7 per cent target was therefore carted off and buried. After that, aid performance actually declined, despite the common observation made in Rio that sustainable development and poverty reduction would require a substantial increase in resources. These were not made available, or were provided only on the condition that developing countries themselves change their governance regime as prescribed by the international institutions. Especially for the poorest countries, for the countries most in debt, without adequate access to commercial lending or foreign trade, countries suffering from violent internal conflict and countries trying to rebuild their society after dictatorship or war, this meant that they were led into a blind alley.

Throughout the 1990s, failure to raise funds paralysed many international talks on global problems, from debt relief to climate change, biodiversity protection, land degradation, peacekeeping, refugee assistance and the fight against AIDS. A promise "to reverse the trend", made during the "UNCED plus five" review in 1997 met with great suspicion and could not save negotiations on how to implement the global sustainable development agenda agreed on in Rio. Moreover, in the 1990s, international talks on trade, capital movements or knowledge-sharing mainly benefited the interests of the North. There were no steps forward whatsoever to enlarge the capacity of developing countries to combat poverty with domestic means. Their utilisation was often curtailed by the demands set by an open economy, the liberalisation of markets and the privatisation of potentially public mandates.

In the meantime two efforts were made to rectify this. Both aimed not so much at increasing resources as at reallocations and increasing coherence among social and economic policies in general. In 1995 poverty was put on the agenda of the Social Summit in Copenhagen. It resulted, among other things, in the "20/20 Initiative". This called for developing countries to increase expenditure on basic social services from the then current average of about 13 per cent of their national budget to 20 per cent. In return, donor countries would increase their aid allocations to basic social services to 20 per cent of their total aid budget. The initiative implied that any developing country reaching this target would have its expenditure matched by aid from the donor countries.

The second initiative was a statement by development ministers in the framework of the OECD Development Assistance Committee proposing 'a global development partnership' to reduce by half the share of the population in extreme poverty by 2015 (OECD 1996: 2). It was the first time that this goal, together with a number of sub-targets, was mentioned. The background of the proposal was the explicit recognition by the ministers during the negotiations on the text that the 0.7 per cent target no longer had a political future. The United States would never agree to such a target, while a number of other countries would never achieve it. They realised that ongoing pressure on the United States to change its stance would be fruitless, but they did not want to let them off the hook. Rather than make developing countries believe or hope that this situation would ever be different, an alternative was chosen: an output target – poverty reduction – instead of an input target – aid volume – together with a firm commitment to get the job done. It was not meant to be a way out, but a new path, leading directly towards the realisation of an appealing objective. The new poverty reduction target was also meant to be acceptable to the United States, which could help to achieve it with other means than just aid, in particular employment-creating trade measures. The message to the developing countries was that there is money for good programmes to combat poverty. Good programmes will be supported with resources and with coherent international economic policies. That was not an expectation, not an expression of belief in the workings of financial systems, but a political commitment.

By unilaterally changing the nature of their commitments to the South, on their own initiative, after 40 years of multilateral talks on the 0.7 per cent, politicians representing the North made themselves more responsible. Without a serious effort to meet the new objective, the initiative would soon be branded as hypocrisy. By agreeing that this originally unilateral initiative should become part of the multilateral Millennium Declaration, the leaders of developing countries declared their commitment to poverty reduction and demonstrated their trust in the commitment from their partners in the North. This once more implied that, politically, there was no way back.

6. ARE WE ON TRACK?

The Millennium Development Goals are not just another set of goals. They are different. They are not the next step, but a final step. At the turn of the millennium, world leaders at the highest political level, looking back on decades of negotiations, policies and targets, declared that from now on the eradication of poverty shall have first and foremost priority, within all countries and worldwide. Any other interpretation of the declaration would be disputable. In 2015, no responsible political leader would be entitled to say, once more, "the goals may not have been reached, but progress has been made and they have helped to keep us alert and the issue alive".

By making them the centrepiece of the Millennium Declaration, the Millennium Development Goals became the responsibility of all countries and all international agencies,. There is no higher or broader forum than a summit. The Declaration is not a legally binding document, but the highest political commitment possible, not made in passing, but with all eyes open. The goals are achievable. All the political leaders concerned have said so, fully aware of both the challenge and the obstacles. But the first signs are not positive. The goals were set at a political summit in 2000. At the Johannesburg Summit in 2002, where decisions ought to have been taken on the implementation machinery, the same political leaders were unable to take any steps other than reconfirming their earlier commitments. No decisions were taken concerning instruments to realise the goals, nor on who should do what, nor on an effective review and feedback mechanism. Since then not much progress has been made. To make further progress six major concerns must be addressed.[18]

The first concern is the lack of a road map. There is much talk of a "development compact", but it does not as yet exist. All references to such a compact in international documents are phrased in terms of "ought to be" and "should be". That is wishful thinking. There is no such a thing as a development compact, with freshly agreed language on how to focus international trade, finance and technology policy on the implementation of the millennium goals rather than the traditional objectives of the previous decades. The recent *Human Development Report* devoted to the Millennium Development Goals is full of questions and recommendations, but has no agreed answers. It is a state-of-the-art document laying out what policies, according to present insights and on the basis of lessons learned, could or should be pursued in trade, debt, industry, science and technology, education, health, water, agriculture, food and nutrition, energy and other resources, the environment and ecosystems. But it cannot point to any new agreement in these fields. It rightly points out the need for country ownership, driven not only by governments but by many actors (local governments, communities and civil society groups) and a need for a comprehensive approach, bottom-up and participatory. It is worded in "correct" development language: gender, empowerment, accountability, partnership, social mobilisation and a cautious approach towards both privatisation and decentralisation.[19] But all these references together do not make a strategy, only proposals, a millennium development compact "proposed" (UNDP 2003: 15).

Second, the Millennium Development Goals are still seen too much as belonging to the traditional field of development policy. It is customary to see development goals as a specific responsibility of developing countries and the specific governmental departments within these countries, together with the corresponding UN agencies, rather than as a common global responsibility that also includes the northern countries and international institutions like World Trade Organization and the International Monetary Fund. After all, heads of state and government committed themselves to deal with global – that is, worldwide – poverty, wherever and with all means, attacking all possible causes, not restricting themselves to foreign aid, but undertaking also to remove national and international constraints: protectionism, monopolistic and discriminatory practices and other external obstacles for developing countries to attack poverty within their borders. Maintaining such obstacles would make foreign assistance a form of compensation for being kept at a distance, rather than a net investment in poverty reduction. So far, however, the agenda for the Doha Round on international trade has not been changed in the light of the commitment to meet the Millennium Development Goals. Nor was poverty on the agenda in Cancun in 2003.

Nor are the Millennium Development Goals a key subject on the agenda of the international financial institutions.[20] Debt rescheduling should focus on poverty alleviation. However, despite the summit agreements, neither the proceedings of the Paris Club nor the debt rescheduling programme for the Highly Indebted Poor Countries (HIPCs) have been redesigned. There can be no debt sustainability without social sustainability. Debt and poverty are related. Debt reduction and poverty reduction ought to be related as well, by making adequate finance available for both, in an integrated policy. The Poverty Reduction Strategy Papers (PRSPs) introduced by the World Bank and serving as the basis for aid allocations from most donor countries, could be the basis of such an integrated approach. However, the concept and procedures of the PRSPs have not yet been systematically reorganised. The World Bank takes as given current budgets and foreign resource levels and has not yet made the agreed 50 per cent cut in poverty by 2015 the key objective of its strategy. Instead of a single integrated approach to poverty reduction we now have competing schemes and pathetic efforts to coordinate (UNDP 2003: 20–22, 149). All of this illustrates that the millennium goals still are not the once and for all objectives of the countries and institutions with the power to decide.

Third, there is no focus. At the Johannesburg Summit, the secretary-general of the United Nations proposed focusing action on five sectors: water, energy, health, agriculture and biodiversity and giving priority to policies in these sectors which contribute to poverty reduction. It was an attractive proposal because sustainable development in these sectors is a precondition for lasting poverty reduction. However, no agreement could be reached on attaching less priority to other sectors. In principle a comprehensive and simultaneous approach in all sectors would be better than a selective approach. However, doing nothing tends to be the result of aiming to cover everything. If an overall approach overburdens implementation capacity, it paralyses action, as now seems to be the case.

Fourth, there still is a shortage of finance. There is no agreement on the costs of implementation. Meeting the goals on health was estimated to require additional foreign aid of US $30 billion annually (Sachs 2001). In policy terms this is a step forward compared with a World Health Organization report on global health tasks published five years earlier, which – after stating the need to set new priorities based on the conviction that health is a human right – concluded, 'Once this conviction is established, society itself will bring new light to bear on resource issues' (WHO 1997: 38).[21] For drinking water it was calculated that the realisation of the target would require US $180 billion annually (Camdessus 2003). This is considered a gross exaggeration by the World Water and Sanitation Collaborative Council, which presented an annual figure of US $30 billion (United Nations 2002b: 16; WSSCC 2002: 8). It is unclear which parts of these amounts would be additional to current expenditure. The United Nations concludes that 'further work' is required to have a 'more accurate and better understanding of the global financial requirements' to meet the Millennium Development Goals (United Nations 2002b: 16). That is obvious, but it would have been helpful if that work had been carried out before the goal was set. For the other Millennium Development Goals there are no separate cost estimates. Needless to say, none of the goals have a financial plan. Nor is there a financial plan to implement the Millennium Strategy as a whole. There are only some rough estimates of the additional external assistance required to meet global objectives. They range from US $40 to $100 billion a year. One of these is the figure of $50 billion mentioned by a UN commission led by former Mexican president Ernesto Zedillo (United Nations 2001). This latter figure – which was a starting point for the World Summit on Finance and Development in Monterrey, Mexico, in 2002 – is called 'conservative' by the authors of the *Human Development Report* (UNDP 2003: 346). That is understandable in light of the figures for health and water only.[22] However, not even this conservative estimate could persuade the donors and international financial institutions to agree on a common finance strategy. The new aid commitments presented in Monterrey by the United States and the European Union were explicitly not intended to help finance a Johannesburg Plan of Action to implement the Millennium Development Goals.

Fifth, foreign aid is missing the mark. Since the Washington Consensus became the guideline for aid allocations there is so much emphasis on good governance in developing countries as a precondition for receiving aid, that the assistance itself can no longer help improve the situation within these countries. Rather than aid being used as a catalyst, helping to bring about better policies and better governance, the countries themselves are expected to make such improvements before receiving any aid. Quite a few countries are not in a position to help themselves, for reasons mentioned earlier. They are then deprived of foreign aid. Countries that are able to improve policies and governance to the liking of donors receive aid as a reward. For them this aid is either no longer necessary, because good governance can be rewarded by the market, or it comes too late. Regardless, as far as poverty reduction is concerned, present-day donor preferences for so-called "performance-driven" aid allocations are overshooting the mark (Pronk 2001, 2003a).

Sixth, poverty reduction will become increasingly urgent and difficult because of the critical trends in the ecological and physical environment. The loss in global biodiversity, the change in the global climate, the increasing scarcity of basic resources such as water and energy and distortions of the ecosystems are no new phenomena. However, recent studies seem to indicate that these changes are now moving faster and are having a greater impact than in the past. The poorest people are the first victims: they live in the most vulnerable places with the least productive soils, in arid regions, polluted slums, eroded hill slopes and flood-prone coastal areas and river plains. They are the least protected against environmental crime and the whims of nature. Poverty due to environmental deterioration is on the rise. Poverty due to conflicts about increasingly scarce resources will increase as well. In international fora, environmental risks seem to have lost the competition for attention against security risks. Yet this imposes a heavy tax on the attainment of the poverty goal (see further Part 3 of this volume).

I am not concerned that the emphasis on the quantitative character of the Millennium Development Goals will jeopardise their quality. The goals cannot be attained without community action at the local level. Quality should be defined in terms of priorities and objectives set by the population rather than authorities. Meeting the goals requires recognition of the desires, aspirations and initiatives of poor people themselves and thus of local processes. However, it is too late to say that the process is more important than the result. The authors of the *Human Development Report* also warn against taking the goals 'out of context and [seeing] them as ends in themselves rather than as benchmarks of progress towards a broader goal of eradicating human poverty' (UNDP 2003: 30). The Millennium Development Goals indeed have a context: survival and a better quality of life, according to people's own wishes. Poverty has many dimensions: food, health, water, knowledge and the resources to survive and secure a better life. But these dimensions cannot be traded against others. They are each ends in their own right. Too much emphasis on process may result in complacency amongst politicians, who tend to be quickly satisfied when they note improvement; slow improvement, less than had been intended, but improvement nevertheless.

This tendency is manifest in the follow up to the summit meetings in New York and Johannesburg. Instead of putting implementation of the Millennium Development Goals on the agenda of the regular international talks in key institutions such as the WTO, IMF, the World Bank and the multilateral environmental agreements, the emphasis is technocratic and thus a-political. Much thought is given to statistics, benchmarks, performance indicators, monitoring and reporting. However, keeping track of what is going on should not be confused with the action itself. Deliberating on which indicators are best to judge whether progress has been made does not constitute action-oriented implementation. On the contrary, efforts to develop performance indicators can often prompt politicians to adopt a wait and see attitude, pending proposals from the bureaucracy, followed by lengthy disputes, thereby postponing action.

For all these reasons, it is no surprise that we are far behind schedule. As was recently noted, the Millennium Development Goals are 'technically feasible and financially affordable. Yet, the world is off track to meeting them by 2015' (Vandemoortele 2003: 16). The authors of the *Human Development Report*, while declaring it 'beyond doubt' that all countries can meaningfully achieve the Millennium Development Goals, present a timeline showing that in most regions most goals will not be achieved if progress does not accelerate (UNDP 2003: 50 and Figure 2.1). That applies not only to Africa, as some might expect, but to Asia as well (United Nations 2003: Table 1).

A dramatic change of direction is needed, and this requires political leadership. The implementation of the poverty targets should not be left to experts, bureaucrats and diplomats. It should be permanently on the political agenda, within countries and in the international system: the intergovernmental machinery and the relevant global institutions. In these bodies the discussion should not be limited to a review of the state of affairs. All political and institutional power available should be used to apply pressure on authorities and agencies not to shy away from the commitments made by their heads of state and government, but to keep their promises. No second thoughts, as is so often the case in international fora, after the political leaders have spoken and left the scene. And those political leaders themselves must live up to their commitments.

Rather than being confident that the job will be done, they have to stay alert and active themselves. Where commitments are still political, without yet having been enshrined in legally binding treaties with compliance regimes and sanctions and in institutions with enforcement regimes, there are still political possibilities to ensure implementation. But that requires leadership and the willingness to go beyond mandating governance in poor countries only, as has been IMF practice, but the governance of the economically advanced as well. That this is possible has been shown by the WTO and the European Commission. These powers should also be applied to combat poverty, not only to ensure stability and a level playing field.

It may seem strange that while poverty eradication basically requires a bottom-up approach, empowerment of the poor, community initiatives, participation of individuals, capacity building at the level of the household, the attainment of the millennium development and poverty goals should depend so much on political leadership, which always implies a certain top-down policy. Is that an anomaly? No, because there *are* community initiatives, social movements *are* alive, empowerment struggles *are* being fought, social mobilisation and emancipation *are* taking place. These do not have to be manufactured from above. They exist. But they are threatened by other forces: the strength of the non-poor and the middle class in a scramble for the fruits of globalisation. After all, poverty reduction is also a matter of redistribution and sharing. That is where political leadership is required: to give room, living space and perspective to poor people trying to take their fate in their own hands and secure a fair redistribution.

7. THE POLITICS OF GLOBALISATION

Earlier, I distinguished between globalisation after 1945 and that since 1989. Modernisation and growth tended to pass by the poor during the first period. The poor were neglected because economic growth did not need them, neither their labour nor their purchasing power. The emerging global economy was similar to that of the western economies, which enjoyed economic growth after the Industrial Revolution. In these countries poverty had been appalling and exploitation harsh. It took some time before the new capitalists and the emerging middle class realised that further economic growth would not only depend on supply factors, but that growing demand would be crucial. This meant that higher incomes would be required for hitherto poor labourers. The way was thus paved for a more positive response to the claims set by the poor and their labour unions and social movements. It resulted in less exploitation, higher wages, better labour conditions, Keynesian economic policies and social welfare systems, based not only on notions of solidarity, but also on enlightened self-interest.

It is no different in the world economy today. In the emerging world economic system the rich did not need the poor. After World War II, modernisation created its own demands. Poverty reduction was not necessary: low-cost labour was amply available and there was much purchasing power around, which could be tapped even if it was far away, because globalisation brought those who could afford to buy and pay closer together. There was no compelling need to sustain demand by raising the incomes of the poor. The poor were dispensable and could be neglected. That is the picture of the second half of the twentieth century: poverty as the collateral damage of globalisation.

Clearly, this could not last. After a while the world economy would experience limitations to further growth as had earlier been the case at the national level. At the end of the 1980s, when adjustment policies had taken their toll everywhere, the moment of change seemed to come a little bit closer. Awareness grew that poverty reduction was not only socially desirable, but also an economic necessity.

But then there was the quantum leap in globalisation during the early 1990s. The fall of the Berlin Wall and the eclipse of frontiers between East and West, together with the breakthrough in information and communication technology, created an unrestricted world market. Neither geographical distance nor time constraints stood in the way of the growth of economic opportunities. The sky was the limit. However, with the growth of the middle class, its better access to resources and public services, and political power, the gap with the poor grew rapidly. The poor were driven away from scarce resources and denied opportunities to strengthen their capacities. Globalisation become occupation, or 'Global Apartheid' in the words of President Thabo M'beki of South Africa, addressing the Johannesburg Summit (Pronk 2003b).

What did governments do? Mostly they were loyal to the new global middle class, modern, embracing the market everywhere, preaching the wonders of liberalisation, skimming the public sector where it felt like a straightjacket to the middle class. Governments skilfully facilitated globalisation by strengthening those institutions and treaties which would ensure a widely open, free world market, unrestrained by considerations of social equity or ecological sustainability. Before 1989 globalisation had been a process. In the 1990s it became a project. Present-day globalisation produces poverty and inequality. The continuation of poverty and inequality is no longer collateral damage. It has become an intended result of policymaking, "calculated default".

A pregnant woman is 100 times more likely to die in pregnancy and childbirth in sub-Saharan Africa than in an OECD country. The poorest 20 per cent of the world's population, who have a greater chance of being affected by poverty-related diseases, receive less than the richest 20 per cent of the benefits of public-health spending. The same is true for public spending on education and other basic human needs. The average land per capita of the rural poor is steadily declining (from less than four hectares in the 1970s to less than half a hectare in the 1990s), and the decline continues.

Poor people pay sky-high prices for life-saving medication, such as drugs to combat HIV/AIDS. Global middle class-oriented companies block lower prices on the world market and resist domestic production of antiretroviral drugs in developing countries. Poor people in slums pay up to ten times more for drinking water than people connected to municipal supplies. They even pay much more, not only as part of their daily budget, but also in absolute terms, than the much richer consumers in the cities of the North.

All this means that the poor are poor because their assets decline and public expenditure in their benefit is declining too, in some countries in both absolute terms and relative to other classes. The better-off block reallocation of public budgets to redress this situation. And when governments want to make an effort 'it is next to impossible politically to shift funds to… basic social services, without incurring the wrath of those better off' (UNDP 2003: 108).

So it is a matter of political power and distribution. In the current phase of globalisation, poverty is no longer a side effect but an intended product of globalisation. Its continuation is seen as beneficial for the middle class, which will continue to resist change and redistribution. That is the basic reason why progress has turned into stagnation and regress in the past decade. It is the main reason to be pessimistic about the possibility of achieving the Millennium Development Goals, unless the dramatic change of direction which was called for really takes place. But, again, that would require a new political approach to globalisation. Not a process to be left to run its course, with some collateral damage to be taken care of on the wayside. Not a project which becomes an end itself, deliberately victimising those who stand in its way. But a project serving the political objective of cutting poverty by half, soon, and then again and again, in a sustainable manner.

What will be the future? It is difficult to predict. Poor people do not resort to violence only because they are poor. But poor people who experience exclusion, who see no perspective whatsoever and who feel treated as less than human beings, may become convinced that there will never be a place for them or for their children in the world system resulting from globalisation. People who feel that the system is turning against them may turn their backs to the system and develop an aversion to that system and its values. Clearly, it is not their system anyway and its values – freedom, justice, solidarity, welfare, modernity – were clearly never meant to be extended to them. Aversion can turn into hatred, and hatred into violence. It is not so difficult to recruit people willing to use violence against a system from a large crowd that is excluded from that system. Nor is it difficult to recruit people within the global system who consider themselves legitimised to act on the behalf of the poor and excluded. Poverty is not a root cause of terrorism. But it can lead to violent action against a system that is believed to be a root cause of poverty. Poverty is a political concept and can provide political rationalisations. To squeeze poverty is an end in itself, but it is also a must if we want to eliminate the possible foundations of violence and insecurity. That is the dual mandate for the politics of globalisation.

NOTES

1. This text is a shortened and edited version of his inaugural speech as professor of the theory and practice of international development at ISS (11 December 2003). Pronk received an honorary doctorate from ISS during the Fiftieth Anniversary *Dies Natalis*, following the conference *Globalisation, Poverty and Conflict*.

2. In the early 1990s the Dutch Government did not hesitate to speak about 'a crisis in the development process, in development thinking, in development policy (and) in "the system"' (Dutch Ministry of Foreign Affairs 1991: 35) in its white paper on development cooperation *A World of Difference*.

3. The figures are 1,292 million for 1990 and 1,169 million for 1999.

4. Falling into poverty does not imply that all entitlements are lost at the same time. Regress is a process whereby losing one entitlement will result in the loss of others. Cernea (1995) has described such a process, starting with landlessness. When there is a parallel loss of access to common property and no access to an income-earning job, homelessness may result. This leads to marginalisation, food

insecurity and increased morbidity, which results in both social disarticulation and a short life. Progress out of poverty, on the other hand, can also start with basic entitlements. When these ensure a minimum level of living within a community and guarantee that the fruits of one's efforts to broaden the entitlements or to use them in a more productive manner without diminishing their future capacity stay with the person – or the household – concerned, that community can embark upon a path towards sustainable human development, provided that the future capacity of these entitlements will not diminish due to their present use. This was the core message of the UNCED as elaborated in the principles laid down in Agenda 21 (United Nations 1992).

5. A better yardstick than an index of income poverty is the Human Development Index, developed by UNDP. This composite index is based on four indicators: per capita income, life expectancy, literacy and school enrolment. The Human Poverty Index, also developed by UNDP, is based on four poverty indicators: access to water, birth weight, life expectancy and adult illiteracy. All these indicators are related to the Millennium Development Goals. Both indices include a measure of the level of education, but the Human Development Index also refers to changes in this level, by explicitly including enrolment data. An index of sustainable development would have to be based on indicators related to energy and the environment (natural resources, biodiversity). As an index of human deprivation the Human Poverty Index is better than the Human Development Index, because of the inclusion of water and nutrition indicators. However, there are no time series available. Half of all developing countries for which it was possible to calculate the Human Poverty Index (47 out of 94) had an index of over 30, meaning that more than 30 per cent of the people of these countries are considered to be very poor. India, South Africa and Nigeria all had an index of slightly more than 30. For most African countries the index lies between 40 and 60 (UNDP 2003: 245–247).

6. Even the trend based on the dollar a day is less promising than politicians are made to believe. Of 67 countries for which data are available, 37 experienced an increase in poverty in the 1990s, measured according to this shallow yardstick (UNDP 2003: 41). Measurement using the Human Development Index reveals a similar pattern: while this index improved in most countries during the last quarter of the previous century, more than 20 countries experienced a drop in this index in the 1990s (ibid.: 40).

7. According to a rough estimate by the World Bank, 2.5 to 3.0 billion people currently live on less than two dollars a day (World Bank 2003a).

8. With a Gini coefficient of 0.66.

9. Global quintile disparity, which was 30 to 1 in 1960, rose to 60 to 1 in 1990. Disparity between the richest 5 per cent and the poorest 5 per cent was even wider. In the 1990s it was, according to an estimate by Milanovic, 114 to 1, while the richest 1 per cent received as much as the poorest 57 per cent (Milanovic 2002).

10. See Pyatt (1999), Ricupero (1999), De Gaay Fortman (2002), Opschoor (2003), Murshed (2003) and Vos (2003).

11. De Gaay Fortman introduced a useful distinction between "growth pro-poor" (growth with redistribution to such an extent that together it will lead to less inequality and poverty reduction) and "pro-poor growth", based on income generation by the poor themselves (De Gaay Fortman 2002). Pro-poor growth does not only rely on redistribution after growth, but also on direct increases in the productivity and incomes of the poor themselves. That means pro-poor growth goes beyond a combination of safety nets and adequate expenditure on social sectors and public services. Pro-poor growth implies all of this, but in addition changes the structure of production and employment as well as the allocation and use of capital, land and technology that would directly increase the income-earning capacity of the poor. This would come close to the distinction between redistribution (of income) after and (of assets) before growth. The latter was strongly advocated by Adelman (1979: 165), 'For equitable growth, at each stage of the development process, access to the critical factor of production should be redistributed before its productivity is improved.' Defining *growth pro-poor* as *growth-cum-redistribution* and *pro-poor growth* as *redistribution before growth* would make more sense than distinguishing between high, moderate and less inequality in conjunction with high or (s)low growth, leading to less or more poverty alleviation, whereby everything is called pro-poor, but some approaches "highly pro-poor" and others just "pro-poor", even when inequality is worsening. As indicated by De Gaay Fortman (2003: 152), the only reason to introduce such a confusing and

euphemistic distinction is political: a stricter definition would reveal that no Asian country could be qualified as "pro-poor". Indeed, '[the] ideal strategy [which] would boost growth, reduce inequality and significantly reduce poverty… has not been seen in any of the spells recorded for Asia and the Pacific in the 1990s' (United Nations 2003: 44).

12. A rights-based approach to poverty is usually juxtaposed against the needs-based approach, which dominated the debate in the 1970s. Klein Goldewijk (2003: 182) rightly pointed out that such a juxtaposition is questionable, 'not all needs justify rights and it is not need alone which justifies a right'. However, she continued by juxtaposing a rights-based approach and a development approach: 'A rights-based approach… recognises that all people are subjects of rights, rather than objects of development.… In practice, a development approach is predominantly resource-based.… A rights-based approach… is primarily access-based' (ibid.). This seems to imply a definition of the concept of development in macro terms, whereby the development approach would have to be complemented, corrected or substituted by a rights-based approach. However, if development itself is defined in terms of people (sustainable human development) a rights-based approach would not have to be juxtaposed against a development approach. On the contrary, it would become the highest form of a development approach. However, we should beware both of artificial distinctions and of wishful conceptualising. Whatever definition is chosen, what matters is the right to take care of one's own needs, together with the right to be treated equally.

13. In nearly all developing countries income redistribution, affirmative action and demand-driven approaches have been accepted and applied more or less, but not the redistribution of assets such as land and other resources. This also applies to development policies pursued by international development institutions. In a review of World Bank policies to redress poverty amongst indigenous groups in Latin America, Van den Berg (2002) observed that the World Bank promotes all means short of a more equitable distribution of land (for instance, with the help of funds to enable small indigenous farmers to purchase land). Van den Berg concluded that the range of demands which the World Bank is prepared to consider is limited, aiming at best at achieving equal opportunities, not at overcoming social inequality.

14. The 1 per cent aid target was not an arbitrary choice, nor a figure solely resulting from an enquiry into the willingness of rich countries to provide aid. The 1 per cent figure resulted from model calculations: an increase of the growth rate of the developing countries to the level of 5 per cent would require a specific increase in the investments in developing countries, on the basis of specific assumptions concerning investment/growth relations. Part of these could be financed out of domestic savings, the remainder had to be provided from outside. That remainder could be calculated as a percentage of the total income of the wealthy nations. That calculation resulted in the 1 per cent figure. So the aid target was a political target to the extent that it implied a promise, a commitment. But it was also a technical-economic target. It resulted from development theory: the two-gap theory (trade gap and savings gap) concerning foreign exchange limitations to a policy to step up productive investments. It was also a sophisticated target in political terms. The theory implied that the faster the income of developing countries increased, the sooner their capacity to raise domestic savings would grow and, so, the quicker their need for additional foreign aid to complement these savings in order to sustain the growth rate. This meant that more aid in the early years would result in less demand for aid later on. This was politically attractive for donor countries. It would be less costly, given specific discount rates. The accumulated aid over a longer period would have to be smaller if donors were generous in the beginning. So the First Development Decade could be seen in more than one respect as a win-win approach.

15. During the First Development Decade, difficult discussions emerged with regard to the commitments of the donor countries. During UNCTAD I and II, in 1964 and 1968, aid commitments were modified. The preparations for the Second Development Decade resulted in a controversy between two approaches: the World Bank-based Pearson Committee, less ambitious, and the UN-based Committee for Development Planning led by Jan Tinbergen, aiming at a higher growth target and higher levels of official development assistance. The dispute circled around a number of issues, which have remained dominant. First is the growth objective: should a more or a less ambitious target be selected? Second is whether the resulting necessary foreign exchange should be brought together through official channels or whether a part should be left to the private sector, even if in a free market system no firm commitments could be expected from the private sector. In addition to this a change in the statistical base was agreed. Instead of net national income (NNI), gross domestic product (GDP) was chosen,

which was roughly – too roughly – estimated to be about 30 per cent higher. So, the aid target – 1 per cent of NNI – was recalculated (and rounded off downwards) at 0.7 per cent of the GDP of the donor countries. All this resulted in less firm aid commitments for the Second Development Decade (1971–1980). The first review and appraisal of performance during that decade, in 1972, made clear that the targets would not be met. That also became clear during UNCTAD III in Santiago. Shortly afterwards the oil crisis broke out, resulting in an instant fourfold increase in world oil prices. Developing countries importing oil were also hit, but oil-exporting developing countries were able to keep the ranks closed within the group of developing countries (G-77) by promising compensation. Moreover, the developing countries as a whole were disillusioned by the meagre results of international trade negotiations and by the low aid performance. Therefore they stuck together during UN negotiations on the so-called New International Economic Order in 1974 and 1975. The negotiations resulted in consensus on an action programme aimed at establishing a new order, including the reconfirmation of aid commitments. However, all this remained on paper only (OPEC's promise also resulted in fewer disbursements than had been expected. Efforts to link oil revenue with investments in food security in poor countries did not materialise. Instead, oil revenue with investments in food security in poor countries did not materialise. Instead, oil revenue became an important source of finance for arms imports). The review and appraisal of the Second Development Decade became a purely statistical exercise. The Third Development Decade (1981–1990) did not draw political attention. The negotiations only served the needs of the UN bureaucracy and the diplomats. The outcome was a strategy without teeth, without any political commitment. Hardly anybody outside New York was aware that, within the framework of the United Nations, the 1990s were proclaimed the Fourth Development Decade.

16. Some goals have been achieved wholly (the eradication of smallpox). Others were achieved partly (the immunisation of children) or better in some parts of the world than in others, whereby in global terms at least there was substantial improvement (primary school enrolment and nutrition). Most others failed completely. This has led to criticism of goal setting as such: goals are seen as inflexible, over-ambitious, too precise, too detailed and therefore deceptive. Against this criticism the authors of the *Human Development Report 2003* argue that 'whether the numerical target of a global goal was achieved is an important but inadequate measure of success, because it does not indicate whether setting the goal made a difference. In many cases enormous progress has been made even though numerical targets have not been reached… Thus global goals can raise ambitions and spur efforts… They are intended to mobilise action' (UNDP 2003: 30–31). Also Jolly (2003), in an overview of all relevant goals set and agreed by the world community in the second half of the previous century, is rather generous when claiming that on the whole many goals have been achieved in a considerable number of countries by the target date or soon thereafter. He pleads for a shift in the emphasis, from achievement or non-achievement to the extent to which the goals have functioned and for an analysis of the specific reasons why specific goals in specific countries have not been achieved: internal reasons (economic constraints, lack of political commitment, violent conflict) or external (the international economic environment or lack of international support). Such an analysis would be useful. However, goals and targets are more than catalysts in a process of awareness building. They are not only meant to mobilise action by others. They imply also a commitment to act by the goal setters themselves. All goals concerned were agreed worldwide and embedded in global political declarations, after full scrutiny – 'carefully constructed, word by word, often syllable by syllable' (ibid.: 4) – and hardly any of these implied a 100 per cent eradication of a deficit in a relatively short period. It was always a decrease of an incidence with a specific percentage, say, a third, within a period of about 15 years. The targets always concerned basic issues. Their achievement was considered fundamental for human welfare. Their non-achievement is failure. The reasons should be analysed, but such an analysis would not clear in advance political leaders from their accountability.

17. The World Employment Programme was under the leadership of Louis Emmerij, former Rector of ISS.

18. With thanks to Richard Jolly and Kitty van der Heijden.

19. See for the development compact UNDP (2003: 15–25). See for proposed elements of such a compact concerning debt, trade, technology and AIDS policy, respectively, pages 152–153, 154–156, 157–159 and 158–159. Privatisation and its limitations in various sectors are discussed on pages 111–121, decentralisation on pages 135–140, social mobilisation on pages 140–143. Desirable policies

concerning food, education, health and water and sanitation are discussed in Chapter 4 (UNDP 2003: 85–110).

20. Richard Jolly, though rather kind in his judgement concerning the non-fulfilment of the various development targets of the decades behind us, is quite outspoken in his verdict on the Bretton Woods institutions, whose 'single-minded focus... [and]... narrow view... often diverted attention from the social dimensions of adjustment, set back progress in the social sectors, and worked against the achievement of global goals in education, health, water and nutrition' (Jolly 2003: 4).

21. Fairness requires that I note I was a member of this task force.

22. Moreover, as recently indicated, for half of the world's poor, meeting only the one dollar a day target would require a redistribution of at least this figure (Dietz 2003: 14). This should mostly be additional to current aid flows, because reallocation would also harm the capacity to meet the other targets.

REFERENCES

Adelman, I. (1979) 'Redistribution before growth: A strategy for developing countries', in *Development of Societies: The Next Twenty-Five Years*. Proceedings of the Institute of Social Studies 25th Anniversary Conference, ISS, pp. 160–76. The Hague: Martinus Nijhoff.

Baker, Ramond W. and Jennifer Nordin (1999) 'A 150-to-1 ratio is far too lopsided for comfort', *International Herald Tribune*, 5 February.

Berg, Maarten, H.J. van den (2002) 'Mainstreaming ethnodevelopment: Poverty and ethnicity in World Bank policy'. Paper presented at ISS, The Hague, 12 November. Amsterdam: RISQ.

Camdessus, Michel, et al. (2003) 'Financing water for all'. Report of the World Panel on Financing Water Infrastructure. Marseilles: World Water Council and Global Water Partnership.

Cernea, Michael M. (1995) 'Understanding and preventing impoverishment from displacement', *Journal of Refugee Studies* 8 (3): 251–252.

Chenery, Hollis B. (1967) 'The effectiveness of foreign assistance', in H.B. Chenery et al. *Towards a Strategy for Development Co-operation*, pp. 9–18. Rotterdam: Universitaire Pers Rotterdam.

Chenery, Hollis B. and Nicholas G. Carter (1973) 'Foreign assistance and development performance, 1960–1970', *American Economic Review* LXIII (2): 459–468.

Dietz, Ton (2003) 'The global social problem'. Inaugural Lecture. Utrecht: University of Utrecht, CERES.

Dutch Ministry of Foreign Affairs (1991) *A World of Difference. A New Framework for Development Cooperation in the 1990s*. The Hague: Directorate General for International Cooperation, Ministry of Foreign Affairs.

Friedman, Thomas L. (1999) *The Lexus and the Olive Tree. Understanding Globalisation.* New York: Farrar, Straus, Giroux.

Brandt, Willy et al. (1980) 'North-South: A programme for survival'. Report of the Independent Commission on International Development Issues. London, Sydney: Pan Books.

Brandt, Willy et al. (1983) 'Common crisis: North-South co-operation for world recovery'. The Brandt Commission. London: Pan Books.

Corea, Gamani (1985) 'The challenge of change', in *The History of UNCTAD 1964–1984*. United Nations publication E.85.II.D.6. New York: United Nations.

Gaay Fortman, Bas de (2002) 'Power and protection, productivism and the poor'. Valedictory Address, Institute of Social Studies, The Hague.

Gaay Fortman, Bas de (2003) 'Persistent poverty and inequality in an era of globalisation: Opportunities and limitations of a rights approach', in Paul van Seters, Bas de Gaay Fortman and Arie de Ruyter (Eds), *Globalisation and Its New Divides: Malcontents, Recipes and Reform*, pp. 149–166. Amsterdam: Dutch University Press.

Hueting, R. (1974) *Nieuwe Schaarste en Economische Groei*. Amsterdam/Brussel: Agon Elsevier.

Hueting, R. (2001) 'Environmental valuation and sustainable national income accounting' (with Bart de Boer) and 'Three persistent myths in the environmental debate', in Ekko C. van Ierland, Jan van der Straaten and Herman R. J. Vollebergh (Eds), *Economic Growth and Valuation of the Environment: A Debate*, pp. 17–77 and 78–89. Cheltenham Glos: Edgar Elgar.

Jolly, Richard (2003) 'Global goals: The UN experience'. Background paper for the *Human Development Report 2003*. New York: United Nations.

Klein Goldewijk, Berma (2002) Dignity and Human Rights: The Implementation of Economic, Social and Cultural Rights. Ardsley, New York: Transnational Publishers.

Klein Goldewijk, Berma (2003) 'Justice for just us? Access to justice and legal empowerment', in Paul van Seters, Bas de Gaay Fortman and Arie de Ruyter (Eds) *Globalisation and Its New Divides: Malcontents, Recipes and Reform.* Amsterdam: Dutch University Press.

Milanovic, Branko (2002) 'True world income distribution, 1988 and 1993: First calculation based on household surveys alone', *Economic Journal* 112 (476): 51–92.

Milanovic, Branko (2003) *Worlds Apart: Global and International Inequality, 1950–2000.* Washington, D.C.: World Bank.

Murshed, S. Mansoob (2003) 'The decline of the development contract and the development of violent internal conflict'. Inaugural Lecture Prins Claus Chair, Utrecht University, Utrecht.

OECD (1996) 'Shaping the 21st Century: The contribution of development cooperation'. Thirty-fourth High-Level Meeting of the Development Assistance Committee, 6–7 May, Paris.

Ohlin, Goran (1966) *Foreign Aid Policies Reconsidered.* Paris: OECD Development Centre.

Opschoor, Hans (2002) 'Sustainable human development and the North-South dialogue', in B. K. Michael Darkoh and Apollo Rwomire (Eds), *Human Impact on Environmental and Sustainable Development in Africa*, pp. 495–518. Oxford: Ashgate.

Opschoor, Hans (2003) 'Sustainability: A robust bridge over intertemporal divides?' in Paul van Seters, Bas de Gaay Fortman and Arie de Ruyter (Eds) *Globalisation and Its New Divides: Malcontents, Recipes and Reform*, pp. 70–100. Amsterdam: Dutch University Press.

Pronk, Jan (1986) 'Markt- en machtsmechanismen na Breton Woods', in W.P.S. Fortuyn (Ed.) *Voor en Tegen de Markt*, pp. 123–143. Deventer: Kluwer.

Pronk, Jan (1987) 'Pleidooi voor een mondiale publieke sector', in L.W. Nauta and J.P. Koenis (Eds) *Een Toekomst voor het Socialisme?* pp. 112–128. Amsterdam: Van Gennep.

Pronk, Jan (1994) *De Kritische Grens. Beschouwingen over Tweespalt en Orde.* Amsterdam: Prometheus.

Pronk, Jan (1997) 'Development cooperation: Out of date?'. Public Address ISS, 12 November 1992, The Hague.

Pronk, Jan (2000) 'Globalisation: A developmental approach', in Jan Nederveen Pieterse (Ed.) *Global Futures: Shaping Globalisation*, pp. 40–52. London, New York: Zed Books.

Pronk, Jan (2001) 'Aid as a catalyst', *Development and Change* 32 (4): 611–629.

Pronk, Jan (2003a) 'Aid as a catalyst: A rejoinder', *Development and Change* 34 (3): 383–400.

Pronk, Jan (2003b) 'Security and sustainability', in Paul van Seters, Bas de Gaay Fortman and Arie de Ruyter (Eds) *Globalisation and Its New Divides: Malcontents, Recipes and Reform*, pp. 25–34. Amsterdam: Dutch University Press.

Pyatt, Graham (1999) 'Poverty versus the poor'. Dies Natalis Address, ISS, The Hague.

Reddy, Sanjay and Thomas Pogge (2002) 'How not to count the poor', Columbia University, Department of Economics, New York, March.

Ricupero, Rubens (1999) 'The challenge of development in the knowledge era'. Public address. ISS, The Hague.

Sachs, Jeffrey et al. (2001) *Macroeconomics and Health: Investing in Health for Economic Development.* Commission on Macroeconomics and Health. Geneva: World Health Organisation.

Schrijver, Nico (2001) 'On the eve of Rio + 10: Development – the neglected dimension in the international law of sustainable development'. Dies Natalis Address, ISS, The Hague.

Schrijver, Nico (2003) 'De verankering van de betekenis van duurzame ontwikkeling in het internationale recht', in N.J. Schrijver and E. Hey, *Volkenrecht en Duurzame Ontwikkeling*, pp. 5–92. Mededelingen van de Nederlandse Vereniging voor Internationaal Recht, Nr. 127. The Hague: T.M.C. Asser Press.

Seters, Paul van, Bas de Gaay Fortman and Arie de Ruyter (Eds) (2003) *Globalisation and Its New Divides: Malcontents, Recipes and Reform.* Amsterdam: Dutch University Press.

Sideri, Sandro (1999) 'Globalisation, the role of the state and human rights', Valedictory Lecture ISS, The Hague.

Tinbergen, Jan et al. (1962) *Shaping the World Economy. Suggestions for an International Economic Policy.* New York: The Twentieth Century Fund, Inc.

Tinbergen, Jan (1990) *World Security and Equity.* Aldershot: Edward Elgar.

United Nations (1961) General Assembly, Sixteenth Session, Resolution 1715 (XVI), 19 December. New York: United Nations.

United Nations (1962) *The United Nations Development Decade. Proposals for Action.* Document E/3613. New York: United Nations.

United Nations (1966) *International Covenant on Economic, Social and Cultural Rights.* General Assembly, Twenty-First Session, Resolution 2200 A (XXI), 16 December. New York: United Nations.

United Nations (1992) *Rio Declaration on Environment and Development. Agenda 21.* June. New York: United Nations.

United Nations (2000a) General Assembly, Fifty-Fourth Session, Document A/54/2000, 27 March. New York: United Nations.

United Nations (2000b) *Millennium Declaration,* General Assembly, Fifty-Fifth Session, Document A/RES/55/2, 18 September. New York: United Nations.

United Nations (2001) 'Report of the High-Level Panel on Financing for Development', General Assembly, 28 June. New York: United Nations.

United Nations (2002a) *World Summit on Sustainable Development: Plan of Implementation,* Johannesburg, 4 September. New York: United Nations.

United Nations (2002b) 'A framework for action on water and sanitation'. WEHAB Working Group, WSSD, August. New York: United Nations.

United Nations (2002c) 'A framework for action on health and the environment'. WEHAB Working Group, WSSD, August. New York: United Nations.

United Nations (2003) *'Promoting the Millennium Development Goals in Asia and the Pacific. Meeting the Challenges of Poverty reduction,* ST/ESCAP/2253, Bangkok.

UN-Habitat (2003) *The Challenge of Slums. Global Report on Human Settlements 2003.* London: United Nations Human Settlements Programme & Earthscan.

UNDP (2003) *Human Development Report 2003.* New York: Oxford University Press.

Vandemoortele, Jan (2003) 'Are the MDGs feasible?', *Development Policy Journal* 3 (April): 1–21.

Vos, R.P. (2003) 'Macro-economisch beleid en armoedebestrijding'. Inaugural Address, Free University, Amsterdam.

Water Supply and Sanitation Collaborative Council (2002) *Kyoto: The Agenda Has Changed.* Geneva: WSSCC.

World Health Organisation (1997) *Health: The Courage to Care.* Task Force on Health in Development, Chaired by Brainford M. Tait. Geneva: WHO.

World Bank (2003a) *Global Economic Progress and the Developing Countries 2002.* Washington, D.C.: World Bank.

World Bank (2003b) *Sustainable Development in a Dynamic World: World Development Report 2003.* Washington/Oxford: Oxford University Press.

CHAPTER 3

CAN IMPROVED HUMAN DEVELOPMENT POLICIES BREAK THE CYCLE OF POVERTY?

JOZEF M. RITZEN[1]

1. INTRODUCTION

If development is complicated, then development cooperation is at least as complex. This chapter considers a number of lessons we have learned about development cooperation. This reflection uses the Millennium Development Goals as benchmarks since they reflect an unprecedented agreement by the development community about the aims of development cooperation (Box 1). The targets and indicators of the Millennium Development Goals range from income poverty (living on less than one dollar a day), to child mortality and primary education, to gender equality, maternal mortality and safe water and sanitation. The goals provide measures by which the development community can monitor its progress towards development in general and to the different dimensions of poverty reduction in particular.

The focus here is on human development: one of the dimensions of development and a prerequisite for overall development. This world has a rich set of "social experiments" in terms of individual countries' policies on human development. These provide evidence of which policies work and which are unlikely to work. This chapter addresses the question of whether improved development policies can break the cycle of poverty. We know what good practice is. For example, money for education should reach the schools, teachers should be paid if they teach (and ghost teachers should not be paid), parents should be able to release their children from the chores of caring for the household, like carrying water or herding cattle, if this prevents the children from attending school. Why then is this level of good practice – no rocket science – not reached everywhere? The reason is because good policies need to be implemented, embedded and then safeguarded against the interests of elites to maintain the status quo.

The introduction of good policies is a matter of political economy. Personally, I have advanced, time and again, the role of transparency and accountability as a mighty force

Max Spoor (ed.), Globalisation, Poverty and Conflict, 35–45.

Box 1. *Millennium Development Goals*

Each goal is to be achieved by 2015, compared to 1990 levels.

1. Eradicate extreme poverty and hunger.
   - Halve the proportion of people living on less than one dollar a day.
   - Halve the proportion of people who suffer from hunger.

2. Achieve universal primary education.
   - Ensure that boys and girls alike complete primary schooling.

3. Promote gender equality and empower women.
   - Eliminate gender disparity at all levels of education.

4. Reduce child mortality.
   - Reduce by two-thirds the under-five mortality rate.

5. Improve maternal health.
   - Combat HIV/AIDS, malaria and other diseases.

6. Combat HIV/AIDS, malaria and other diseases.
   - Reverse the spread of HIV/AIDS.

7. Ensure environmental sustainability.
   - Integrate sustainable development into country policies and reverse loss of environmental resources.
   - Halve the proportion of people without access to potable water.
   - Significantly improve the lives of at least 100 million slum dwellers.

8. Develop a global partnership for development.
   - Raise official development assistance.
   - Expand market access.
   - Encourage debt sustainability.

---

behind the implementation of good policies. Show the poor that the money meant for the poor also reaches them. And – if that is not the case – give them voice in changing the situation. This simple recipe (also now advanced in the *World Development Report 2004*) requires a firm policy line, namely, to create and maintain transparency.[2]

This reasoning boils down to a general conclusion for development cooperation. Development cooperation should contribute to increase the room for developing-country governments to manoeuvre to implement "good policies", in particular, with regard to transparency.

While this chapter was written to debate the question of why poverty still endures, an alternative and less pessimistic question could be posed: "Will poverty endure?" Evidence from the past shows that persistent poverty in a country is not a fate that will forever remain unchanged. Let us briefly examine the worldwide trends in poverty in the recent past and consider the near future.

2. POVERTY AND DEVELOPMENT TRENDS

The number of poor people (those living on less than one dollar a day) has hardly decreased in the past decade, and continues to hover around 1.2 billion, or more than one in every five inhabitants of the world. So poverty certainly continues to be entrenched in many poor countries. At the same time we have seen some significant changes. For example, over the past thirty to forty years East Asia has experienced the fastest reduction in poverty in human history. Also, the trend is to some extent encouraging in South Asia.

For the near future, of the six regions of the world, all except Africa are expected to achieve the Millennium Development Goal of reducing income poverty by half between 1990 and 2015, if current growth rates continue (World Bank 2003). Africa is the exception, however. The continent could reach the goal only if it were to double per capita growth forecasts.

Many lessons can be learned from the past. Some countries have done well and their populations have prospered; others have been unable to fulfil the expectations and desires of their people and the international community. The difference between the two groups seems to lie in their *governance and the quality of their policies*. An extreme expression of lack of governance and social cohesion is internal civil conflict or civil war. The conclusion that civil conflict and war are a major factor in the endurance of poverty is thus not a surprising one. Less extreme, but equally frustrating for improving the position of poor people, is the failure of the rule of law. Such failure drives out the basic economic incentives that lead to better living conditions. People seek shelter for their savings outside their country, rather then investing in their own environment. Worldwide this is a common phenomenon. To exemplify, 40 per cent of Africa's private wealth (US $360 billion) is held outside the region, compared to 6 per cent of East Asia's and 10 per cent of Latin America's (World Bank 2003). More broadly, the rule of law and the quality of governance affect the investment climate of a country and the resulting net inflow of capital (the difference between the local capital seeking shelter elsewhere and the foreign capital entering the country).

Investment climate surveys show the tremendous impact of governance on operating conditions for industries. In a recent survey in India, it turned out that on all accounts (costs of capital, costs of utilities, costs of taxation) business costs were higher in India than in China, often due to corruption. This may be one explanation for the growth differential between India and China or for the difference in foreign domestic investment between the countries (a factor of around ten in the past ten years).

Governance and the rule of law not only contribute to a net inflow or outflow of physical capital. Brain drain is also often the result of a push out of countries due to corruption, unemployment and adverse working conditions, combined with the pull of the international labour market. Sound macroeconomic policies are a necessary condition for development. Fortunately, in the 1990s inflation rates have been cut in half and, more in general, macro conditions in developing countries have improved. In Africa, the New Partnership for Africa's Development (NEPAD) has taken a stronger stand on corruption than ever before.

External conditions count as well. To focus on just one point, rich countries may preach about the power of trade to help developing countries grow, yet these same countries have erected powerful trade barriers in agricultural goods against their poorer trading partners in Africa and elsewhere. Development cooperation can be a powerful tool for development. However, we know it is more effective when coordinated and harmonised and if money for cooperation is not tied to a specific purchases of foreign consultants or foreign goods. Yet only a small part of the US $50 billion of official development assistance (ODA) is harmonised or untied.

These are but a few, but important, issues in development. They focus on developing countries, but could easily be translated to the developed countries. In the developed world, the quality of governance and the rule of law also play vital roles in economic growth (compare the Worldcom and Enron scandals).

3. HUMAN DEVELOPMENT

The title of this chapter implies that human development and economic development can be untied. Moreover, it suggests that human development can be proactive (rather than merely dovetailing economic development) and that such a course could raise the prospects for poverty reduction. Let us consider these two steps one by one. First, human development often as much follows economic development as it leads. However, this relation is not cast in stone. Countries with the same per capita income can have widely differing levels of human development. In other words, human development can indeed be untied from economic development to some extent and be proactive towards economic development. Second, not every input for development leads to the same output in terms of actual human development that uplifts the poor. Much depends on the effectiveness of policies in reaching the poor or their children. Yet, as an example, Figure 1 is a well-established way of assessing the path of education's impact on human development and ultimately on economic development.

We may ask whether the human development objectives set out in the Millennium Development Goals will be reached. As mentioned, five of the six developing regions are on course to reach the Millennium Development Goal on income poverty. Along with income poverty, the millennium goals reflect the international consensus that health, education, gender equality and a sustainable environment are all integral elements of human well-being. Yet in general progress in human development has been conspicuously slow, even though developing countries have achieved major gains in life expectancy and literacy.

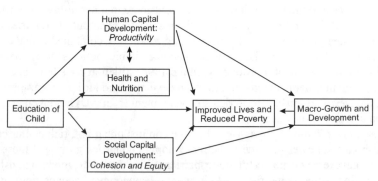

Figure 1. *Education and Macro Growth*

Source: Adapted from World Bank (1999).

Growth is a necessary condition for each type of poverty reduction. But economic growth is not enough (see Table 1). This is particularly true for the health and education dimensions of poverty reduction. The table shows that only three of the six developing regions is set to reach the primary school enrolment goal. If one chooses (more properly) primary school completion as the goal, only two regions are on course to reach the goal, assuming the continuation of present growth rates. For infant mortality, prospects for reaching the Millennium Development Goal are even worse: if present growth rates continue, none of the six developing regions will reach the goal.

Since growth alone is insufficient, many observers call for increased financial resources for health and education in order to accelerate progress towards those goals. Under current patterns of allocation and use, however, additional resources may also fail to improve outcomes in these sectors. In fact, comparisons among countries suggest that public expenditures have no significant effect on health and education outcomes (covering five of the Millennium Development Goals), when countries' per capita GDP and a small number of socioeconomic variables are taken into account (Filmer et al. 1999, 2000).

Financing human development is a problem. The developed world recognised this at the Monterrey conference in 2001 and promised to increase financing by at least 20 per cent. However, with current patterns of public expenditure productivity, the additional financing promised in Monterrey is unlikely to make a difference in the rate at which the human development Millennium Development Goals are being achieved.

4. THE QUALITY OF HUMAN DEVELOPMENT POLICIES

The limited impact of policies on human development in many countries can be explained by several factors. *First*, governments may be spending on the wrong goods or the wrong people. For instance, they may be spending on the military rather than on social services. Governments should intervene only if there is a market failure or

Table 1. *Growth Is Not Enough*

	Poverty Headcount		Primary Education Enrolment		Infant Mortality per 1,000	
	Target	2015 Growth Alone	Target	2015 Growth Alone	Target	2015 Growth Alone
East Asia	14	3	100	100	14	33
Europe and Central Asia	1	1	100	100	9	22
Latin America and Caribbean	8	7	100	100	14	30
Middle East and North Africa	21	1	100	92	20	46
South Asia	22	18	100	87	29	70
Africa*	24	40	100	64	33	87

Note: * More recent figures indicate that South Asia might well reach the income poverty Millennium Development Goal.
Source: World Bank (2001).

redistributive rationale. Note that this is an argument for *government financing* of these activities – not necessarily *for* government *provision*. In India where 80 per cent of health expenditures are private, the poor often bypass free, public clinics to use fee-based private services because the latter are better equipped and medical personnel better trained (Peters et al. 2003). Most studies of public spending in health and education show that the benefits accrue largely to the rich and middle class. The share going to the poorest 20 per cent of the population is often less than 20 per cent (see Table 2).

In several countries in Africa, budget management institutions themselves contribute to inefficiency and wastage in public spending and poor service delivery. In Zambia, for example, the budgeting system for monthly cash flows to sectors was effectively negotiated, which meant that those sub-sectors with "voice" within the public financing circles had access to resources, regardless of whether these delivered services to the poor. Recent budgetary reforms in Zambia and elsewhere, supported by the international community, have extended the planning horizon for cash budgeting and linked budgets to agreed medium-term expenditure frameworks which are publicly known and whose execution is tracked.

A *second* factor limiting the impact of policies on human development is that even when governments spend on the right goods or the right people, the money fails to reach the frontline service provider. The provision of public services like health and education involves substantial non-wage recurrent expenditures, such as textbooks, chalk, medicines and syringes. Yet these are the very expenditures that "leak" from initial central government allocations to service provider or delivery agency. Also, within recurrent expenditures in education and health service sectors, the lion's share goes to wages and salaries. With their political power, teachers and doctors are able to protect their incomes when there is pressure for budget cuts. The only thing left to cut, therefore, is non-wage operations and maintenance expenditures.

Systematic comparative data, when available, can support public feedback; it is difficult to ignore and can spark public action. This happened in Uganda as a result of a public expenditure tracking survey in the education sector. This case demonstrated that the political power of teachers' unions or the local (bureaucratic or political) capture of

Table 2. *Benefit Incidence of Public Spending*

Country/Year	Health		Education	
	Poorest 20% of Population	*Richest 20% of Population*	*Poorest 20% of Population*	*Richest 20% of Population*
Côte d'Ivoire (1995)	11	32	13	35
Ghana (1992)	12	33	16	21
Guinea (1994)	4	48	5	44
Kenya (1992)	14	24	17	21
Madagascar (1993)	12	30	8	41
South Africa (1994)	16	17	14	35
Tanzania (1992/93)	17	29	14	37

Source: Castro-Leal et al. (2000).

public funds can be countered if citizens have the information and "voice" to lobby for a better mix of inputs or for making sure that budget allocations are actually received. A study of Uganda in the early 1990s showed that only 13 per cent of non-wage recurrent expenditures for primary education actually reached the primary schools (Reinikka et al. 2001). These findings are comparable to those of similar studies in Ghana and Tanzania. As evidence of the degree of leakage became public in Uganda, the central government enacted a number of changes: it began publishing the monthly transfers of public funds in the district newspapers, broadcasting information on the transfers on radio and requiring primary schools to post information on inflows of funds. The objective of the "information campaign" was to promote transparency and increase public-sector accountability. The central government signalled that, as a result of the reforms, it had more information on local government actions than previously was the case. Further, by giving citizens access to information on the grants for primary schools, they empowered both schools and citizens to monitor and challenge abuses of the system. Initial assessments of these reforms a few years later showed dramatic improvement in the flow of non-wage recurrent funds to schools (ibid. 2001)

Third, even when the money reaches the primary school or health clinic, the incentives to provide the service are often weak. Indeed, many of the problems confronting service delivery have to do with the incentives facing service providers. First, these services require skilled personnel, such as qualified doctors or trained teachers. Not only are many countries short of these skills, but the pay and working conditions, especially in remote rural areas, makes it difficult to get qualified people to work there. The staff vacancy rate for clinics in Kalimantan, Indonesia is almost 80 per cent, whereas it is less than 2 per cent in Bali or Jakarta (Chomitz et al. 1998).

Effective service delivery involves more than just being able to hire skilled staff. It requires that employees show up for work and perform their duties in a reasonable fashion. This is probably the biggest constraint to making services work for poor people. Since the delivery of the service is poorly monitored (if at all), service providers do not feel accountable. They are typically employees of the central government, which issues pay checks regardless of how the functions are performed. In Bangladesh, for example, the central government provides a monthly subvention equivalent to 90 per cent of salary to secondary school teachers (World Bank 2003). Yet, the subvention is not linked to the education performance of the schools. The symptoms of such a lack of performance-based financing are everywhere (absenteeism in schools, nurses hitting mothers during childbirth, drunk teachers) (PROBE 1999).

Many governments have tried to address this problem by contracting out services to NGOs or private organisations, which are then held accountable for performing the contract. In Cambodia, where 30 years of civil war had depleted the country's physical and human health infrastructure, the government introduced a programme of contracting NGOs to manage district health services. They introduced two types of contracting: "contracting out", where the NGO can hire and fire staff, set wages and procure drugs on its own; and "contracting in", where the NGO manages a district from within the Ministry of Health. In this latter construction, the NGO does not hire or fire staff (though they may transfer employees) and they obtain medicines from the government. A before and after survey of households and health facilities carried out by a third party revealed a clear pattern: districts that contracted out registered the biggest improvement, with those contracting in being next best and the control group last.

Yet the success of the contracting out arrangement can be elusive, depending on the particular sequencing and other design details of such programmes. Performance contracts initiated by governments to address problems of chronic inefficiency and poor service delivery in municipality-run water and sanitation services shared many features in Cartagena, Colombia, and Cochabamba, Bolivia. While the first contract led to significant improvements and coverage (new connections in poor areas almost doubled), the Cochabamba experiment had less favourable results. Recent dramatic events in Bolivia are related to these outcomes.

Fourth, even if services are effectively provided, households may not take advantage of them. For economic and other reasons, parents pull their children out of school or fail to take them to a clinic. This "demand-side" failure combined with supply-side failures is often responsible for the low level of public services and human development outcomes among the poor.

Government can intervene on the demand side. One example is the Female Secondary Scholarship Programme in Bangladesh. Under this programme, unmarried girls who attend secondary school and maintain passing grades receive a stipend, deposited directly into a bank account in their name (World Bank 2003). The school in turn receives an amount proportional to the number of girls enrolled. While the programme has yet to be rigorously evaluated, preliminary estimates indicate that it has had a sizeable impact on secondary school enrolment rates among girls.

Another example is the PROGRESA programme in Mexico, which has been extensively analysed. This programme gives cash payments to families provided they regularly visit a clinic and send their children to school (ibid.). Most of the evaluations show significant improvements in the health and education status of these poor families. Still unknown, however, is whether these improvements are due to the cash transfers or to the clinic visits and school attendance.

Fifth, aid modalities sometimes exacerbate weaknesses in service delivery systems. When donors provide financing for development, they tend to set requirements on how the money is to be used. This is of course part of any contract. It is perhaps less appreciated that the nature of the contract between donor and policymaker affects the delivery of services between front-line provider and clients. Specifically, the requirement by some donors that resources provided be used only for capital investments results in under-financing of recurrent expenditure, especially non-wage operations and maintenance, which, especially in the health and education sectors, strongly influences expenditure outcomes. When in Uganda the World Bank and other donors switched their assistance from specific investment projects to untied budgetary support, the recurrent budget became able to accommodate an increase in the teaching force to lower the student-teacher ratio, which had reached 100 to 1 (World Bank 2003).

A second phenomenon is donor insistence that the recipient country follow donor procedures in financial management and procurement. As a result, some countries have to comply with many different sets of procedures (typically one for each donor), exhausting their scarce administrative talent. In other cases, donors may set up their own project implementation units, which draws scarce manpower away from government, undermining the rest of the government. An important feature of the World Bank's Poverty Reduction Support Credit (PRSC) introduced in 2001, is that the World Bank and other co-financing donors follow a common set of disbursement procedures – the recipient government's own – rather than each establishing their own standards and systems. There are high risks in this approach. Everyone recognises that Uganda's own systems, for example, need to be strengthened in terms of accountability and monitoring. Nonetheless, the reform programme associated with the Uganda PRSC aims to strengthen the national systems rather than bypass them to establish enclaves of excellent accounting in a traditional project lending mode.

As we look to the dimensions of the Millennium Development Goal challenge for the international community, it is clear that development efforts must be scaled up if the Millennium Development Goals are to be met. What is required is new thinking on aid delivery mechanisms, to enable external financing to support well-structured national public budgets. Among other things, external financiers must be prepared to strengthen the effectiveness of the public budget as a whole. Countries will be unable to scale up through a traditional, narrowly defined donor projects. The PRSC is a step in the right direction in this respect.

5. ACCELERATING HUMAN DEVELOPMENT

The preceding sections have tried to capture elements of "good" and "bad" human development policies. The *World Development Report 2004* presumably will be more extensive on many of the notions mentioned. This may, first of all, be important reading for any government, whether from a developing or a developed country. Notice that the lessons are more generally applicable than just to developing countries. Developed nations also suffer some deficiencies in applying the concepts of "good policy".

A constant aim of good policy is to promote transparency and voice. Developed nations have noticed how powerful this is. Germany, for example, benefited from increased transparency provided by its Project International Student Assessment (PISA). Information from this project enables international comparisons to be made of educational achievements of 14-year-olds. The German government used the results, which showed German pupils' achievements to be among the lowest in Europe, to rally for educational reform. In the same vein, developing countries benefit from political fora which are based on transparency, with Uganda's education financing as a prime example.

Yet, experience shows that governments cannot just introduce transparency as a technical tool. Those who have an interest in the status quo will fight transparency, often on grounds that no proper comparisons can be made, that benchmarks are inaccurate and so on. This happened, for example, in the Netherlands with respect to

44 JOZEF M. RITZEN

transparency mechanisms related to school performance. It is also the reason why the Moroccan government has not (yet) been able to publish figures on school attendance in villages and neighbourhoods.

In other words, introducing transparency is as much a matter of political economy, of room for manoeuvre of governments, as any other line of policy.

6. CONCLUSION

Poverty can be reduced by relinquishing old and inadequate policies and embracing those that have been effective in improving poor people's physical and mental well-being. Questions remain about the appropriate design of such policies and programmes. Still other questions relate to the way to achieve a transition to a new governance system within a given country. Lastly we may ask how the international community can best help countries in the midst of this transition to identify and build stakeholder coalitions for change.

Poverty reduction has not yet been sufficiently achieved, not so much because we do not know what constitutes good practice, but because governments' room for manoeuvre to change policy is limited. Old policies may be inadequate for poverty reduction, but at the same time, the authorising environment for changing policies may be missing or too small. The dominant voices in society may not perceive these policy changes as holding sufficient benefit for them, and many may view them as threatening their privileges. Popular trust in government may be too weak to lend broad-based support to policy changes that might involve short-term losses for certain groups and yet deliver substantial benefits in the long run.

Increasing levels of trust is a slow process of climbing upwards, rung by rung. Trust evolves as people experience consistently fair behaviour from their key leaders and institutions. Where citizens see policy changes working to their benefit over time, they will gradually become more supportive and involved in the policy reform needed for a more rapid pace in poverty reduction. Vietnam is a case in point. That country seems to be in an upward spiral with a pace of poverty reduction that is – as far as we know – unique in history.

Development cooperation aims to create more room for manoeuvre for governments in developing countries, by focusing on ownership, but also by generating and communicating information which can give rise to a stronger platform for policy change. Such information can also lead to pressure from citizens for reform. Financial support for good policies is an additional incentive for governments to engage in policy reforms.

NOTES

1. I am most grateful for the support of Triny Haque and Santa Devarajan (both at the World Bank). They were the masters of the literature on which this contribution is based. Of course, all errors remain mine.

2. This is not an easy task, as I recall from my own experience as Minister of Education in the Netherlands. Interest groups may fight policy decisions on transparency.

REFERENCES

Castro-Leal, Florencia, Julie Dayton, Lionel Demery and Kalpana Mehra (2000) 'Public Spending on Health Care in Africa: Do the Poor Benefit?' *Bulletin of the World Health Organisation* 78 (1): 66–74.

Chomitz, Kenneth, Gunawan Setiadi, Azrul Azwar and Nusye Ismail Widiyarti (1998) 'What Do Doctors Want? Developing Incentives for Doctors to Serve in Indonesia's Rural and Remote Areas', Policy Research Working Paper No. 1888. Washington, D.C.: World Bank.

Collier, Paul and Jan Willem Gunning (1999) 'Explaining African Economic Performance', *Journal of Economic Literature* 37 (1): 64–111.

Devarajan, Shantayanan (2002) *Growth Is Not Enough*. Washington, D.C.: World Bank (In Mimeo).

Filmer, Deon, Jeffrey S. Hammer and Lant H. Pritchett (2000) 'Weak links in the chain: A diagnosis of health policy in poor countries', *World Bank Research Observer* 15 (2): 199–224.

Filmer, Deon and Lant H. Pritchett (1999) 'The effect of household wealth on educational attainment: Evidence from 35 countries', *Population and Development Review* 25 (1): 85–120.

Peters, David H., Abdo S. Yasbeck, Adam Wagstaff, G.N.V. Ramana, Lant Pritchett and Rashmi R. Sharma (2003) *Better Health Systems for India's Poor: Findings, Analysis and Options*. Washington, D.C.: World Bank.

PROBE (1999) *Public Report on Basic Education in India*. PROBE team in association with the Center for Development Economics. New Delhi: Oxford University Press.

Reinikka, Ritva and Jakob Svensson (2001) 'Explaining leakage in public funds', Policy Research Working Paper No. 2709. Washington, D.C.: World Bank.

World Bank (1999) *Education Sector Strategy*. Washington, D.C.: World Bank.

World Bank (2001) *Global Economic Prospects and the Developing Countries*. Washington, D.C.: World Bank.

World Bank (2003) *Making Services Work for Poor People: World Development Report 2004*. Oxford: Oxford University Press, published for the World Bank.

CHAPTER 4

INEQUALITY, POVERTY AND CONFLICT IN TRANSITION ECONOMIES

MAX SPOOR

1. INTRODUCTION

This chapter is not about the success of transition, a process which has profoundly transformed many societies behind the Iron Curtain in the short time-span since the late 1980s and early 1990s. Instead it focuses on how transition has influenced income distribution and social differentiation. Indeed, increased inequality, poverty and conflict have clearly been inter-linked phenomena in the transition processes that took place in Eastern Europe. The economic, social and political transformation of the countries of Central and Eastern Europe (CEE), the Baltic states and the Commonwealth of Independent States (CIS) was unprecedented in scale.[1] Moreover, this complex process led to highly differentiated outcomes. Eight of the formerly socialist countries will soon graduate to the European Union; but transition has also caused widespread human suffering in many countries and regions. Strong economic contraction in most of the transition countries during the first half of the 1990s left millions of people without jobs or sufficient income during a period of galloping inflation which eroded real wages and pensions.

After economic recovery resumed in the mid-1990s, growth was accompanied by increasing inequality and marginalisation of large sections of the population. This affected not only traditionally vulnerable groups, such as the elderly and children, but also created a new class of "working poor". Armed conflicts, mostly fought under ethnic banners, caused millions of internally displaced persons, refugees and hundreds of thousands of deaths and wounded, alongside the enormous destruction of infrastructure and losses of human capital. It is no surprise that internally displaced persons and refugees are amongst the poorest of society. Furthermore, in contrast to the urbanised transition countries in Central Europe, Ukraine and the western parts of the Russian Federation, much of South-eastern Europe (SEE) and the southern states

47

Max Spoor (ed.), Globalisation, Poverty and Conflict, 47–65.
© 2004 Kluwer Academic Publishers. Printed in the Netherlands.

of the former Soviet Union are still largely rural with an accompanying higher relative risk of poverty.

While the more successful cases of transition, China and Vietnam, are rapidly reducing their absolute and relative poverty rates, in some parts of Eastern Europe, poverty, rural poverty in particular, was still increasing towards the end of the 1990s. A new periphery seems to be emerging, formed by SEE, including parts of the former Balkans and some of the heirs of former Yugoslavia. This is comparable to the periphery of the former Soviet Union formed by the states of the Caucasus and Central Asia and including some of the conflict-ridden Russian republics (such as Dagestan, North Ossetia and Chechnya) and the Republic of Moldova. With few exceptions, these countries have large (often majority) rural populations, and poverty is mostly more acute in the rural areas.

The poor countries of the CIS are Armenia, Azerbaijan, Georgia, Kyrgyzstan, Moldova, Tajikistan and Uzbekistan. These were recently grouped in the so-called "CIS-7" initiative of the World Bank and IMF which aims to resolve their development problems of high human insecurity and poverty. A similar grouping could be formed in SEE followed possibly by similar special attention and assistance. This chapter makes a plea to set up such a SEE-7 initiative, which would include Albania, Bosnia and Herzegovina, Bulgaria, Kosovo, FYR of Macedonia, Romania and Serbia and Montenegro. Though this grouping is justified by comparable levels of (under)development, there are nonetheless some significant differences.[2]

First, Bulgaria and Romania are in a special position. Earmarked to enter the European Union in 2007 they currently benefit from a whole spectrum of pre-accession investment funds (ISPA) and strong incentives to change their institutional framework in line with the EU *acquis communitaire*. Second, in the heir states of former Yugoslavia, such as Bosnia and Herzegovina, Serbia and Montenegro and the FYR of Macedonia, there is still a high level of human insecurity caused by the most destructive and massive conflict that Europe has seen since World War II. Third, one member of the proposed SEE-7, Kosovo, is not an independent country; it is actually governed by a UN-led administration and is still an integral part of Serbia and Montenegro.

Differences are large in the CIS-7 as well. Uzbekistan has been reluctant to implement large-scale market reforms, but has done remarkably well in avoiding the contraction that characterised the rest of the CIS. Tajikistan is slowly recovering from a devastating civil war, but remains extremely poor, with Moldova a close second. Other states, such as Armenia, Azerbaijan and Georgia, still have enormous problems resulting from armed conflicts.

This chapter emphasises the dangers of developing (one or more) peripheries at Europe's borders, pointing amongst others to a growing gap between these seven CIS states and Europe and the rest of the CIS. Emergence of such new peripheries will mean continued human suffering, social unrest and resumption of conflicts. The current CIS-7 and the proposed SEE-7 initiatives should lead to economic growth that reduces inter- and intrastate inequality and thus help to avert further marginalisation of large parts of the populations of the respective countries. The current tendency in transition countries, in particular under influence of the (post) Washington Consensus, is to focus on growth *per se*, which is a necessary but insufficient condition for reducing poverty. It cannot be expected that any growth will be "good for the poor". Without broad-based growth, particularly in the sectors which employ most of the poor, growth will not "trickle down".

This analysis pays particular attention to the locational disadvantage of the rural poor. In many, though not all, of the peripheral transition countries rural poverty is more severe than urban poverty, and the relative poverty risk is higher in the countryside. The rural poor seem to be an ignored section of the population (see EBRD 2002). Only a growing agricultural sector can tackle this important problem. Agricultural sector growth can also provide a foundation for a more balanced, broad-based and sustainable development model.

This chapter proceeds in the following manner: The second section provides a brief overview of transition, using broad strokes to compare developments in the CEE (and SEE) countries with the former Soviet Union (or better, CIS, as the Baltic states rapidly moved towards the CEE grouping). A dualistic development is emerging within the groupings of countries. A number of the CEE countries is moving towards EU membership in May 2004. Others lag behind, in particular those in SEE, with some, such as the heirs of former Yugoslavia, still struggling with the aftermath of violent conflicts that caused human suffering and widespread poverty. Indeed, these form a new periphery of Europe, sometimes even defined as the New Developing Countries in Europe (NDCEs). At an even lower level of human development, as evidenced by their Human Development Index (HDI) rankings, is the group of poor countries that forms the periphery of the former Soviet Union.

The section analyses reform and performance in these countries, taking into account initial conditions and the consequences of conflict, that last having been a major determinant of relative underdevelopment. It is striking that the transition peripheries show clear signs of traditional development problems, having a large informal and grey economy, a largely backward agricultural sector (with only some modern sub-sectors), increasing rural-to-urban migration and high incidence of poverty, which is usually more acute in rural areas. As agricultural reform and agriculture-based growth have been mostly slow, this enormous potential for rural poverty reduction and more balanced and broad-based growth has yet to be sufficiently tapped.

The third section of the chapter analyses poverty, in particular that in the SEE-7 and CIS-7 groupings. It shows poverty incidence to be very high, particularly when absolute poverty is estimated, as relative poverty can easily underestimate the severity of the problem where large sections of the population have subsistence or below subsistence incomes. Rising inequality is an important phenomenon underlying poverty, sometimes at levels close to those in Latin American developing countries. At the outset of transition (and structural adjustment) it was *en vogue* to see this phenomenon as the temporary price of adjustment; it would not impede growth and would be redressed after reaching a certain developmental stage. That thesis is now increasingly being questioned, and the investment-led "growth with equity" strategy is proposed instead as a necessary (and sufficient) condition for sustainable growth and poverty reduction (McKinley 2001, Chang 2003).

It will become clear that while all of the countries analysed have resumed positive growth, in some cases already in the mid-1990s, poverty has been increasing for a number of years. Though growth trends have been positive since the recovery from the 1998 Russian crisis, growth is seldom in the sectors that employ the "working poor". Furthermore, there is insufficient access to education and health services, and the initial advantages of the socialist heritage, namely a high level of human development, will rapidly erode if this is not corrected. Some positive exceptions can be identified in terms of growth with poverty reduction. A strong decrease in poverty rates (albeit from very high levels) in the rural areas of Albania and Kyrgyzstan was clearly caused by agricultural sector recovery and growth. Rural poverty remains a serious threat to sustainable growth however, and to blame for this is insufficient attention to the agricultural sector, through land reform, institution building and access to markets (see Spoor 2003).

The fourth and final section of this chapter discusses the previous and current transition strategies being implemented in most countries of the SEE-7 and CIS-7, including the implementation of Poverty Reduction Strategy Papers (PRSPs). It calls for renewed attention to the dangers of forming one or more peripheries in Europe and the CIS, with high incidence of poverty and human insecurity, social instability and therefore future conflicts. This could possibly be counteracted by adding to the existing CIS-7 initiative a parallel SEE-7 programme through which regional coordination of assistance, economic cooperation and trade could be promoted. Quite a lot of donor coordination (WB, IMF, EU, UNDP) is already undertaken in the regions, in particular as there are huge UN-led peacekeeping operations active. However, a more strongly based regional approach is needed to bridge the growing gap between the CEE (plus Croatia and the Baltic states) and the SEE-7.

Returning to the main line of argument, this chapter contends that the current growth strategies will be unable to reduce poverty sufficiently and under them high inequality and poverty levels will hamper sustainable growth. Instead, broad-based growth, based on investment in (human) capital formation is called for in order to reach the goal of growth with poverty reduction. In the peripheral countries, where poverty is largely a rural phenomenon, this calls for an agricultural (and rural non-farm economy) growth strategy, which can positively feed back into overall macroeconomic growth and linkages with other economic sectors. Finally, such a growth strategy can be successful only if the major (ethnic and territorial) conflicts are resolved in a sustainable manner, which is not yet the case.

2. GROWTH WITH INEQUALITY: ACCESSION COUNTRIES AND EMERGING PERIPHERIES

Transition is currently taking place in three large blocks of formerly socialist countries: Central and South-eastern Europe (CEE and SEE) and the Baltic states of the former Soviet Union; the Commonwealth of Independent States (CIS); and China, Vietnam and Mongolia (the Asian transition economies). The initial conditions were quite different among these blocks (World Bank 2000, Swinnen 2003: 31), while also the policies of transformation differed widely (Kolodko 2000). At the outset of transition in 1989 the CEE, SEE and CIS (in 1991) had very different levels of human development with in the latter two groupings large differences not only between the countries but also within them.[3]

After *Die Wende* several countries of CEE, such as Poland, Czecho-Slovakia (soon after to become the Czech Republic and Slovakia), Hungary and the Baltic states, Estonia, Latvia and Lithuania, moved rapidly with the introduction of market reforms. Poland, with its "shock therapy"-type strategy, was the most far-reaching in this respect (Kolodko 2000). In SEE only Albania followed such a strategy, albeit within a much weaker institutional framework than in the CEE countries. All suffered a steep economic decline (in particular the Baltic states, partly caused by the break in highly dependent economic relations with the Russian Federation), but CEE and the Baltic states soon showed economic recovery.

From the start substantial differences were evident in this process (Table 1). The heirs of the former Yugoslavia that were heavily affected by the 1992–95 armed conflict (Bosnia and Herzegovina, Serbia and Montenegro, and Croatia) saw their economies

Table 1. Growth in Real GDP in Central and Eastern Europe and the CIS

	1990	1991	1992	1993	1994	1995	1996	1997	1998	1999	2000	2001	2002	GDP#
SEE-7														
Albania	−10.0	−28.0	−7.2	9.6	8.3	13.3	9.1	−7.0	8.0	7.3	7.8	6.5	6.5	116
Bosnia & Herzegovina	−23.2	−12.1	−30.0	−40.0	−40.0	20.8	86.0	37.0	9.9	10.6	4.5	2.3	3.0	70
Bulgaria	−9.1	−11.7	−7.3	−1.5	1.8	2.9	−9.4	−5.6	4.0	2.3	5.4	4.0	4.0	80
Kosovo*	−5.6	−11.5	−11.5	−11.5	−11.5	−11.5	—	—	—	—	—	—	—	—
FYR Macedonia	−9.9	−7.0	−8.0	−9.1	−1.8	−1.2	1.2	1.4	3.4	4.3	4.6	−4.1	2.0	77
Romania	−5.6	−12.9	−8.8	1.5	3.9	7.1	3.9	−6.1	−5.4	−3.2	1.8	5.3	3.5	84
Serbia and Montenegro	−7.9	−11.6	−27.9	−30.8	2.5	6.1	7.8	10.1	1.9	−18.0	5.0	5.5	3.0	50
*SEE**	−7.3	−14.8	−9.6	−2.4	3.0	6.4	3.5	−0.5	−0.7	−3.4	3.6	4.5	3.6	79
CEE and Baltic states	−6.6	−10.3	−2.2	0.3	3.9	5.4	4.7	5.0	3.6	2.8	4.0	2.5	2.3	113
CIS-7														
Armenia	−7.4	−11.7	−41.8	−8.8	5.4	6.9	5.9	3.3	7.3	3.3	6.0	9.6	8.0	74
Azerbaijan	−11.7	−0.7	−22.6	−23.1	−19.7	−11.8	1.3	5.8	10.0	7.4	11.1	9.9	8.8	62
Georgia	−12.4	−20.6	−44.8	−25.4	−11.4	2.4	10.5	10.8	2.9	3.0	2.0	4.5	3.5	37
Kyrgyzstan	3.0	−5.0	−19.0	−16.0	−20.1	−5.4	7.1	9.9	2.1	3.7	5.1	5.3	2.0	71
Moldova	−2.4	−17.5	−29.1	−1.2	−31.2	−1.4	−5.9	1.6	−6.5	−3.4	2.1	6.1	3.5	37
Tajikistan	−1.6	−7.1	−29.0	−11.0	−18.9	−12.5	−4.4	1.7	5.3	3.7	8.3	10.3	7.0	56
Uzbekistan	1.6	−0.5	−11.1	−2.3	−4.2	−0.9	1.6	2.5	4.4	4.1	4.0	4.5	2.5	105
*CIS***	−3.3	−8.1	−11.0	−6.9	−6.1	−0.2	0.1	2.3	−1.0	3.0	5.5	4.2	3.4	76

Sources: EBRD 2002, UNDP 2002.

Notes: * Kosovo is still officially part of Serbia and Montenegro (FR Yugoslavia), although it is currently UN-governed and functioning as a quasi-independent country.

** Kosovo is not included in the SEE weighted average.

*** The CIS-average is weighted according to nominal dollar-GDP (lagged by one year) and includes all 12 countries (FSU minus the Baltic states).

1989=100.

crash. If the Baltic states are left out of the comparative data presented in Table 1, the contrast in the first years of transition between SEE and CEE is even sharper. Estonia, Latvia and Lithuania suffered deep contractions, though they recovered rapidly thereafter. During the first four years after 1989, GDP fell to a level of 82.2 (1989 = 100) for CEE and the Baltic states, while in SEE GDP dropped to 69.7. After the Bosnia war, there were still several economic downturns in SEE, such as in Albania, with its "pyramid crisis", and in Romania and Bulgaria because of incomplete reforms and political instability. More recently, Serbia and Montenegro and Macedonia experienced economic contractions because of the Kosovo and Macedonia conflicts.

Overall, the CEE countries did quite well, and the Baltic states, after an initial huge economic crisis, also recovered rapidly, resulting in the pre-accession status of the Czech Republic, Estonia, Hungary, Latvia, Lithuania, Poland, Slovakia and Slovenia and their imminent membership to the European Union in May 2004. SEE faired much worse, for a variety of reasons, namely insufficient reforms, weak institutional framework and political governance and violent social, ethnic and territorial conflicts. This is particularly true of the three countries most involved in the war on Bosnia: Croatia, Bosnia and Herzegovina (including the FbiH and the Republica Serbska), and Serbia and Montenegro. Croatia does not, however, form part of our suggested SEE-7 grouping, since its income per capita is substantially higher than the other members, and the country suffered much less from the war in terms of destruction of human capital and material infrastructure.

Transformation in the CIS is a more complicated and lengthy process. Belarus, Ukraine, Turkmenistan and Uzbekistan were slow or non-reformers during most of the 1990s, while in the other newly independent states the economic and political reforms were mostly inconsistent or incomplete. In particular, privatisation was effected in a non-transparent and chaotic manner, which in the case of Russia led to the formation of huge private financial and industrial conglomerates. These are run by a small group of omnipotent oligarchs (Goldman 2003). Land reform was mostly very partial and restructuring of the farm sector slow, with a continued overall predominance of large-scale enterprises that still benefit from "soft budget" constraints (Lerman 2003, Spoor 2003).

For the CIS and in particular most of the CIS-7, the transition period was also marked by violent conflicts, contracting economies, insufficient reforms and continued authoritarian regimes in power. Also notable in the CIS are the dual development paths that had already become visible in the early 1990s. Even before transition, the CIS-7 belonged to the Soviet periphery, but the gap has since quickly grown, as all of the members of this grouping, except for Uzbekistan, had a lower GDP in 2000 (compared with 1989) than the weighted average for the CIS as a whole. Their average GDP was only around 40 per cent the level of the other five CIS countries. The worst performers were Georgia and Moldova, followed by Tajikistan, Azerbaijan, Armenia and Kyrgyzstan (Table 2).

This peripheral development is more sharply evident from a comparison of GDP per capita for the SEE-7 and CIS-7 with that in CEE and the remaining CIS countries (using exchange rates in US dollars rather than purchasing power parity). Using non-weighted averages, the SEE-7 had an average per capita GDP of US $1,258 in

Table 2. SEE-7 and CIS-7: Gross Domestic Product (US dollars)

	2000		2000
SEE-7		*CIS-7*	
Albania	1,094	Armenia	614
Bosnia and Herzegovina	1,031	Azerbaijan	653
Bulgaria	1,548	Georgia	562
Kosovo	764	Kyrgyzstan	289
FYR Macedonia	1,792	Moldova	398
Romania	1,636	Tajikistan	160
Serbia and Montenegro	942	Uzbekistan	264
SEE-7	*1,258*	*CIS-7*	*420*
CEE	*4,526**	*CIS (5)*	*1,079*
CEE+SEE	*2,653***	*CIS (12)*	*695*

Sources: EBRD 2002, UNDP 2002.
Notes: These figures exhibit some differences with World Bank (2000) which presents GDP/capita as measured by the Atlas method. Nevertheless, the relative difference in wealth between the SEE-7 and CIS-7 remains similar, with differences being only within the groupings.
* Without Slovenia (which is the positive outlier), the average drops to US $3,954 per annum.
**Without Slovenia (which is the positive outlier), the average drops to US $2,222 per annum.

2000 (EBRD 2002). Albania, Bosnia and Herzegovina, Kosovo, and Serbia and Montenegro are below this mean, while Bulgaria, the FYR of Macedonia and Romania are above it. In the CIS-7, average per capita GDP is only $420 (exactly one-third the level of the SEE-7 grouping), with Kyrgyzstan, Moldova, Tajikistan and Uzbekistan even below this mean and the three Caucasus countries, Armenia, Azerbaijan and Georgia, above it (Table 2).

The accession countries to the European Union have a GDP per capita that is more than tenfold higher than that in the poor CIS countries, while compared with the SEE-7 this gap is 3.6: 1 (see Table 2). This is already indicative of a pronounced dual development, although a more extensive comparative exercise based on the HDI confirms this picture (see Table 3).

The 2001 GDP per capita, calculated with PPP (purchasing power parity) in US dollars (see UNDP 2003), shows the eight accession countries to be in the second half of the 55 countries that the UNDP considers representing high human development (HDI from 0.800 to 1.000). In descending order these are Slovenia, the Czech Republic, Poland, Hungary, Slovakia, Estonia, Lithuania and Latvia. For comparative (and possibly sentimental) reasons, Table 3 also includes the Netherlands, while the CIS-7 is compared with the Russian Federation. Croatia is also included in that group, as well as Belarus, although it is generally agreed that GDP growth in Belarus has been overestimated during the past decade by 1–2 per cent per annum. In the category of medium human development, in which many of the medium- and high-income developing countries are represented, are firstly the countries in our proposed SEE-7 grouping (again in descending order Bulgaria, FYR Macedonia, Bosnia and Herzegovina, Romania, Kosovo and Albania). With only Georgia and Azerbaijan

MAX SPOOR

Table 3. Relative Human Development Rankings, SEE-7 and CIS-7, 2001

HDI Rank	Country	Group	GDP/Capita (PPP US$)	HDI
High human development				
5	The Netherlands	EU	27,190	0.938
29	Slovenia	CEE	17,130	0.881
32	Czech Republic	CEE	14,720	0.861
35	Poland	CEE	9,450	0.841
38	Hungary	CEE	12,340	0.837
39	Slovakia	CEE	11,960	0.836
41	Estonia	Baltic	10,170	0.833
45	Lithuania	Baltic	8,470	0.824
47	Croatia	SEE	9,170	0.818
50	Latvia	Baltic	7,730	0.811
53	Belarus*	CIS	7,620	0.804
Medium human development				
57	Bulgaria	SEE-7	6,890	0.795
60	FYR Macedonia	SEE-7	6,110	0.784
63	*Russian Federation*	*CIS*	*7,100*	*0779*
66	Bosnia and Herzegovina	SEE-7	5,970	0.777
72	Romania	SEE-7	5,030	0.773
±82	Kosovo	SEE-7	2,712	0.733
88	Georgia	CIS-7	2,560	0.746
89	Azerbaijan	CIS-7	3,090	0.744
95	Albania	SEE-7	3,680	0.735
100	Armenia	CIS-7	2,650	0.729
101	Uzbekistan	CIS-7	2,460	0.729
102	Kyrgyzstan	CIS-7	2,750	0.727
108	Moldova	CIS-7	2,150	0.700
113	Tajikistan	CIS-7	1,170	0.677

Sources: UNDP 2002, 2003.
Note: It is mostly assumed that Belarus is overstating by its annual growth rates by 1–2 per cent, which would likely place the country in the medium human development group, rather than its current place. Kosovo is not represented in UNDP (2003), as it is not an independent country. Strangely, Serbia and Montenegro is also absent.

preceding Albania, the rest of the CIS-7 then follows suit (Armenia, Uzbekistan, Kyrgyzstan, Moldova and Tajikistan). For comparison, China (GDP/capita US $4,020 PPP and HDI = 0.721) and Vietnam (GDP/capita $2,070 PPP and HDI = 0.688) are between Kyrgyzstan and Tajikistan, though their starting points in the 1980s were much lower than even the poorest areas of the former Soviet Union. The Russian Federation is near the top of this group (close to Brazil and Colombia, with GDP/capita $7,100 PPP and HDI = 0.779).

Three main factors explain the current highly differentiated development levels of the former socialist economies. First, as mentioned, initial conditions were very different, with some peripheral regions in these parts of the former Soviet Union and Balkans amongst the poorest in the 1980s. China and Vietnam were substantially poorer, but are now rapidly overtaking parts of the former Soviet Union and Balkans in standard of living, after nearly two decades of fast growth and substantial poverty reduction. Geographical position and closeness to (or remoteness from) mainly EU markets should also be counted as part of initial conditions. Second, the new regimes were very diverse in their determination to fundamentally transform their economic and political systems. In some cases, such as with the CEE countries and Baltic states, the option to closely link with the West European economic system and, somewhat later, entry into the European Union (in combination with membership of NATO), was an enormous stimulus for change. Thirdly, those countries that lagged in development are largely the ones that had (or even emerged from) major armed conflicts, which caused enormous human suffering and destruction of infrastructure and human capital.

Table 4 (*a* and *b*) shows the vast scale of the armed conflicts and civil wars fought in a number of the transition countries that (except Russia) now form part of the SEE-7 and CIS-7 groupings of poor, peripheral countries. Particularly troubling amongst the new poor within these countries are the millions of internally displaced persons and refugees from neighbouring states, since they lack assets, rights and human security (World Bank 2000).

3. INCREASED INCOME INEQUALITY AND HIGHER RURAL POVERTY RISKS

Though authors and international institutions differ about the precise causes of impoverishment in these regions, and possible solutions to it, there is tacit agreement about the seriousness of the problem and the vastness of its size. Milanovic (1998), UNDP (1998, 1999), Ellman (2000), the World Bank (2000), Hutton and Redmond (2000) and Falkingham (2003) have produced similar analyses of increased income inequality and poverty in the transition economies. Next to the attention given to the alarming re-emergence of poverty (Spoor 2002), other indicators of human development and security are often added, such as the deterioration of access to health services and education and decreased life expectancy at birth. Perhaps the most disturbing indicator of serious social problems is the substantial increase in child poverty and infant mortality (UNICEF 2003). The importance of rural poverty, particularly in the poor SEE and CIS countries with their substantial and often even majority rural populations, is however mostly ignored. Only recently, with the publication of two major studies (EBRD 2002 and IFAD 2002) has a finger been pointed to this important aspect of poverty and the growing inter-sectoral gap in incomes.

Although the data on increased poverty rates and the millions of poor in many previously socialist countries are already striking, the personal accounts of families or individuals who have fallen into poverty are even more telling. The "Voices of the Poor" project (Narayan and Petesch 2002) gave a face to the large populations that underwent dramatic change in their social status, often moving from a solid

Table 4. The Human Costs of Conflicts in CEE and CIS

a. Major Armed Conflicts: Deaths, Internally Displaced Persons (IDPs) and Refugees (1989–2001)

Area of Conflict	Period	Related Deaths	Internally Displaced Persons (IDPs) and Refugees
CIS			
Uzbekistan (Ferghana Valley)	1989	—	—
Kyrgyzstan (Osh and Djalalabad)	1990	300–1,000	—
Moldova (Transdnestria)	1992	2,000	IDPs: 60,000
Armenia-Azerbaijan (Nagorno-Karabakh)	1988–94	67,000	IDPs: 616,000; refugees: 233,000 Azerbaijanees, 325,000 Armenian.
Russia (North Ossetia/Ingushetia)	1992 and 1997	583	IDPs: 40,300; refugees: 37,700 (in North Ossetia)
Tajikistan	1992–97	157,000	IDPs: 600,000; refugees: 50,000
Russia (Chechnya)*	1994–96, 1998 to present	—	IDPs: 76,000–207,000; refugees: 7,000
Georgia (Abchasia, South Ossetia)	1992–93	—	—
SEE			
Former Yugoslavia (Bosnia & Herzegovina)	1992–95	278,000	IDPs: 1,300,000; refugees: 900,000
Former Yugoslavia (Croatia)	1992–95	20,000	IDPs: 34,134; refugees: 323,784
Serbia Montenegro (Kosovo)	1999	10,000 (in Kosovo)	Serbia: IDPs: 230,000 (mostly Serbs from Kosovo); Kosovo: IDPs: 590,000 (during war); 228,500 (2001)
Macedonia	2001	50	IDPs: 25,000; refugees: 6,000

b. Internally Displaced Persons, Asylum Seekers and Refugees (x 1,000)

	1998	2002		1998	2002
SEE-7			*CIS-7*		
Albania	—	—	Armenia	310.0	247.6
Bosnia & Herzegovina	876.4	396.0	Azerbaijan	798.2	585.8
Bulgaria	—	4.8	Georgia	277.0	265.8
Kosovo**	—	228.5	Kyrgyzstan	14.9	8.4
FYR Macedonia	1.7	13.6	Moldova	—	1.2
Romania	—	—	Tajikistan	5.4	3.8
Serbia and Montenegro	733.0	616.3	Uzbekistan	3.8	46.0

Sources: UNHCR 2002, cited in UNICEF 2003; UNHCR 2003.
Notes: *Van Tongeren et al. (2002), p. 355, state that the reported number of conflict-related deaths was 3,384 (2000), but they also note that this total is likely much higher, in particular amongst the civilian population.
**UNHCR (2003) does not separately mention Kosovo (it not being an independent country). We therefore use data from Van Tongeren et al. (2002) for the post-war (2001) situation.

middle-class position to the lowest classes of society. They suffer not only from insufficient income, but also from inadequate access to health services, education and social transfers. In particular the elderly, youths and children from these families are in danger of further impoverishment, which for the latter categories sometimes means that they resort to crime and prostitution, lacking any form of future social perspective.

In several of the poorer SEE and CIS countries extremely worsened income inequality has meant the disappearance of most of the middle class. For example, in Kyrgyzstan, Voices of the Poor concluded that "self-observed poverty" increased dramatically in just a decade. The change in welfare status in Kok Yangak was particularly striking, although being a town in which many were dependent on a mine it may not be representative. Those who were considered wealthy dropped from 25 per cent to 8 per cent and the middle class fell from 60 to 8 per cent since the start of transition. The poor increased from 10 to 76 per cent (Narayan and Petesch 2002: 280). Income inequality had increased as well, though the socialist countries were very egalitarian at the outset of transition.

In the CIS-7 grouping income distribution has become highly unequal. Most countries have a Gini coefficient for income distribution close to or even greater than 0.50, comparable with Latin American developing countries such as Brazil and Colombia. These would possibly be even higher if the full scale of the informal and grey economy were taken into account (Table 5).

Table 5. *Inequality in the Distribution of Income, Gini coefficients for SEE-7 and CIS-7*

	1989–90	2000		1989–90	2000
SEE-7			*CIS-7*		
Albania	—	—	Armenia	0.251	0.590
Bosnia & Herzegovina	—	—	Azerbaijan	0.275	0.506
Bulgaria*	0.233	0.332	Georgia	0.270	0.503#
Kosovo	—	—	Kyrgyzstan	0.270	0.414
FYR Macedonia*	—	0.346	Moldova	0.251	0.437
Romania	0.237	0.310	Tajikistan	0.281	0.470
Serbia and Montenegro*	—	0.373	Uzbekistan	0.280	0.475**

Source: UNICEF 2003: 93.
Notes: * Data is from the UNICEF MONEE database (2003). The data differ from that in UNICEF 2003.
** Gini coefficient of income distribution was estimated at 0.45–0.50 for 2000 (Cornea 2003).
# 1998.

High inequality in the distribution of incomes has also translated into increased incidence of poverty. On the basis of the authoritative study of Milanovic (1998), and further analysed by UNDP (1998, 1999, 2000) and Ellman (2000), alarming absolute numbers of the "new poor" in the CIS, CEE and SEE were given. It was estimated that the total number of poor increased from 14 to 147 million from 1989 to 1995. Relative poverty rates, mostly defined as those below 50 per cent of the median income, are often used to compare poverty between countries (or regions). However, when a country has a vast majority of very low incomes, as do Moldova and Tajikistan, the threshold of 50 percent of the median might well underestimate the seriousness of the problem. Therefore, as there are by now internationally comparable data on absolute poverty rates, these are used in this chapter to stress the size of the poverty problem, in particular in the SEE-7 and CIS-7. While concentrating on these, it should not be forgotten that the majority of the poor are actually living in the larger transition countries such as the Russian Federation, Ukraine and Romania (World Bank 2000: 3) with several regional pockets of poverty.

Table 6 displays absolute poverty rates for the countries in the two groupings, first based on the internationally comparable poverty lines of US $4.30 (1996 PPP) for poverty incidence and $2.15 (1996 PPP) for extreme poverty. Second, additional data are given, following the study of Falkingham (2003), for the CIS-7 which compare these data with poverty rates when a nationally defined poverty line is used. Because the national poverty lines are defined using different standards and methodologies they are difficult to compare. Nevertheless, Table 6 shows poverty is a serious problem in the countries in the two mentioned "peripheral" groupings, and the situation in the CIS-7 countries (except possibly Uzbekistan) is highly problematic.

Table 6. Absolute Poverty in the SEE-7 and CIS-7

	Extreme Poverty (%) < 2.15 US$/day	Poverty (%) < 4.30 US$/day	Poverty (%) < national poverty line
SEE-7			
Albania	—	—	46.6 (2001)
Bosnia & Herzegovina	—	—	64.0 (1998)
Bulgaria	3.1 (1995)	18.2 (1995)	—
Kosovo	—	—	50.3 (2001)*
FYR Macedonia	6.7 (1996)	43.9 (1996)	20.7 (1998)
Romania	6.8 (1998)	44.5 (1998)	—
Serbia and Montenegro	—	—	20 and 27 (2001)
CIS-7			
Armenia	43.5 (1999)	86.2 (1999)	47.4 (2001)
Azerbaijan	23.5 (1999)	64.2 (1999)	49.6 (2001)
Georgia	18.9 (1999)	54.2 (1999)	51.4 (2000)
Kyrgyzstan	49.1 (1998)	84.1 (1998)	56.4 (2001)
Moldova	55.4 (1999)	84.6 (1999)	62.3 (2001)
Tajikistan	68.3 (1999)	95.8 (1999)	95.7 (1999)
Uzbekistan	—	—	27.5 (2001)

Sources: Falkingham 2003; World Bank 2000; UNDP 1998, 2002; FRY 2002; IFAD 2002; and Republic of Albania 2001.
Notes: *The full poverty line is defined as $1,534 per day, while extreme poverty is measured at consumption levels below $0.813 per day (11.9 per cent), see UNDP 2002.

Rural poverty incidence is higher in the countries of the SEE-7 and CIS-7 if one uses the relative rural poverty index developed by the EBRD in its special report on agriculture and rural transition (EBRD 2002: 91). The EBRD notes that the report might have overestimated rural poverty incidence, as the data were a combination of an absolute poverty line (for national poverty) and a relative line for the contribution of rural poverty in total poverty (see Table 7). Indeed, Falkingham (2003: 15), who compared headcount poverty rates for urban and rural areas, concluded that in the more urbanised societies of the Caucasus (Armenia, Azerbaijan and Georgia), poverty had become increasingly an urban problem. However, in the other CIS-7 countries (Kyrgyzstan, Moldova, Tajikistan and Uzbekistan), poverty was indeed much more a rural (and small town) phenomenon.

Table 7. *Relative Rural Poverty and Poverty Risk for SEE-7 and CIS-7 (2001)*

	Share of Rural Population	RRP Index	RRP Risk		Share of Rural Population	RRP Index	RRP Risk
SEE-7				*CIS-7*			
Albania	57.1	—*	1.4	Armenia	32.7	1.3	0.9
Bosnia and Herzegovina	56.6	—	—	Azerbaijan	48.1	1.6	—
Bulgaria	32.5	2.0	1.3	Georgia	43.5	1.5	1.1
Kosovo	65.0	—	—	Kyrgyzstan	65.6	6.3	1.3
FYR Macedonia	40.5	3.6	1.3	Moldova	58.3	1.7	—
Romania	44.7	2.6	1.5	Tajikistan	72.4	1.3	1.0
Serbia and Montenegro	—	—	—	Uzbekistan	63.3	—	—

Sources: EBRD 2002: 91. World Bank 2000; UNDP 2003.
Notes: RRP Index = relative rural poverty index or the share of rural people in poverty relative to the share of urban people in poverty. RRP Risk = relative rural poverty risk or the rural poverty rate relative to the national poverty rate.
*IFAD (2002: vi) states that 90 per cent of the poor live in rural areas.

Table 7 also shows that the two groupings of countries still have substantial rural populations, with the exception of Bulgaria (32.5 per cent) and Armenia (32.7 per cent). Albania, Bosnia and Herzegovina, Kosovo, Kyrgyzstan, Moldova, Tajikistan and Uzbekistan are predominantly rural economies. In nearly all cases, except for Armenia, the relative rural poverty (RRP) risk (i.e. relative to the national poverty rate) is higher than the relative urban poverty risk. The RRP risk in the SEE-7 countries is also generally higher than in the CIS-7 countries. The World Bank (2000) explained the difference by pointing to the also very high urban poverty rates and the consumption buffer that was created for the rural poor with the distribution (and expansion) of subsidiary household plots.

In IFAD's (2002) extensive study on rural poverty, including Albania, Armenia, Azerbaijan, Bosnia and Herzegovina, Georgia, the Republic of Moldova, Romania and the FYR of Macedonia, the most vulnerable rural dwellers were listed as farmers in upland and mountainous areas, rural wage earners, rural women, the elderly, ethnic minorities and internally displaced persons and refugees. Apart from income poverty, often due to the dramatic changes in the (often still stagnating) agricultural sector, the report also points to other factors:

> The situation is especially bad in some rural areas where rural health centres, schools and clean water supplies were mainly developed to suit the needs of the former state and collective farms and are not always appropriate to the requirements of today's more dispersed population (IFAD 2002: 21).

Finally, the report gave some rough estimates of the absolute numbers of rural poor in the eight countries that were studied (all of them members of our SEE-7 and CIS-7 groupings). Based on World Bank data (originating in the 1996–99 period), of an estimated total population of 53.3 million and a rural population of 23.2 million, 12.3 million would form the rural poor (income < US $4.30 per day) of which 4.0 million would be categorised as extremely poor (< $2.15 per day). If other countries in our groupings that have majority rural populations were added, such as Kyrgyzstan, Tajikistan and Uzbekistan, this picture would substantially worsen.

4. CONCLUSIONS

Transition as a concept suggests a smooth transformation of one system into another. Unfortunately this has not been the case, and transition countries have developed in a highly unequal manner. Unequal because they show, after more than a decade of transformation, a wide disparity of economic and human development. Of the eight newcomers to the European Union, Slovenia has a per capita GDP of PPP US $17,130 (and a HDI of 0.881), while at the other end of the spectrum Tajikistan has a per capita GDP of PPP $1,170 and a HDI of 0.677. This chapter has shown a number of phenomena. Firstly, unequal development has expressed itself in the formation of one or more *peripheries*, such as that of the poor countries of the CIS and SEE. The former group has now been labelled the "CIS-7" (Armenia, Azerbaijan, Georgia, Kyrgyzstan, Moldova, Tajikistan and Uzbekistan) from an initiative undertaken by the World Bank and the IMF. This chapter proposed a similar initiative for the latter group (Albania, Bosnia and Herzegovina, Bulgaria, Kosovo, FYR of Macedonia, Romania and Serbia and Montenegro). Like the CIS-7, this SEE-7 is a far from homogeneous group. Bulgaria and Romania are working hard to comply with the requirements to enter the European Union in 2007. The heirs of the former Yugoslav Republic have a high degree of externally supported (and enforced) peacekeeping and development assistance and large numbers of internally displaced persons and refugees, while many conflict areas remain unresolved. Finally, Kosovo is governed as an independent economy, but is formally still an integral part of Serbia and Montenegro. Still, though politically complex, they form a region which is an emerging periphery in danger of being marginalised from the CEE countries now entering the European Union. In view of the danger of donor fatigue, the current coordination between the World Bank, IMF, the European Union and UNDP could be continued and even strengthened within the proposed SEE-7 initiative.

Secondly, unequal development is a problem within the countries in transition as well. In some cases, the Gini coefficient of income distribution has doubled, reaching the levels of very unequal Latin American developing countries. This means that while in the above-defined periphery of the CIS there are relatively high levels of poverty, most of the poor actually live in the largest countries of the former Soviet Union, namely the Russian Federation and Ukraine. Also, within the SEE-7 and CIS-7, poverty is often concentrated in certain (mostly peripheral) regions. Furthermore, rural poverty predominates in most of these countries, and the relative rural poverty risk is higher than that of urban poverty, despite the consumption buffer that was created by the distribution of small household plots. In these countries, poverty struck those who got unemployed, often hitting traditionally vulnerable groups, such as the elderly, youths and children and the less educated. Transition has created a new and substantial category of poor people as well, namely the working poor: labourers with bad jobs, poorly paid (if paid at all) and working under bad conditions. This pertains particularly to agricultural workers, as many of the still dominant agricultural enterprises are in dire straits. Yet poverty cannot be measured solely by income and expenditures; it is also expressed in the loss of human security and access to education and health services, all of which tend to be worse in rural areas which already lacked sufficient infrastructure and investments at the onset of transition (EBRD 2002, IFAD 2002, Falkingham 2003).

The impoverishment of livelihood conditions for such large sections of the populations in the more peripheral transition countries and regions, can be attributed to at least four important factors. First, initial conditions were important: the level of industrialisation, distance to international markets, existence of mineral wealth and the overall development level. Although poverty was not recognised to exist, low income groups were large in countries such as Albania, Bulgaria, Romania and some parts of former Yugoslavia (SEE) and in all of the current members of the CIS-7. Unsurprisingly, levels of industrialisation were much lower than, for example, in some of the Central European transition economies, and the importance of the agricultural sector was much larger.

Second, governance was influential, since in many of the countries political instability was problematic, corruption levels were (and are) high and economic reforms were sometimes implemented in highly erratic and non-transparent ways.

Third, there is an unfortunate correlation between the low level of human development (and a corresponding high level of human insecurity) and the armed conflicts that have occurred in most of the countries. In SEE armed conflicts in Bosnia and Herzegovina, Serbia and Montenegro, the FYR of Macedonia (which was affected by "ethnic" conflict in 2001) and Kosovo (since the war of 1999) were linked with the most violent war on the European continent since World War II. The continuing presence in these countries of nearly one million internally displaced persons and refugees, who are the poorest of the poor, is indicative (see Table 4b). In the CIS most conflicts were in the CIS-7 (except the Russian war in Chechnya), the most violent having been wars between Armenia and Azerbaijan (on Nagorno Karabakh), the war in Tajikistan, and the conflict in Abkhasia and South Ossetia in Georgia. Apart from these wars, several local conflicts have taken place, and the countries have received refugees from other conflict areas. Again, a high incidence of poverty coincides with the occurrence of conflicts, in particular where human costs and destruction of infrastructure were substantial.

Fourth, in many of the peripheral countries identified in this chapter, rural poverty is a severe problem. In some more urbanised countries, such as Bulgaria, Romania and the countries of the Caucasus, urban poverty is higher. But in all cases except one (Armenia), the rural population has a higher risk of becoming poor. Within the rural poor, the old *kolkhoz* workers are now the worst off (Spoor and Khaitov 2003, World Bank 2000). There are few cases in which rural poverty is decreasing, since rural poverty tends to accompany poor agricultural growth, low rural investment, lacking institutions and missing markets (EBRD 2002, IFAD 2002). In Kyrgyzstan and Albania, the recovery of the agricultural sector, jointly with a land reform, has reduced rural poverty. However, in most countries, the urban-rural gap is growing and the process of marginalisation of rural areas has led to migration, dead villages and social disintegration in the countryside.

This is the more serious problem, as in the two groupings of countries the population share of the rural areas is between 40 and 60 per cent and even higher in the Central Asian states (again except Bulgaria and Armenia, with their rural population share in the low thirties). While there is growth in all of these countries, it is seldom in the sectors where the poor are employed, and in particular agricultural growth has been poor (with only some exceptions). As the EBRD (2002: 91) acknowledged, 'The gap between rural and urban poverty is probably due, at least partially, to the neglect of rural issues during the early years of transition.' Land reform has been implemented in most countries, giving many rural dwellers access to subsidiary household plots, guaranteeing a minimum of food consumption and starting a rapid process of "individualisation" of agricultural production (Lerman 2003).

However, the accompanying efforts of institution building (land markets, rural finance, property rights and legislation) and investment in infrastructure, extension and research services have been weak, making it difficult for private farmers or commercial enterprises to modernise and produce with profits (IFAD 2002, Spoor 2003). Agriculture-led growth (and investment-led growth) is lacking, which has a large impact on the reduction of rural poverty and the ability to attain a more balanced and sustainable growth model. In the end such a model will be more poverty-reduction-oriented than the current "growth that will trickle down" model (Cornea 2003, McKinley 2001). There is a non-linear relation between inequality and growth, in which low Gini coefficient and very high ones become an impediment for sustainable growth.

The latter might well contribute to erosion of social cohesion, as is clearly shown by the "Voices of the Poor" project and social conflicts. Finally, rural poverty can also become a serious strain to development in these peripheral transition countries. The industrial and trade sectors cannot yet absorb sufficient surplus rural labour, while a stagnating agricultural sector will continue to expel labour. In the absence of employment opportunities and with diminishing access to social services, rural poverty could become more severe. Currently most small farmers are producing foodstuffs, often for own consumption, while production of export crops, which require more investment and technology and therefore entail higher financial risk, has contracted enormously. Incomes therefore remain low, translating into insufficient rural demand for locally produced manufactured goods, restraining broad-based growth. Emphasis on the recovery and growth of the agricultural sector, based on land reform, the building of markets for land, labour, inputs and outputs, the improvement of infrastructure and increased access to services is thus a vital priority for these countries. Additionally, as China's success illustrates, the development of a viable rural non-farm economy is crucial to increase rural incomes (Bezemer and Davis 2003).

NOTES

1. CEE is formed by the Czech Republic, Hungary, Poland, Slovenia and Slovakia. More recently the Baltic states are also included: Estonia, Lithuania and Latvia (sometimes changing the abbreviation in CEB). South and Eastern Europe (SEE) consists of Albania, Bosnia and Herzegovina, Bulgaria, Croatia, the Former Yugoslav Republic of Macedonia, Kosovo, Serbia and Montenegro, and Romania. The CIS has the following members: Armenia, Azerbaijan, Belarus, Georgia, Kazakhstan, Kyrgyzstan, Moldova, the Russian Federation, Tajikistan, Turkmenistan, Ukraine and Uzbekistan.

2. Croatia (also in SEE) is substantially more developed than the other countries and suffered less from the war. Its GDP/capita was US $4,206 in 2000, comparable with that in the CEE group that are to become EU members in May 2004. Croatia filed for EU membership in February 2003.

3. Human development can be measured by the composite Human Development Index (HDI) as used by UNDP in their global and national human development reports. The HDI takes into account GPD/capita, life expectancy at birth, adult literacy rate and combined gross enrolment in school.

REFERENCES

Bezemer, Dirk and Junior Davis (2003) 'The rural nonfarm economy in transition countries', in: Max Spoor (Ed.), *Transition, Institutions and the Rural Sector*. Lanham and Oxford: Lexington Books, Rowman and Littlefield Publishers, pp. 163–84.

Chang, Ha-Joon (Ed.) (2003) *Rethinking Development Economics*. London: Anthem Press.

Cornea, Andrea Giovanni (2003) 'An overall strategy for pro-poor growth', in: *Macroeconomic Policy and Poverty Reduction: The Case of Uzbekistan*. Consultancy Report for UNDP. Tashkent.

Dudwick Nora, Elizabeth Gomart, and Alexandre Marc, with Kathleen Kuehnast (Eds) (2003), *When Things Fall Apart: Qualitative Studies of Poverty in the Former Soviet Union*. Washington, D.C.: World Bank.

EBRD (2002) *Transition Report 2002: Agriculture and Rural Transition*. London: European Bank for Reconstruction and Development.

Ellman, Michael (2000) 'The social costs and consequences of the transformation process', in: *Economic Survey of Europe*, No. 2/3, pp. 125–45.

Falkingham, Jane (2003) 'Inequality & poverty in the CIS-7 1989–2002'. Paper presented at the Lucerne Conference of the CIS-7 Initiative, January 20–22.

Federal Republic of Yugoslavia (2002) Interim Poverty Reduction Strategy Paper. Belgrad.

Goldman, Marshall I. (2003) *The Privatization of Russia: Russian Reform Goes Awry*. New York: Routledge.

Government of Georgia (2000) Poverty Reduction and Economic Growth Program. Intermediate Document. Tiblisi.

Government of Moldova (2002) *Interim Poverty Reduction Strategy Paper*. Chisinau.

Government of the Republic of Macedonia (2000) *Poverty Reduction Strategy Paper* (Interim Version). Skopje.
Government of the Republic of Montenegro (2002) *Interim Poverty Reduction Strategy Paper*. Podgorica.
Government of the Republic of Tajikistan (2002) *Poverty Reduction Strategy Paper*. Dushanbe.
Hutton, Sandra, and Gerry Redmond (Eds) (2000) *Poverty in Transition Economies*. Routledge Studies of Societies in Transition. London and New York: Routledge.
IFAD (2002) *Assessment of Rural Poverty: Central and Eastern Europe and the Newly Independent States*. Rome: International Fund for Agricultural Development.
Kolodko, Grzegorz W. (2000) *The Political Economy of Postsocialist Transformation*. WIDER Studies in Development Economics. Oxford: Oxford University Press.
Kyrgyz Republic (2002) *National Poverty Reduction Strategy 2003–2005 'Expanding the Country's Capacities'*. Comprehensive Development Framework 2010 Document. Bishkek.
Lerman, Zvi (2003) 'A decade of transition in Europe and Central Asia: Design and impact of land reform', in: Max Spoor (Ed.), *Transition, Institutions and the Rural Sector*. Lanham and Oxford: Lexington Books, Rowman and Littlefield Publishers, pp. 5-26.
Mckinley, Terry (Ed.) (2001) *Macroeconomic Policy, Growth and Poverty Reduction*. Basingstoke and New York: Palgrave.
Milanovic, Branko (1998) *Income, Inequality, and Poverty during the Transition from Planned to Market Economy*. World Bank Regional and Sectoral Studies. Washington: The World Bank.
Narayan, Deepa and Patti Petesch (2002) *Voices of the Poor: From Many Lands*. New York: Oxford University Press, published for the World Bank.
Republic of Albania (2001) *National Strategy for Socio-Economic Development. Medium-term Program of the Albanian Government 'Growth and Poverty Reduction Strategy'*. Tirana.
Republic of Armenia (2001) *Interim Poverty Reduction Strategy Paper*. Yerevan.
Republic of Azerbaijan (2003) *State Programme on Poverty Reduction and Economic Development 2003–2005*. Baku.
Spoor, Max (2001) 'Central Asian transition: Dual economies and growing inequality'. Paper read at the Centenary Conference of the Royal Society for Asian Affairs (RSAA), London, 1–2 November 2001.
Spoor, Max (2002) 'The [re]emergence of poverty in transition economies'. Paper presented at the 50th Anniversary Conference of the Institute of Social Studies 'Globalisation, Conflict and Poverty'. The Hague, 7–9 October 2002.
Spoor, Max (Ed.) (2003) *Transition, Institutions and the Rural Sector*. Lanham and Oxford: Lexington Books, Rowman and Littlefield Publishers.
Spoor, Max and Aktam Khaitov (2003) 'Agriculture, rural development and poverty', in: *Macroeconomic Policy and Poverty Reduction: The Case of Uzbekistan. Consultancy Report for UNDP*. Tashkent.
Swinnen, Johan (2003) 'Lessons from Ten Years of Rural Transition', in: Max Spoor (Ed.), *Transition, Institutions and the Rural Sector*. Lanham and Oxford: Lexington Books, Rowman and Littefield Publishers, pp. 27–46.
Tongeren, Paul van, Hans van de Veen and Juliette Verhoeven (Eds) (2002) *Searching for Peace in Europe and Eurasia*. The European Centre for Conflict Prevention. Boulder and London: Lynne Reinner Publications.
UNDP (1998) *Poverty in Transition?* New York: UNDP/Regional Bureau for Europe and the CIS.
UNDP (1999) *Transition 1999: Human Development Report for Europe and the CIS*. New York: UNDP/Regional Bureau for Europe and the CIS.
UNDP (2000) *Beyond Transition: Ten Years after the Fall of the Berlin Wall*. M. Spoor (Ed.). New York and Bratislava: UNDP/RBEC in co-operation with ISS and the Centre for the Study of Transition and Development (CESTRAD), The Hague, The Netherlands.
UNDP (2002) *Human Development Report Kosovo 2002*. Pristina: UNDP.
UNDP (2003) *Human Development Report 2003*. New York: UNDP.
UNECE (2003) *Economic Survey of Europe, No. 1*. Geneva: United Nations Economic Commission for Europe.
UNHCR (2003) *Statistical Yearbook 2002*, Paul van Tongeren et al. (Eds), Searching for Peace in Europe and Eurasia, European Centre for Conflict Prevention. Boulder and London: Lynne Reinner Publishers.

UNICEF (2001) *A Decade of Transition. Regional Monitoring Report, No. 8.* The MONEE Project CEE/CIS/Baltics. Florence: UNICEF Innocenti Research Centre.

UNICEF (2002) *Social Monitor 2002.* The MONEE Project CEE/CIS/Baltics. Florence: UNICEF Innocenti Research Centre.

UNICEF (2003) *Social Monitor 2003.* The MONEE Project CEE/CIS/Baltics. Florence: UNICEF Innocenti Research Centre.

World Bank (2000) *Making Transition Work for Everyone: Poverty and Inequality in Europe and Central Asia.* Washington, D.C.: The World Bank.

World Bank (2001) *World Development Report 2000/2001: Attacking Poverty.* Oxford and New York: Oxford University Press for the World Bank.

CHAPTER 5

GLOBALISATION, MARGINALISATION AND CONFLICT

S. MANSOOB MURSHED[1]

1. INTRODUCTION

Globalisation has marginalised many nations in the South since 1980, and there are systematic mechanisms via which this occurs. Globalisation has also increased inequality between nations and peoples and polarised the world into rich and poor nations. Nineteenth-century colonialism had a similar impact, and the world experienced another truly globalised era in the 1870–1914 period. This chapter begins with a discussion of globalisation and marginalisation. It then addresses the dispensation with the earlier North-South development contract as evidenced by the structural adjustment policies that have been inflicted on low-income developing countries, the debt crises that drain resources from the poorer South to the affluent North and the protectionism that prevents competitive goods from the South from reaching Northern markets. Accompanying international economic marginalisation has been a simultaneous growth in violent internal conflict in developing countries. The conclusion argues the need to restore the development contract. But this will necessitate root and branch reform of the international institutions that govern globalisation, with a greater say for the South.

2. GLOBALISATION AND MARGINALISATION

2.1 Globalisation and Growing North-South Disparities

It is a truism to state that we live in an era of globalisation. Indeed, globalisation is a much used and abused term in contemporary social science. In economics it is used to indicate accelerated economic integration: greater trade and financial flows across nation-states. This phenomenon has continued unabated since 1980. It cannot be said to have begun in earnest before that date, as at least a quarter of the world's population

Max Spoor (ed.), Globalisation, Poverty and Conflict, 67–80.
© *2004 Kluwer Academic Publishers. Printed in the Netherlands.*

then lived under socialist systems. The 1980 date approximates when China began to liberalise its economy. Globalisation thus implies systemic hegemony, British in the nineteenth century, American at present. It also connotes powerlessness on the part of nations, societies and groups to shape their own destiny in the face of the "silent takeover" by the forces of globalisation (see Hertz 2002). Furthermore, economic globalisation is not a new phenomenon. As many authors such as Murshed (2002b) and Williamson (2002) point out, the late nineteenth century up to World War I marks another historical episode of globalisation.

Globalisation is meant to be beneficial for the world's poorer nations, the proponents of globalisation or unfettered capitalism would have us believe. Participation in international trade and policies aimed at attracting foreign finance is meant to narrow the gap between rich and poor nations and to pull the world's chronically poor up by their bootstraps. This process is known as "convergence", meaning that the real incomes per capita of richer and poorer nations move closer to one another. International trade is meant to be the engine that achieves this. My argument is that it has not, and does not do so, because of the unequal nature of North-South trade. But first some stylised facts.

Globalisation marginalises much of the Third World and low-income developing countries. Table 1 attests to that fact. Apart from East and South Asia, all of the world's

Table 1a. GDP Per Capita (1995 Constant US$) Growth Rates

Region/Countries	Annual Average GDP Growth (%)			
	1960–1970	*1970–1980*	*1980–1990*	*1990–2000*
Low and middle income	3.1	3.3	1.2	1.9
East Asia and Pacific	2.9	4.5	5.9	6.0
South Asia	1.8	0.7	3.5	3.2
Latin America and the Caribbean	2.6	3.4	−0.8	1.7
Sub-Saharan Africa	2.6	0.8	−1.1	−0.4
High-income OECD	4.4	2.6	2.4	1.7

Source: World Bank (2002).

Table 1b. GDP Per Capita (1995 Constant US$)

Region/Countries	GDP per Capita				
	1960	*1970*	*1980*	*1990*	*2000*
Low and middle income	535	725	999	1129	1,356
East Asia and Pacific	194	256	396	705	1,252
South Asia	186	221	236	332	456
Latin America and the Caribbean	1,983	2,549	3,548	3,275	3,856
Sub-Saharan Africa	473	609	658	587	564
High-income OECD	10,324	15,885	20,525	26,080	30,757

Source: World Bank (2002).

less-developed regions grew faster during the less globalised era prior to 1980. Yet all developing regions have expanded their exposure to international trade in recent years (Table 2). While it is true that much of the developing world continues to participate in (and even seeks to expand exposure to) international trade, in a sense it is effectively decoupled from meaningful participation in the world economic system. Some middle-income developing countries as well as the most populous developing countries, India and China, are doing well from globalisation, but the benefits of globalisation are far from widespread in the South (Murshed 2002b). The best example of this is the pattern of foreign direct investment (FDI) flows. Three nations, China, Mexico and Brazil, receive half of FDI flows to the developing world. Some eleven nations account for two-thirds of all exports from developing countries. Low-income countries account for only 2.5 per cent of world merchandise exports and 1.4 per cent of FDI inflows. The corresponding figures for all developing countries are 19.7 and 21.6 per cent, respectively.

Data is more scarce for the nineteenth century. But it can be argued based on the figures suggested by Angus Maddison in UNDP (1999), that the North-South gap has continually widened since the dawn of modern capitalism and colonialism, but particularly so during the two periods of accelerated international economic integration that we refer to as globalisation (Table 3). The great first wave of globalisation ending in 1914 produced for the first time a sizeable North-South income

Table 2. *Sum of Exports and Imports as a Percentage of National Income (GDP), Selected Regions*

Region/Countries	1980	1985	1990	1995	2000
Low-income countries	34.4	29.2	37.6	47.6	55.9
UN least-developed countries	42.2	37.5	36.5	44.8	52.9
East Asia and the Pacific	44.7	43.9	51.7	63.5	79.7
South Asia	21.6	19.2	22.0	29.5	33.4
Latin America and the Caribbean	26.5	26.5	26.1	30.8	35.7
Sub-Saharan Africa	62.7	54.2	52.8	58.7	64.3
High-income OECD	37.4	37.9	35.7	37.0	40.3*

Note: * 1999 (data for 2000 not yet available).
Source: World Bank (2002).

Table 3. *Historical Gaps Between Rich and Poor Nations*

Year	Gap Between Richest and Poorest Nations
1820	3:1
1913	11:1
1950	35:1
1973	44:1
1992	72:1

Source: UNDP (1999).

gap, creating the present third world. This gap has widened in the more recent globalisation episode, which has cemented the unequal status of the Third World that was first created a century ago.

2.2 Globalisation and Increasing Inter-Nation Inequality

Three different means may be used to measure inter-country inequality (see Milanovic 2003). All three methods arrive at a Gini coefficient, the most commonly used measure of income inequality. Gini coefficients range from perfect equality (0) to perfect inequality (1), or in percentage terms from 0 to 100. The first means to measure international inequality treats all countries, large or small, equally. This is known as category 1 inequality. All countries for which data are available are arranged according to per capita income in comparable purchasing power parity (PPP) dollars. Thus each nation is characterised by a representative individual, whose income is that country's average. If, for example, there were 150 nations, we would be effectively measuring the inequality across 150 different individuals. Category 2 inequality is the same as category 1 except that each national per capita income is weighted by its relative population size (national population relative to the world). Thus, China is given a weight of about 0.2 since it accounts for a fifth of humanity. This may appear to be a reasonable procedure, but a serious flaw in category 2 inequality is that most changes in indices are accounted for by the alterations in populous countries, such as China and India. These downwardly bias results elsewhere, such as for poor but small African nations. Moreover, when nation-states are the unit of analysis, each state should be treated equally, as each nation is an independent entity, representing a unique policy experiment. Equal treatment of all nations means that each unit's income should be weighted equally, which implies no weighting at all. Category 3 inequality focuses on individuals rather than nations as the unit of analysis. This, however, represents a tall order in terms of comparable international data collection as it requires standardised household income surveys across the globe. Nonetheless, household surveys have become more common and streamlined in coverage and scope since the late 1980s.

What do the three types of international inequality measures show us? Category 1 measures indicate an increase in inequality in the globalised era. From about 46 in 1978, the Gini coefficient had risen to over 54 by 2000. In contrast, during the less globalised period of 1960 to 1978 the Gini remained fairly stable, between 46 and 48 (Milanovic 2002). Clearly, this shows that globalisation produces winners and losers and does the converse of achieving income convergence. Recall that category 2 inequality is the same as category 1 except that it is population weighted. Category 2 measures show a fall in international inequality in the highly globalised era because of the impressive growth in China's real per capita income. Using this method, the Gini for the world declines from 54.4 in 1978 to 50.1 in 2000 (Milanovic 2003), indicating a decline in inter-country inequality. But not only are category 2 indicators biased by what happens in large countries like China and India, but any country's income growth success in overall terms masks changes in income inequality along spatial or socio-economic lines within that country. The category 3 measure, based on household surveys, shows a rise in inequality (Gini coefficient) from 62.4 in 1988 to 66.0 in 1993, falling back slightly to 64.6 in 1998. Moreover, these figures would be higher, implying greater inequality, if ordinary dollars based on market exchange rates were used instead of PPP dollars. It therefore captures the huge rise in inequality amongst citizens within the former communist bloc. The urban-rural divide in inequality within China and India are also captured by this method.

Is rising global inequality, or inequality for that matter, a cause for concern? Or should we only worry ourselves with absolute levels of poverty, whether based on national standards or the international dollar-a-day measure of abject poverty. Clearly this depends on our notion of justice. The current development-donor focus is on poverty alleviation. Poverty is an absolute concept at any point in time, and those in poverty are individuals who fall below the poverty line measured in monetary terms. Inequality is a relative concept, measuring the differences in standard of living across groups who may or may not be poor. While poverty reduction is a lofty ideal, citizens of the globe, including those residing in poor nations are more aware of the differences in their own circumstances and capabilities compared to those of fellow human beings in rich nations. This gap needs to be addressed. The share of the poorest five per cent of the world's population is 0.3 per cent of world income using PPP dollars. The share of the richest five per cent is about a third or 33.3 per cent of world income. Thus the gap between the rich and the poor is 100:1. The important point to bear in mind is that the inequality between nations is greater than the inequality within any nation-state, and such inequality could not be tolerated in any democracy.

2.3 Openness and North-South Trade: Is it Beneficial for the South?

Let us move on to consider what the theories of international trade say about increased trade and relative incomes between trading partners, considering neo-classical theories first.[2] The Ricardian model of international trade, based on the notion of comparative advantage, has little to say about the convergence or divergence of relative wages in countries after trade is opened up. The Heckscher-Ohlin-Samuelson (HOS) paradigm, however, reveals much more on the matter. Under certain assumptions free trade will lead to the equalisation of factor prices, including wages. Free trade is therefore similar to regional integration. The restrictive conditions necessary for complete factor price equalisation are the absence of trade barriers, competitive pricing, non-sector-specific factors of production and incomplete specialisation in the countries in question. As with the per capita income convergence literature, in growth theory one might expect a tendency towards partial but not total factor price equalisation after episodes of trade expansion.

The model described above is static. Trade in the HOS paradigm is driven by differences in factor endowments: countries with more capital abundance would export capital-intensive goods and so on. Clarida and Findlay (1992) presented a model in which comparative advantage and endowments are endogenous and policy induced. These are proffered via a public knowledge-based input (non-rivalled and non-excludable) that lowers production costs. This input will not be provided by the private sector and is therefore a purely public good. One can also think of this input as human capital or infrastructural investment. Further, there are two sectors in the economy. One of these is akin to a natural resource sector or agriculture, in which the benefit from the publicly financed input in terms of reduced production costs is relatively lower. The other sector may be likened to manufacturing, which derives greater benefit from the publicly provided input. Capital is a specific factor in manufacturing, whereas land is specific to the resource sector. All sectors require labour input. In these circumstances, a commodity boom will induce a lower optimal supply of the publicly financed input, as the resource sector returns a proportionately smaller benefit from this input. This could be imagined to occur episodically in the South. Consequently, both sectors will be less productive over the course of time. The expansion of international trade will enable countries in the North with their greater capital endowments to gain absolute advantage in all sectors, as exports of manufacturing increase, inducing greater provision of the cost-reducing public good. Factor price equalisation therefore may not result from an increase in North-South trade.

The new theories on international trade which emphasise increasing returns to scale and product diversity offer more insight into the unequalising effects of North-South trade. Krugman (1981) presented a model of uneven development between North and South. Let us say there are two sectors: manufacturing and agriculture. Manufacturing or another dynamic sector is subject to increasing returns to scale, this benefit being related to initial endowments of capital, which we can interpret as know-how or primitive capital accumulation. At some point in the early nineteenth century, the North (say Great Britain), as a result of the extraction of colonial surplus, had a greater capital stock. After the introduction of free trade, its manufacturing sector expanded and the South's manufacturing sector contracted. At some point manufacturing in the South vanished, and the North's manufacturing could not increase due to the exhaustion of labour supply released from agriculture. This model explains the historical de-industrialisation of, say, India, the mechanism being "free trade". Eventually, the North's manufacturing sector can expand only if it exports capital to the South or imports labour from the South. Free trade does not produce factor price equalisation because the South is completely specialised in agriculture. Rather, the North will have a high-wage labour aristocracy. The equalisation of profit rates may, however, take place if the North invests in the South. This model aptly describes what Hobson (1902) and Lenin (1917) had in mind in their seminal works on imperialism. Moreover, Northern capital flows may occur towards a relatively more developed semi-periphery rather than the really under-developed periphery. It is this semi-periphery that moves closer to the North, their modern counterparts being the successful globalisers.

Then there are the product cycle models of North-South trade. Here again, Krugman (1979) did the pioneering work, though there are many extensions in the literature. In this model there is a continuum of goods, and new products are constantly being invented in the North. Eventually, they become old goods and the technology gradually moves to the South because of the South's cost advantage. Due to the North's temporary monopoly over new goods, and because it is the sole innovating region, the wage rate is always higher in the North compared to the South, which is the imitating region. This wage gap can be narrowed only by accelerating the pace of technological diffusion to the South. At present, technological diffusion is greatly hampered by arrangements in the World Trade Organization (WTO) governing knowledge transfer, particularly TRIPs agreements which regulate trade in intellectual property rights. The South can achieve parity with the North only when it too innovates new goods.

In summary, there are a variety of means by which free trade with the North might disadvantage the South and cause its average income to continue to diverge from that in the richer developed regions. The North has the larger stock of know-how and is where new goods are innovated. The important point here is that the developing world no longer has the choice between an open or closed trade regime; openness is compulsory.

3. SUSPENSION OF THE DEVELOPMENT CONTRACT

3.1 Structural Adjustment

The history of structural adjustment programmes can be traced back to the two oil shocks of the 1970s, particularly the second oil shock of 1978. Increases in energy prices hit non-oil-producing developing countries hard, and these countries soon ran into balance of payments difficulties. These and other macroeconomic difficulties compelled many developing countries to seek financial assistance from international financial institutions such as the International Monetary Fund (IMF) and the World Bank. Structural adjustment was not confined to macroeconomic issues, but a whole host of reforms were imposed and instituted under its umbrella. Measures included unilateral trade liberalisation and the removal of subsidies and fiscal exigency amounting to reductions of social-sector expenditures. While the need for a degree of fiscal prudence could not be denied, a major casualty of structural adjustment was human capital forming expenditure on education and health in many low-income developing countries.

Structural adjustment programmes provided an opportunity to force the tenets of the "Washington Consensus" (economic conservatism) on hard-pressed developing countries in dire need of financial assistance. The Washington Consensus can be traced back to the work of Milton Friedman (1968). According to his view, the primary goal of economic policy is to contain inflation, which in a developing country context means not only a brake on money supply growth but fiscal restraint as well. The secondary set of policy goals is the removal of distortions, mainly related to trade taxes, capital controls and domestic subsidies. The removal of trade taxes, however, has often had long-lasting deleterious effects on government revenues. Furthermore, hasty financial sector liberalisation has helped to spread disastrous financial contagion, as in East Asia in 1997. In general, the elimination of subsidies and reduced social-sector expenditure halts the accumulation of growth-enhancing human capital. Public expenditure is generally credited with helping to reduce inequality by allowing the poor to accumulate assets and skills with which they can break out of the vicious cycle of poverty and eventually catch up with the rich.[3] It is also worth remarking that policies associated with a major reduction in social protection would be politically untenable in most OECD countries, where social protection is mandatory and a human right.

Structural adjustment policies and the Washington Consensus are based on an economic philosophy of continuously clearing markets with flexible prices. Adherents of these views virtually rule out the possibility of market imperfections or temporary price rigidities (except in the very short term). Countries that have fared well in East Asia and elsewhere have not subjected themselves to the Washington Consensus orthodoxy, at least not until recently. Furthermore, structural adjustment programmes and what followed have set developing countries well and truly on the route to indebtedness to multilateral and bilateral donors. They have also inculcated the culture of dependence on aid for budgetary support. The era of structural adjustment is associated with the ideological hegemony of the Washington-based international financial institutions reflective of American interests. Moreover, their philosophy and approach has been adhered to by other major bilateral donors in Europe and Japan, in contrast to the earlier era of more independent European aid policies.

Allied to the ideology of structural adjustment was the (mis)management of the transition in the former communist bloc in Eastern Europe and the former Soviet Union (Stiglitz 2002). Once again obsession with price stability and macroeconomic balances under any circumstances proved costly in terms of vast income compression, comparable only to the devastating effects of a long war. More damaging was the approach to privatisation, which resulted not only in corruption but also in asset stripping which further reduced output. This helped to produce huge, historically unprecedented growth in poverty and inequality in those regions.

Structural adjustment has been replaced by the allegedly pro-poor Poverty Reduction Strategy Paper (PRSP) process. While concern with poverty is laudable, as alluded to earlier, true long-term development requires reduction of global inequalities and the North-South income gap. Moreover, PRSPs can be viewed as a cynical repatriation of conditionality: the recipient country must demonstrate its worthiness to receive aid by signing *ex ante* its agreement with the donor's priorities, usually with the donor's connivance.

3.2 Debt Crises

Few developments are more indicative of the abrogation of the development contract than the various Third World debt crises that have been with us since the Mexican default of 1982. The debt crises of the 1980s mainly afflicted middle-income countries in Latin America and elsewhere. Today the focus is on poorer (low-income) nations, mainly in Africa. Initially, concern was with private debt, as the lending was by commercial banks to middle-income developing countries; now we speak of debt owed by low-income nations to multilateral agencies and bilateral donors.

A great deal has been written on the origins of the 1980s debt crisis, emphasising the profligacy and untrustworthiness of sovereign debtor governments in the South. However, as Murshed (1992) pointed out, the developments that converted indebtedness into a crisis were a result of macroeconomic policies pursued by Northern governments motivated by the need to conquer their own domestic inflation, through contractionary monetary and expansionary fiscal policies in the United States and contractionary monetary policies elsewhere in the OECD, all of which served to raise world interest rates and create payment problems for debtor nations in the South.

In dealing with the debt crisis of the 1980s the international financial institutions were more concerned with the interests of Northern creditors, western banks. A truly developmental agenda would have balanced the needs of creditors with those of debtors. In fact, it is universally true that domestic bankruptcy laws do not allow creditors to let debtors sink to such a low degree of capability and the types of misery associated with the lost decade in Latin America during the 1980s. Table 1a indicates the negative growth rates associated with the debt crisis of the 1980s in Latin America. These rates may not have been so low or negative if sufficient debt forgiveness had been provided.

Moreover, putting humanitarian considerations aside, punitive levels of debt servicing actually damages the debtors' ability to repay loans due to the debilitating effect of excess debt servicing on investment, a phenomenon known as debt overhang. As the 1980s progressed much of the debt owed to private banks in the North was converted to public debt, owed to international financial institutions, something that was achieved by discounted buybacks financed by the taxpayer in the North. This amounted to a taxpayer bail-out of their privately owned banking sector, an indicator of the financial sector's influence over OECD governments.

The various debt crises represented crises in solvency, not short-term liquidity problems that could be overcome with increased lending. Given the nature of this insolvency, debt relief measures are needed. True, at present there is the Highly Indebted Poor Country (HIPC) initiative for countries, mainly in Africa, whose debt is primarily with multilateral and bilateral official donors, but progress is slow. Such debt should be easy to forgive, as it is paltry for creditors and is mainly a government-to-government or multilateral agency obligation. Furthermore, part of this debt is odious,[4] as it was incurred by unrepresentative leaders.

3.3 Protectionism in the North Towards the South

The mercantilist colonial contract in the nineteenth century had an important trade policy component. It compelled or induced colonies to purchase manufactured goods from the mother country. This certainly helped the development of manufacturing industries in the North and sustained their international competitiveness. As a result one would expect markets in the North to be open to goods from the South in the post-colonial phase as part and parcel of a development contract. Yet at present the vast majority of protectionist measures in the world are instituted by the North against goods from the South in areas in which the South has competitive advantage, such as agriculture and textiles. This too has a modern-day mercantilist motivation, bearing in mind that the South is too weak to retaliate with protectionism of its own. Protectionism helps to boost output and employment in flagging import-competing sectors in the North, sectors that would be assigned to the dustbin if the tenets of capitalism were followed to their logical conclusion. Trade policy in the North has become a substitute for traditional Keynesian demand-management policy instruments, such as monetary and fiscal policy. These instruments are increasingly rules-based and beyond the scope of political discretion.

Developments in the arrangements governing multilateral trade and technology transfers have left nations in the South more vulnerable than in the past. First, the WTO negotiating process excludes countries in the South from meaningful participation, even though the WTO is based on the one-country one-vote principle. Second, there is a greater reluctance at present to grant non-reciprocal special and differential treatment to less-developed countries compared to thirty years ago. Third, WTO rules on trade in intellectual property rights (the so-called TRIPs) are especially inimical towards the process of technological diffusion in the South. The experience of, say, South Korea in developing indigenous technical capacity is vastly more difficult to emulate in an era in which TRIPs regulations are in force. The TRIPs regime also prevents affordable solutions to humanitarian crises, such as the production of cheaper generic drugs to combat AIDS. At a more fundamental level, authors such as Bhagwati (1994) have indicated a growing concern in the North (particularly in the United States) for fair as opposed to free trade. The current regime also leads, as Nayyar (1996) pointed out, to increasing asymmetry in the application of the principle of free trade. Free trade is advocated when it is in conformity with American interests, for example, with regard to the entry rights of American industries in other countries' markets. The principle of free trade is dispensed with, however, when it is in conflict with American and European interests in import-competing sectors such as textiles and agriculture.

4. THE DEVELOPMENT OF VIOLENT INTERNAL CONFLICT

We are used to viewing war as something that happens between nation-states. Today's wars mostly occur between groups within the same country, and in the developing world.[5] Are these civil wars fundamentally irrational, and could the differences underlying these disputes be settled peacefully? Sadly, conflict may be the product of rational decisions, even if the rational choice is only of a bounded or myopic variety. Even a terrible genocide, as in Rwanda, is often planned well in advance and carried out to meet a well-defined objective. Since the end of the Cold War conflict in developing countries has cost five million civilian lives (85% to 90% of total casualties) and displaced 50 million people from their homes.

The new rational choice literature on conflict often distinguishes between "grievance", a motivation based on a sense of injustice in the way a social group is or has been treated, and "greed", an acquisitive desire similar to banditry, albeit on a much larger scale. In many ways the former refers to intrinsic motivation and the latter to an extrinsic or pecuniary incentive to go to war (see Murshed 2002a). These motives are not entirely separate in practice and evolve as conflict progresses.

Grievances include systematic economic discrimination against groups based on ethno-linguistic or religious differences. Extreme poverty and poor social conditions, including refugee camps, also facilitate conflict by making soldiering less unattractive. Many of today's civil wars have an ethnic or nationalist dimension. Ethnicity, whether based on language, religion or other distinctions, is often a superior basis for collective action in poorer countries than other social divisions, such as class. In coalescing groups, therefore, current and historical grievances play a crucial part. This is all the more possible when there are inequalities across a small number of clearly identifiable groups. More often than not, these take the form of high asset inequality, discriminatory public spending across groups and unequal access to the benefits of state patronage, such as government jobs. Furthermore, state failure to provide security and a minimal level of public goods often force individuals to rely on kinship ties for support and security, as in the former Soviet Union. Ethnicity, however, must be treated with caution. Indeed, where ethnicity is very diverse conflict is less common.

Greed as a motive for conflict arises mainly in the context of natural resource endowments in Africa (see Collier and Hoeffler 2001). Natural resources, such as alluvial diamonds in Angola and Sierra Leone, can lead to contests for control. Such contests can take the form of warfare, as well as criminality and corruption. Furthermore, there may be attempts to obstruct the flow of natural resources with a view to extracting tribute, such as in the case of oil pipelines. This too can produce conflict. Even where greed or the desire to control valuable resource rents is the primary cause of conflict, as in some mineral-rich countries, poverty and grievance are important in fuelling conflict, especially in providing ready recruits for armed struggle.

The greed versus grievance dichotomy provides a useful analytical starting point for discussing the causes of conflict. But for these forces to take the form of large-scale violence there must be other factors at work, specifically, a failing social contract and conflict triggers (see Addison and Murshed 2002a, Murshed 2002a). A functioning social contract and the concomitant institutions that distribute income and resolve disputes can prevent violent expression of greed or grievance. Therefore, violent internal conflict is a consequence of development failure and the repudiation of the development contract rather than a symptom of development or state failure. Furthermore, the outbreak of conflict always requires triggers, both internal and external. External triggers involve support and succour from an outside power; internal triggers are events that induce parties to abandon peaceful negotiation in favour of outright war.

Conflict-affected nations have histories of weak social contracts, or a once strong social contract that has degenerated. This weakness is often a legacy of colonialism, which institutionalised mechanisms favouring one group over another. The risk of conflict is greatest when societies are in transition from autocracy to democracy, because it is precisely then that state failure is most acute and the social contract weakest (see Hegre et al. 2001). In a number of countries with new democracies, governance indicators have worsened in recent years while ostensible multiparty electoral competition has gained ground.

Domestic conflicts or civil wars are not the only new form of war. Transnational terrorism and, to combat it, the strategy of war on terrorism is another new form of war. Here intrinsic motivation, often in the form of a collective sense of humiliation, plays a greater role. Deterrence against terrorists therefore may backfire if it hardens the resolve to resist. Insights from behavioural and experimental economics also tell us that individual agents dislike threats and reject humiliating offers in ultimatum games, even when accepting these offers is the rational course of action. This type of decision-making may also apply to "terrorists". This means that political solutions are necessary. In terms of rational choice theory the "terrorists" need to be involved in the design of mechanisms to resolve conflict.

The other type of war is that associated with aggressive unilateralism on the part of the United States and other regional powers such as Israel. Unilateralism allows them to pursue strategic aims through force in a manner unthinkable in the days of superpower rivalry, when they were compelled to take into account other views and reactions.

Conflict resolution is more difficult when the intrinsic motivation to fight is strong, as in secessionist wars driven by historical grievances and certain types of terrorism. It is also difficult to sustain peace when parties feel tempted to return to war so they can continue looting valuable resources as in many African countries and in Central Asia (see Addison and Murshed 2002b). To be successful, however, peacemaking must reconstitute and refashion the social contract. That means broad-based reconstruction and a solution that does not leave any of the belligerents worse off than they were prior to war. Conflict-ridden countries usually have highly distorted economies, favouring services and trading over production (see Addison and Murshed 2002a). More often than not, this form of resource allocation is not pro-poor. Post-conflict recovery along such a skewed path exacerbates poverty, even if output rapidly recovers to pre-conflict levels. It is far better to have a long-term growth strategy that yields results gradually but is based on rebuilding shattered infrastructure with an emphasis on relatively pro-poor production activities, such as agriculture. Moreover, in countries rich in natural resources, such as oil or diamonds, output recovery and poverty reduction is insufficient to bring about enduring peace. For that to occur relative income differences, those inequalities that produced conflict in the first place, must be narrowed. Again, donors must not confine their attention solely to alleviating poverty, but inequality must be addressed as well. Lastly, peace can only exist within a secure environment; this includes external security as well as domestic law and order.

5. CONCLUSION

The period since 1980 has been one of accelerating globalisation, which has resulted in the marginalisation of vast swathes of the South. Evidence of this includes declining economic growth rates in Africa and Latin America and rising income inequality between rich and poor nations. In fact, the middle class in the international community of nations has shrunk in the past forty years. While concern with poverty reduction is laudable, true development also necessitates narrowing the North-South income gap. This is all the more true in a digital age when information disseminates rapidly. The past two decades have also seen the rise in violent internal conflict, civil wars, international crime, terrorism and aggressive unilateralism on the part of some great powers, despite the growth of multiparty electoral competition. This too is a symptom of development failure and of malfunctioning institutions of conflict management, both domestic and international.

Yet the outlook is not entirely gloomy. The early version of the Washington Consensus (crude monetarism) is now discredited. A dialogue appears to be at hand between neo-classical economists on the one hand and development practitioners such as the ubiquitous NGOs on the other regarding their different approaches to development and development outcomes (see Kanbur 2001 on the nature of these differences).

I would like to conclude this chapter by recalling the pressing need for reform of our outmoded international institutions (see Nayyar 2002). New specialised institutions are needed that deal with specific international issues, such as the environment, brought closer to those affected by the problem. This would also result in greater influence for the South. In the Bretton Woods organisations in particular, the balance of power between creditor and debtor nations needs to be redressed in favour of the latter. Debtor nations are mainly from the South. Debt servicing adds considerably to the share capital of the IMF and the World Bank. Consequently debtor nations should have greater voting rights. The United Nations system is the world's truly democratic international institution, one that can contain aggressive unilateralism and bring about a renaissance of the development contract. Ultimately, however, the United Nations is what its member states make it. In the final analysis, the rationale for a global developmental contract is rooted not in security considerations or economic self-interest, but in our universal sense of common humanity.

NOTES

1. Based on the author's presentations at the ISS tenth lustrum conference, 7 October 2002 and his inaugural lecture as the first holder of the Prince Claus Chair in Development and Equity, University of Utrecht, 12 May 2003.

2. Here I am not considering the famous Prebisch-Singer macroeconomic hypothesis regarding the secular tendency of primary goods prices to decline vis-à-vis manufactured goods prices due to a low income-elasticity of demand for primary goods.

3. Free education, including free higher education did much to reduce income inequality in OECD Europe.

4. The term odious debt was employed by the United States during the Spanish-American war of 1898 to refer to the debt owed to Spain of the Spanish colonies taken over by America. Clearly, the United States felt at that time, unlike now, that such odious debt should be repudiated.

5. There were more than forty internal conflicts that led to at least 1,000 deaths in any single year in the 1990s.

REFERENCES

Addison Tony and S. Mansoob Murshed (2002a) 'On the economic causes of contemporary civil wars', in S. Mansoob Murshed (Ed.), *Issues in Positive Political Economy*. London: Routledge.

Addison Tony and S. Mansoob Murshed (2002b) 'Credibility and reputation in peacemaking', *Journal of Peace Research*, 39: 487–501.

Bhagwati, Jagdish (1994) 'Free trade: Old and new challenges', *Economic Journal*, 104: 231–246.

Clarida, Richard H and Ronald Findlay (1992) 'Government, trade and comparative advantage', *American Economic Review*, 82: 122–127.

Collier, Paul and Anke Hoeffler (2001) 'Greed and grievance in civil wars'. Internet: www.worldbank.org.

Friedman, Milton (1968) 'The role of monetary policy', *American Economic Review*, 68: 1–17.

Hegre, Håvard, Tanja Ellingsen, Scott Gates and Nils Petter Gleditsch (2001) 'Towards a democratic civil peace? Democracy, civil change, and civil war 1816–1992', *American Political Science Review*, 95: 17–33.

Hertz, Noreena (2002) *The Silent Takeover: Global Capitalism and the Death of Democracy*, London: Arrow Books.

Hobson, John A. (1902) *Imperialism: A Study*, London: Nisbet.

Kanbur, Ravi (2001) 'Economic policy, distribution and poverty: The nature of disagreements', *World Development*, 29: 1,083–1,094.

Krugman, Paul (1979) 'A model of innovation, technology transfer, and the world distribution of income', *Journal of Political Economy*, 87: 253–266.

Krugman, Paul (1981) 'Trade, accumulation and uneven development', *Journal of Development Economics*, 8: 149–161.

Lenin, Vladimir I. (1917) *Imperialism: The Highest Stage of Capitalism*, Moscow: Foreign Language Publishing House.

Milanovic, Branko (2002) 'The two faces of globalization: Against globalization as we know it'. Internet www.worldbank.org/research/inequality.

Milanovic, Branko (2003) Notes on inter-national inequality (In Mimeo).

Murshed, S. Mansoob (1992) *Economic Aspects of North-South Interaction*, London: Academic Press.

Murshed, S. Mansoob (2002a) 'Conflict, civil war and underdevelopment', *Journal of Peace Research*, 39: 387–393.

Murshed, S. Mansoob (2002b) 'Perspectives on two phases of globalization' in S. Mansoob Murshed (Ed.), *Globalization, Marginalization and Development*, London: Routledge.

Nayyar, Deepak (1996) 'Free trade: Why, when and for whom?', *Banca Nazionale Lavoro Quarterly Review*, 198: 333–350.

Nayyar, Deepak (2002) (Ed.) *Governing Globalization: Issues and Institutions*, Oxford: University Press for UNU/WIDER.

Stiglitz, Joseph (2002) *Globalization and its Discontents*, London: Allen Lane.

UNDP (1999) *Human Development Report 1999*, New York: United Nations Development Programme.

Williamson, Jeffrey G (2002) 'Winners and losers over two centuries of globalization', 2002 WIDER Annual Lecture. Internet www.wider.unu.edu.

World Bank (2002) *World Development Indicators*, Washington, D.C.: World Bank.

CHAPTER 6

THE SLOW PROGRESS OF INTERNATIONAL FINANCIAL REFORM

STEPHANY GRIFFITH-JONES

1. INTRODUCTION

The wave of currency and banking crises that began in 1997 in East Asia, then spread to Russia and other emerging markets, and even threatened to spill over to the United States, generated broad consensus that fundamental reforms were required in the international financial system. Particularly during 1997 and 1998, the view emerged that existing institutions and mechanisms, designed in the mid-1940s, were inadequate for preventing and managing crises in the twenty-first century. Significant reform – as well as strengthening – of global financial governance was seen as urgent. Besides the objective of achieving international financial stability, an equally important aim, to which insufficient attention has been given, is the provision of adequate capital flows, both private and public, to different categories of developing economies. Such flows can complement domestic savings and provide additional foreign exchange and technology to these economies. This does not imply a return to the excessive levels of easily reversible private lending that characterised the first half of the 1990s, but rather, sufficient levels of stable private and official flows that contribute to higher growth of both low- and middle-income countries.

Thus, from a development perspective, the two major goals for a new international financial architecture are (i) to prevent currency and banking crises and better manage them when they do occur and (ii) to support the adequate provision of private and public investment flows to developing countries, including particularly to the lowest income countries.

This chapter assesses progress on international financial reform in relation to these two goals (see also Griffith-Jones and Ocampo 2003). In this sense, this contribution is broader than most of the literature on the subject, which has focused on achieving international financial stability and avoiding contagion. It should be stressed that a development-oriented international financial architecture would benefit not only

81

Max Spoor (ed.), Globalisation, Poverty and Conflict, 81–96.
© 2004 Kluwer Academic Publishers. Printed in the Netherlands.

developing countries. Stable growth in these countries provides growing markets for developed-country exporters and profitable opportunities for investors. More generally, avoidance of crises in developing countries reduces the risk of crises spilling over to the developed economies. Though small, this risk is significant, as the Latin American debt crises of the 1980s and the combined effect of the Asian and Russian crises of 1997–98 showed.

Though changes have taken place, deep crises continue to occur, most recently in Turkey and Argentina (and the risk of crises severely restricts growth-orientated policies, such as recently in Brazil). These indicate that the international financial system in place needs further changes in the area of crisis prevention and management, in parallel with further improvements in domestic economic policies in developing countries. On top of these issues, the availability of sufficient external finance has emerged as a particularly urgent issue in recent years, given that net private capital flows both to emerging economies and to low-income countries have fallen sharply since 1997.

According to the World Economic and Social Survey (UN 2003: 43), net private financial flows to developing economies peaked at almost US $208 billion in 1996, more than halved to $97 billion in 1997, and continued to fall rapidly (with a small upsurge in 1999) to only $18 billion in 2001. If foreign direct investment (FDI) is excluded, net private flows were $127 billion in 2001. Although there was some improvement in 2002, since 1997 emerging economies have practically become 'net exporters of capital to the developed world', as the 2003 *Global Development Finance Report* puts it (World Bank 2003).

Whilst the initial decline of capital flows to emerging economies after 1997 mainly affected East Asia, more recently flows to Latin America have fallen sharply. Data from the Inter-American Development Bank (IADB) show net private capital flows to Latin America fell from around five per cent of GDP in 1996 to around zero per cent of GDP in 2002. Indeed, according to the Economic Commission for Latin America and the Caribbean (ECLAC), negative net transfers – US $39 billion from Latin America – in 2002 were the highest in nominal terms since statistics are available, more negative than even the worst years of the early 1980s (albeit as a percentage of GDP they are somewhat lower).

At the time of this writing (mid-2003), there was some recovery of private flows to emerging economies, especially bond flows. However, it is too early to know whether these flows are sustainable. It is important to note that private flows to low-income countries (which were never very high) have also fallen since the Asian crisis, but less sharply than flows to emerging markets (Griffith-Jones and Leape 2002). Two comments seem relevant here. Firstly, low-income countries did not experience the large surges of private flows that went to the emerging markets; they were thus less vulnerable to costly, sudden stops or reversals. Secondly, if attracting private flows to low-income countries is an important policy objective (e.g. to complement low levels of savings, particularly in sub-Saharan Africa), then the decline in private flows to emerging markets would seem to be a strong inhibitor of such flows to low-income countries, especially in sub-Saharan Africa. Some portfolio equity investors in the mid-1990s seemed to view sub-Saharan Africa as the "last frontier" of emerging markets. However, when risk aversion increased and, more specifically, appetite for investing in emerging markets fell due to frequent crises in them, these investors dismissed any possibility of investing in low-income countries, even though those countries did not experience similar crises.

The structure of private flows to developing countries has changed as well, with a rotation from debt to equity. This shift has positive features in that FDI flows are more stable and less prone to reversals. However, the fact that an increasing proportion of FDI has recently gone into sectors such as public utilities, where sales are in local currency, prompts multinational companies to hedge their liability exposures, especially in times when there is fear of a large depreciation. This puts pressure on the exchange rate in difficult times, that is, in a pro-cyclical manner.

One statistical and two policy questions therefore arise. For policy analysis purposes, should FDI in companies producing mainly for home markets be registered separately from FDI for exports? Should policymakers in developing countries consider FDI for home production as potentially having more volatile effects than FDI for exports? Could and should measures be taken to affect the level and timing of hedging by FDI for producing in the home market? One option could be for developing country governments to require multinational companies investing for domestic sales to hedge their exposure at the time of entry. This appears an attractive option, though it may pose additional costs for the multinational company and could be difficult to monitor.

Returning to the previous point, to the extent that private capital flows do not recover sufficiently (either spontaneously or encouraged by government policies), a greater role would need to be played by official liquidity and development finance. A particular source of concern is that an important part of the decline in private flows discussed above may be due to structural reasons and not just to cyclical ones. The structural changes include the fact that banks have crossed borders and will substitute foreign for domestic lending and that there are not many "sufficiently large" companies left for equity investors to buy in developing countries. To the extent that these flows are determined by cyclical factors, an important question is how long is the relevant cycle (see Griffith-Jones 2001, IMF 2001). Net private flows to developing countries could remain very low for a fairly long time period.

2. WHAT PROGRESS TILL NOW?

Almost six years after the Asian crisis and with new crises still unfolding, it is important to evaluate progress achieved in reforming the international financial system. Though some progress has been made, it is clearly insufficient. The mechanisms that existed previously and the adaptations of recent years still do not fully meet the demands created by financial globalisation.

The extensive debates in recent years, and the parallels from national systems, indicate that the international financial architecture must provide four services: (i) appropriate transparency and regulation of international financial loan and capital markets and private capital account flows, as well as mechanisms to encourage flows in times of drought; (ii) provision of sufficient international official liquidity during crises; (iii) mechanisms for standstill and orderly debt work-out at the international level; and (iv) appropriate levels and instruments of development finance.

The first two mechanisms are essential for preventing crises, which have proven to be developmentally, socially and financially costly. The third and fourth mechanisms would help manage crises to lessen their costs and also have preventive effects, since a system better suited to manage crises is less prone to destabilising capital flows. This has indeed been the experience of national financial systems in relation to the lending of last resort by central banks. Finally, it is essential to channel flows of development finance to countries, especially low-income ones, that have insufficient access to private flows. An adequate supply of funds must also be guaranteed to middle-income countries during periods of insufficient private capital flows.

Progress in reforming the international financial system has so far suffered four serious problems. Firstly, there has been no agreed international reform agenda. Furthermore, the process has responded to priorities set by a few industrialised countries. These priorities have not always been explicit and have varied through time. In this regard, the "Monterrey Consensus" of the International Conference on Financing for Development of the United Nations, held in March 2002 (see United Nations 2002), provided for the first time an agreed comprehensive and balanced international agenda. That agenda should be used to guide and evaluate reform efforts.

Secondly, progress has been uneven and asymmetrical in several key aspects. The focus of reforms has been largely on strengthening macroeconomic policies and financial regulation in developing countries – that is, on the national component of the architecture –, while far less progress has been made on the international components. This is a major weakness, since crises have been caused not only by country problems (though these have obviously been important) but also by imperfections in international capital markets, such as herding, leading to rapid surges and reversals of massive private flows and multiple equilibria, which may push countries into self-fulfilling or deeper crises.

Another set of asymmetries relates to the excessive focus of the reform effort on crisis prevention and management, mainly for middle-income countries. Important as this is, it may have led to the neglect of equally if not more important issues of appropriate liquidity and development finance for both low-income and middle-income countries. For several years, the problem has seemed to be one of insufficient private flows. Therefore, an important task is to design measures that both encourage higher levels of private flows (especially long-term ones) and provide counter-cyclical official flows (both for liquidity and for development finance purposes) during periods when private flows are insufficient. These important tasks have been relatively neglected in recent years, certainly in the policy arena and even in academic debate. They now require urgent attention. (Some initial ideas are presented in the conclusions section below.)

Within the realm of crisis prevention and management, progress has also been uneven. To prevent crises, much work has been done to strengthen domestic financial systems in developing countries and to draft international codes and standards for macroeconomic and financial regulation. The review of the Basel accord on international banking regulation also concentrated efforts, but many of the main concerns and possible negative impacts on developing countries have yet to be properly addressed. Also, the drafting of new IMF financing facilities has received much more attention than international debt standstills and work-out procedures. In the IMF financial facilities, the Supplementary Reserve Facility (SRF) was a positive development. But frustration has characterised the design of the new facility to manage contagion, the Contingency Credit Line (CCL). The IMF quota increase and the extension of the arrangements to borrow, which became effective in 1999, also represents an advance. However, several proposals on more active use of Special Drawing Rights (SDRs) as a mechanism of IMF financing have not led to action.

Thirdly, some advances in the international financial architecture run the risk of reversal. There is growing reluctance by developed countries to support large IMF lending (or to contribute bilateral short-term lending) to better manage crises. The main arguments given are that these large packages lead to excessive moral hazard, which implies that both borrowers and lenders behave irresponsibly knowing that they will be "bailed out", and that taxpayer money from industrialised countries should not, in any case, be risked in these operations. These arguments have been vastly overstated, as we will see below, but have been quite influential in recent international action.

Fourthly, as discussed in more detail in Griffith-Jones and Ocampo (2003) the reform process has been characterised by insufficient participation of developing countries in key institutions and fora. As regards the international financial institutions (especially the IMF, World Bank and Bank for International Settlements), more balanced representation needs to be discussed in parallel with a redefinition of functions. It is also urgent that developing countries be fully represented in the Financial Stability Forum and in standard-setting bodies, like the Basel Banking Committee. After all, developing countries will be asked to implement the standards there defined.

3. CRISES PREVENTION: HAVE THE RIGHT MEASURES BEEN TAKEN?

3.1 Codes and Standards

As regards crisis prevention, most emphasis and activity has been in the development and implementation of codes and standards for macroeconomic policy and financial sector regulation in developing countries. Clearly the aims of such efforts, such as strengthening domestic financial systems, are worthy and desirable. A concern, however, is whether implementing existing codes and standards would always be meaningful in helping to prevent crises. Indeed, it could be argued that standards to be implemented by developing countries or which would affect developing countries should more explicitly incorporate criteria for crisis prevention in developing countries. The introduction of explicit elements of counter-cyclicality into banking (and possibly other financial) regulation seems particularly relevant in this context (see Ocampo and Chiappe 2003, BIS 2001). This is both because financial actors seem particularly pro-cyclical in their behaviour within and towards developing economies and because this pro-cyclicality can be especially damaging to these more fragile economies, with their fairly thin financial markets.

A concrete example of such a measure could be the introduction of obligatory forward-looking provisions along the model applied by the Central Bank of Spain. This approach estimates risk based on past experience (to cover at least one business cycle) and creates a mechanism in which provisions increase during economic expansion; provisions are drawn down during slowdowns and recessions (Ocampo and Chiappe, op. cit., Poveda 2001). Such a cycle-neutral approach could be complemented by explicitly counter-cyclical mandatory provisions on rapidly increased bank lending, for example, to sectors characterised by cyclical risk (such as building and real estate), or the growth of foreign currency-denominated loans to non-tradable sectors (as suggested in Ocampo and Chiappe, op. cit.). Such measures could help smooth the link between pro-cyclical behaviour of financial actors and excessive macroeconomic cyclicality.

More generally, the introduction of counter-cyclical or cycle-neutral elements into financial regulation (both domestic and international) could create countervailing forces to dampen the natural tendency of financial markets to pro-cyclicality and short-termism, a tendency that has been accentuated in recent times. A key objective for regulators could be to encourage diversity of risk-management models, to match the diversity of investment objectives as well as of characteristics of lenders and investors. Regulators also should take account of (and attempt to compensate for) the complex and problematic interactions between risks that pro-cyclical and herding behaviour in different actors generate. For example, a downgrade of a sovereign by a rating agency can cause investors to sell bonds immediately; simultaneously domestic counterparties of derivatives may have to meet margin calls (Dodd 2002) and banks may stop lending.

Counter-cyclical and integrated regulation seem to be the way ahead, both within developing countries and for regulation of international actors lending and investing in developing countries. Some aspects of such regulation, however, are technically difficult to implement. For example, it can be difficult to distinguish between cyclical and long-term trends (Goodhart 2002). Whilst wishing to slow booms, regulators should not curb increases in sustained growth, though in practice this can also be difficult to distinguish. A second problem is that most regulators do not normally regulate in a counter-cyclical manner, though such practice needs to become established. Thirdly, there will be political-economic pressures to avoid tightening of regulation in boom times. In spite of the problems and difficulties, more integrated and counter-cyclical regulation would seem to offer large potential gains in curbing damaging boom-bust patterns.

3.2 Basel II

The main regulatory change introduced since the Asian crisis is the proposed Basel Capital Accord. Though implying improvements in some areas, this new accord does not deal with the two problems affecting developing countries in terms of international bank lending going to them: their boom-bust pattern and their reversal since 1997.

Quite the contrary. The impact of Basel II could be to increase the pro-cyclicality of bank lending in general, and to developing countries in particular, and to reduce further and increase the cost of bank lending to developing countries.

The proposed internal rating based (IRB) approach, to be implemented initially by large international banks, would exacerbate pro-cyclical tendencies; the drive for risk weights reflecting probabilities of default (PD) as estimated by banks is inherently pro-cyclical. During an upturn, PD falls, thus implying lower capital requirements. In a downturn, PD grows, as the same portfolio of loans is seen as more risky, thus raising capital requirements. As it is difficult to raise capital in a downturn, the result could be a credit crunch that could downturn into recession. This increase in pro-cyclicality goes against what is increasingly accepted as best practice and, as outlined above, the need to introduce neutral or counter-cyclical elements into regulation.

Equally serious, the IRB approach would further reduce international bank lending and increase the cost of such lending, particularly to developing countries that do not have investment grade (the large majority) (Griffith-Jones, Spratt and Segoviano 2003). Recent research shows that the current Basel proposal would quite significantly overestimate the risk of international bank lending to developing countries. This would increase capital requirements excessively on such lending, leading to a sharp increase in the cost of bank borrowing by developing countries, as well as to a fall in the supply of bank loans. This is particularly serious now because, as discussed above, in the last five years bank lending to the developing world has fallen sharply. Thus, the current proposals are problematic, both in terms of the Basel Committee's own aims (more accurate measurement of risk for determining capital adequacy) and due to their further discouragement of already insufficient bank lending to emerging markets, which damages growth of their economies. The latter impact is manifestly counter to one of the aims of the G-10, which is to actively encourage private flows to developing countries so as to provide an engine for stimulating and funding growth.

One of the reported major benefits of lending to – and investing in – developing countries, is their relatively low correlation with mature markets. The research quoted above carefully tested this hypothesis empirically and found strong evidence, for a variety of variables and over a range of time periods, that correlation between developed and developing countries is significantly lower than correlation amongst developed countries. For example, spreads on syndicated loans, which reflect risks and probability of default, tend to rise and fall together within developed and developing regions more than between developed and developing countries. Similar results are obtained for the correlation of profitability to banks. Furthermore, broader macroeconomic variables (such as GDP growth, interest rates, evolution of bond prices and stock market indexes) show far more correlation within developed economies than between developed and developing ones.

Finance theory and practice tells us that a bank's loan portfolio diversified between developed and developing countries thus has a lower level of risk than one focused exclusively on lending to developed economies. In order to test this more directly we simulated two loan portfolios, one diversified only across developed economies and another diversified across developed and developing regions. The estimated unexpected losses for the portfolio focused only on developed-country borrowers was 23 per cent higher.

Given that the capital requirements determined by Basel regulators are aimed precisely at helping banks cope with unexpected losses, it is extremely unfortunate that the current Basel proposals do not explicitly incorporate the benefits of international diversification. The surprising fact that, at the time of this writing, the Basel proposal did not do so implies that in this aspect, capital requirements will not accurately reflect risk and thus unfairly penalise lending to developing countries.

It is encouraging that there is a clear precedent for change, as the Basel Committee has already done (in a fairly limited way) with respect to lending to small- and medium-sized enterprises (SMEs). After the release of the consultative document in January 2001, there was widespread concern – especially in Germany – that the increased capital requirements would sharply reduce bank lending to SMEs, with negative effects on growth and employment. The technical case was made that the probability of a large number of SMEs simultaneously defaulting was lower than for a smaller group of large borrowers. Intensive lobbying by the German authorities led to this technical argument being recognised, and the Basel Committee agreed to lower capital requirements by 10 per cent on average for smaller companies. Our empirical research (Griffith-Jones, Segoviano and Spratt, op.cit) found that at least as large a modification is justified with respect to international diversification related to lending to developing countries. There are no practical, empirical or theoretical obstacles to such a change, which could potentially benefit the developing world and ensure more precise measurement of risk and capital adequacy requirements. This, after all, is the main aim of the entire process.

It is difficult to persuade the Basel Committee to make changes that are technically and economically correct because of the committee's composition and the dominant influences on it. Developing and transition economies are not represented in the Basel Committee. It is heavily influenced by the large G-10 countries' regulators, and these are strongly influenced by their international banks. The modelling "industry" also seems to wield large influence. The G-10 bank regulators are not particularly democratically accountable even in their own countries, and certainly not to the developing countries. Here there is a clear problem of governance. G-10 bank regulators are mainly concerned with bank stability (a legitimate concern) and with enhancing the competitiveness of their own banks. They also wish to improve the micro-economic efficiency of allocating risk by banks and to discourage what they see as "excessively risky lending"; but they are insufficiently concerned with the macroeconomic effect of such an approach, even within their own countries. They are still less concerned with macroeconomic effects in developing countries, though these have large but indirect effects on the G-10 economies.

More broadly, important differences have arisen between the United States and Europe, with the former wanting to require only the very largest banks to implement the IRB approach and the latter wishing to implement it amongst all banks and even non-financial institutions (see Milne 2003). These divisions could bring about new opportunities for developing country bargaining, though divisions among the G-10 could also detract attention from developing country concerns.

3.3 Contingency Credit Line (CCL) and Access to IMF Lending

During the 1990s, capital account liberalisation and the large scale of private flows greatly increased the need for official liquidity in times of crises. As a result of the Asian crisis, IMF resources were significantly increased, which facilitated the provision of the fairly large financial packages that played a positive role in managing and containing crises.[1] Two new facilities were created as well: the Supplementary Reserve Facility (SRF), which has been successfully used and whose maturity was recently extended, and the preventive Contingency Credit Line (CCL), which remains unused.

Creation of the CCL was a positive initiative, since it could reduce the chances of a country entering into crisis by providing contingency lending agreed in advance. However, at the time of this writing, no country had yet applied for use of the CCL. If no strong case is made and no country applies, the facility will expire in autumn 2003. Though precautionary stand-by agreements can to some extent play a function similar to the CCL, it would be unfortunate if this preventive contingency facility were eliminated rather than improved.

The key problem with the CCL has been that countries with "good" policies, and which are perceived as such, fear that if they apply for a CCL they will be stigmatised by the markets. No country is eager to be the first to apply. To make this facility more attractive and to diminish or eliminate any potential stigma attached to it, modifications have been suggested. For instance, it could be agreed that all countries favourably evaluated by the IMF in their annual Article IV consultations would automatically qualify for a CCL. This would imply that quite a large number of countries – including the developed ones – would qualify, thus eliminating the current stigma on its use. This proposal is similar to one suggested by the UK treasury, whereby after a positive evaluation in Article IV consultations a country would automatically become eligible for a CCL. In this latter variant, the country would still have to apply, but this step would be far easier because the country would already know it was eligible. Being named by the IMF as eligible for a CCL would thus reflect a country's strength (indicating good policies), rather than, as currently feared, being a sign of possible future weakness. The fact that countries could access a CCL would hopefully diminish the likelihood of crises and therefore the need to draw on it.

Another source of possible concern for developing countries is the IMF review of access policies in the context of capital account crises (IMF 2003). The stated purpose of this review is to 'establish a stronger framework for crisis resolution, and provide member countries and financial markets with greater clarity and predictability'. The source of concern is whether this review will lead *de facto* to more restricted and slower access. It is encouraging that no presumptive limit on cumulative exceptional access was introduced – it was discussed and dismissed by the IMF board. The board concluded that 'while some moral hazard is bound to be present in Fund lending, there is little evidence that the use of exceptional access in general has had large effects on moral hazard'. Developing countries may be concerned that they will need to meet tighter criteria for exceptional access in case of capital account crises. The criteria include some straightforward ones, like exceptional balance of payments pressures on the capital account that cannot be met within normal limits and a policy programme that provides reasonably strong prospects of success. Other conditions are more difficult to quantify, however, and may therefore delay or even restrict approval.

Furthermore, certain procedures after the review have been strengthened. While positive in providing additional safeguards and enhancing accountability, this could produce delays and even decrease the likelihood of packages. The problem is the speed with which capital account crises can unfold and deepen, and the need to have responses from the official sector within a timeframe that is not too much longer than that of private behaviour.

Two final issues seem relevant in the broader context of the international financial architecture. One is that the high percentage of quotas for IMF lending in the context of capital account crises is the result of the fact that IMF quotas seem to be insufficient in relation to the needs of a globalised financial economy. Indeed, since the creation of the IMF, quotas have grown less than GDP or trade indicators; they have certainly not expanded to take sufficient account of the larger scale of private flows and their volatility and reversibility. The second issue relates to the speed and ease of access to IMF lending related to liquidity problems (where a combination of adjustment and financing would probably be sufficient to restore countries to a stable path). Experience shows that liquidity crises can deteriorate into solvency ones, especially if the former are mishandled. Difficult or slower access to IMF lending could be damaging to developing economies, because no clear international framework for orderly debt work-out has been agreed (though there is some initial progress on collective-action clauses and in other areas).

4. POLITICAL ECONOMY

Progress on international financial reform has been uneven and asymmetrical. More has been achieved nationally by developing countries (e.g. codes and standards) than in the equally important and complementary international arena (e.g. provision of sufficient official liquidity and development finance and design of international debt work-out procedures). What are the main reasons for this uneven progress? More importantly, what strategy and bargaining tactics could be most productive for achieving a more symmetrical process?

Clearly, the asymmetries in the international financial reform process reflect certain political and economic characteristics of the world. Few powerful governments – or their financial authorities – have thrown their weight consistently behind a deep international financial reform, though they were enthusiastic about such reform after the 1997–98 Asian and Russian crises, largely due to the brief credit crunch they generated in the industrialised world.

One reason for the lack of consistent developed country support for the reform process may be that powerful actors in those countries (e.g. major financial agents) do not see major changes in the international financial architecture as in their interest. Another problem is that those who would benefit most from such changes in developed countries (e.g. shareholders and workers of companies trading and investing long-term in developing economies, along with supporters of development in poor countries) are not represented in financial decision-making.

As a result, it would seem that the main impulse for international financial reform must come from developing countries. However, developing countries have their own restrictions. First, and most importantly, they have limited power, as reflected in their exclusion or limited participation in key bodies. Second, developing countries have seen their ability to generate strong coalitions weakened. This may be linked to "policy competition" to attract foreign capital, and a resulting unwillingness to make or support proposals that could undermine their image as friendly to foreign investors. The fact that some large developing countries (e.g. Brazil) have started to take initiatives in this area is, however, encouraging. Indeed, without conscious and deliberate effort to overcome the basic asymmetries in global power relations and to form international coalitions to compensate for power imbalances, the international financial agenda will continue to be biased towards the views of a limited set of actors in the industrialised countries.

However, two positive elements may be helpful in the process of genuine international financial reform. One is that all the key actors share a common objective: they favour and benefit from sustained growth in developing countries. As shown in Table 1 below, for some actors this is more important than for others, but all share this objective.

Table 1. *Objectives of Key Actors*

Actors	Dominant Objectives	Other Objectives
Developed country governments	– Growth in their own economies – Profits for their financial sectors – Global financial stability – No large bail-outs	– Growth in developing countries – Avoidance of crises
Developing country governments	– Growth in their own economies – Global financial stability – Stable and adequate flows	– Growth in developed countries
Banking and financial markets	– Maximum profits	– Global financial stability – Growth in developed and developing economies

The second positive element is the existence of a set of actors in developed countries that is an important ally of developing countries in building a better international financial system. These include the non-financial parts of governments such as development cooperation ministries, parliamentarians and political parties, and non-financial corporations and NGOs. In different ways, and for different reasons, these actors are supportive of more rapid growth in developing countries. They therefore are or could become very supportive of international financial reforms that help make such growth possible. For this purpose, developing country governments need to maintain an active dialogue on international financial reform, not just with financial authorities in developed countries, market actors and international financial institutions (which are clearly the main actors in the reform process) but also broadly with other actors in the developed world.

Developing countries could attempt to design and offer a "grand bargain" on international and national financial reform that would be attractive to a whole range of developed country actors, both public and private sector, as well as supportive of their own growth and development.

Such a bargain would have two sets of elements. Developing countries could say they would be keen to implement initiatives that are of particular interest to developed economies, such as codes and standards and financial regulations and fuller liberalisation of their capital accounts, if and only if developed countries start reforming the global financial system in ways that would facilitate larger and more stable capital flows to developing countries and would make costly crises in these countries less likely. Without such a reformed international financial system, they would be less able and less willing to fully open their capital accounts, since the potential risks of doing so could outweigh the benefits. Particularly, developing countries could argue that implementing codes and standards and a commitment to proper domestic macroeconomic policies should be explicitly linked to four elements of reform: (i) some regulation of developed countries' financial markets to help avoid surges of potentially reversible capital flows to the developing economies; (ii) mechanisms that encourage long-term flows; (iii) the design of (low-conditionality) international liquidity mechanisms that would significantly protect individual developing countries from crises and stop crises from spreading to other countries; and (iv) fair multilateral debt work-out mechanisms that would be used to manage solvency crises (debt overhangs).

Thus, developing countries that follow good macroeconomic policies and significantly improve their financial regulations (as certified, for example, in their annual Article IV IMF consultations) could have virtually automatic access to sufficient IMF lending if hit by a crisis that was not of its own making, but rather, was due to unexpected changes in international lenders' perceptions of investors or to terms-of-trade shocks. Low-income countries that follow good macroeconomic policies and improve their financial regulations would have sufficient access not just to international liquidity, but also to development finance. Debt work-out mechanisms would be used only when crises faced by developing countries are due to unsustainable debt burdens (and would not be used when they are associated with insufficient international liquidity). Moreover, appropriate mechanisms would be designed to guarantee financing in the post-debt restructuring environment to facilitate reinsertion into private capital markets.

Such a bargain would provide incentives for developed countries to make necessary international changes, since these would ensure the desired changes in developing countries, and vice versa. Collective-action problems could thus be overcome if genuine progress was made simultaneously by developed and developing countries. Most importantly, the result would be of great value, not just to developing countries, but also to developed ones.

Developing countries could draw valuable lessons from both the bargaining tactics used and the vision presented by Keynes in negotiations that led, at Bretton Woods, to the creation of the post-war international financial order (see Skidelsky 2001). Keynes presented two clear alternatives: an "ideal" scheme, with key international elements – such as a large IMF – and a "second best" case, wherein the United Kingdom would reluctantly follow a far more closed approach in trade and the capital account if the international financial system was not properly developed. There was, he argued, no middle way (though in practice he made some important concessions later).

Suitably adapted to the features of the early twenty-first century world economy, developing countries could argue that the same two clear options remain. The first option is an appropriate international financial system that supports development and makes crises far less likely and less costly, not just for developing countries but for the global economy. Developing countries could contribute to this by implementing regulatory standards, adopting good macro-policies and by gradually liberalising their capital account. The second option is for an incomplete and lopsided international financial system that is unable to guarantee support for developing country aims. Developing countries would not be able to fully open their capital accounts; they would regretfully have to protect their interests by having, as a "second best solution", more rather than less national policy autonomy. Similarly, they may be forced to rely on regional institutions and mechanisms even to perform functions that could best be performed globally, given vacuums in the existing global financial architecture.

Developing countries could draw lessons from Keynes' preparation of a clear vision of the key elements which need to be included in a "first-best" international financial system and show how such a superior system would benefit all involved. This system would be superior both because it would support more stable growth in developing countries – of benefit to many actors in the developed world – but perhaps more importantly, because it would increase financial stability globally. Here there is a clear parallel with Keynes' position at Bretton Woods. In defending the interests of the relatively weaker, debtor countries like the United Kingdom, Keynes was at the same time defending global prosperity.

Just as Keynes appealed then to US internationalism and liberalism to help overcome opposition to his proposals, developing countries should now appeal to the US ideals of supporting and deepening the market economy globally. For this, they should stress how a "first-best" international financial system, that would facilitate growth and prosperity for them, would clearly increase their own commitment to the global market economy and their ownership of policies to integrate further into it.

5. CONCLUSIONS AND POLICY SUGGESTIONS FOR ENCOURAGING PRIVATE FLOWS

Recent trends in private capital flows and changes in the international financial system are not particularly encouraging. As regards private flows, there has been a sharp fall in the five years since the Asian crisis, especially in bank lending. Moreover, actions to encourage private flows, especially long-term flows, to developing countries, particularly emerging ones, have been limited. Whilst there has been an improvement in the structure of private flows, with a change from debt to FDI, the favourable impact should not be over-estimated, due to problems such as hedging of FDI for the domestic market and short-termism and reversibility of non-FDI flows.

As regards measures to improve the international financial and development architecture, progress has been insufficient and excessively slow. There have even been some reversals, although these have perhaps not been as major as seemed likely previously (e.g. in terms of formally reducing access to IMF lending during capital account crises). Valuable theoretical work was recently done on the need for counter-cyclical regulation which could help curb boom-bust cycles in financial and banking markets. But there has been as yet little practical implementation of these findings and as a result little is known about how effective such measures would be in practice. It is worrying that the only major regulatory change being discussed internationally, a rather major modification of the Basel Capital Accord, whilst having some positive features, could actually increase the pro-cyclicality of bank lending and discourage international bank lending to developing countries, especially those below investment grade (the majority).

As regards IMF lending to prevent or better manage capital account crises, the CCL remains unused and could expire. There seems to be some tightening of access to IMF lending in times of capital account crises and the possibility of some slowing down of granting such access.

This chapter concludes with some suggestions of measures to encourage private flows in times of drought, which seems particularly relevant at the present time. Such measures could be reduced or reversed if private capital surged; indeed, in such a case, particularly recipient countries would need to discourage flows. To the extent that the new trend towards a drought of capital flows to emerging markets is likely to last longer, the policy agenda needs to shift, at both the national and international levels. The immediate problem is how to encourage sufficient private flows to developing countries. This chapter has focused on measures to be taken internationally, and/or in source countries, though measures in developing countries are also important.

One of the novel problems that has arisen during and in the aftermath of recent crises is that trade credit has dried up. At present, government institutions such as export credit guarantee agencies (ECA) and multilateral development banks limit their activities (providing guarantees and credits) to longer term assets. An important policy question is whether they should extend their activities to also cover short-term assets. In fact, the Inter-American Development Bank (IADB) is currently exploring the creation of a guarantee mechanism tailored specifically to encourage trade finance provided by commercial banks. Such guarantees might be particularly useful for a country like Brazil, which in 2002 experienced difficulties in accessing short-term trade credit but was not in a full crisis. One could of course go one step further, and have an institution like an ECA or the IADB grant trade credit in special circumstances (e.g. if a guarantee programme failed to restore an adequate level of trade credit). Such a programme for either guarantees or the direct provision of trade credits could be temporary and be phased out once full access to trade credit from commercial banks was restored.

In the case of long-term trade credit, ECAs already play a large, if declining, role in guaranteeing credits. An important issue is the extent to which these agencies and development banks should be willing to be counter-cyclical in the guarantees they grant. If it is accepted that international financial markets tend to overestimate risk in difficult times and underestimate it in good times, there is a strong case for introducing an explicit counter-cyclical element into risk evaluations made by ECAs. In times when banks and other creditors lower their exposure, ECAs would increase or at least maintain their levels of guarantees. When the markets view matters as improved, so that banks increase their willingness to lend, then ECAs could decrease their exposure, for example, by selling export credit guarantees in the secondary market. This would avoid a greater counter-cyclicality of guarantees resulting in an increased average level of guarantees. .

One way to increase the effectiveness of multilateral development bank guarantees in inducing private flows might be to guarantee only those risks that the markets are not prepared to cover (e.g. possibly covering only country risk and not commercial risk). It would also be possible to cover only initial maturities, and then roll over the guarantee once these initial payments are made. Other mechanisms include introducing guarantees in local currency instruments. Alternatively, in some cases private actors may be willing to lend for early maturities, and institutions like the IADB and the World Bank may need to guarantee later maturities or provide co-financing for later maturities. This is particularly appropriate for infrastructure investments, which have high initial sunk costs and long gestation periods before projects become profitable (see Gurria and Volcker 2001, Griffith-Jones 1993). Infrastructure projects therefore often need financing for up to 25–30 years, while the private market normally only provides loans with significantly shorter maturities.

Public-sector institutions could play a much more consciously anti-cyclical role than has been customary. Further, a more proactive role could be encouraged for socially responsible investment. Such investment has tended to be negatively slanted, focusing on restrictions on investing in undesirable activities, such as those that employ child labour, do not meet environmental or labour standards, or involve "sins" like tobacco, alcohol and gambling. Yet such restrictions might discourage investment in developing countries. A new definition of socially responsible investment could incorporate a central positive aim: to support long-term private flows to developing countries to help fund pro-poor growth. This would over time help to improve labour standards, not least because socially responsible foreign investors, by being present and engaged in developing countries, could have a positive influence on wages.

A change in the concept of what amounts to socially responsible investment, both by institutional and retail investors (where social responsibility has an important and growing presence), from a negative "anti-bad things" to an emphasis on pro-poor growth in developing countries, could have a positive impact on both the level and stability of private flows to developing countries. In particular, pension funds could provide more stable flows as their liabilities are on average very long term. In the United Kingdom, legislation introduced in 1999 requires that all pension funds set out in their annual report the way that social and environmental factors were taken into account in their investment decisions. This facilitates the examination of investment practices by fund trustees and members, and provides them with a basis on which to lobby for change if they wish. The change in UK regulation was soon replicated in various other European countries. In the United States too a number of large institutions, both pension funds and religious foundations, have a tradition of socially responsible investment; those investments could be in part channelled to emerging markets.

An important challenge is therefore to encourage socially responsible investors to expand their horizons and recognise their responsibility to help promote development. This need not imply an inferior long-run investment performance, for there is evidence that the return/risk ratio of a portfolio that has a part of its assets invested in developing country equities will be higher in the long term than if it invests only in developed countries (see e.g. Armendariz et al. 2002). The potential is large, given the rapidly growing scale of socially responsible investment assets. A global estimate of such assets calculated a total of US $2,700 billion for 2001 (Persaud 2003).

NOTE

1.   The conditionalities attached to these packages have been controversial.

REFERENCES

Armendariz, E., R. Gottschalk, S. Griffith-Jones and J. Kimmis (2002) 'Making the case for UK pension fund investment in development country assets', Institute of Development Studies, July (In Mimeo).
BIS (2001) *The New Basel Capital Accord, Basel Committee on Banking Supervision*, May. Basel: Bank for International Settlements.
Dodd, R. (2002) 'Improving financial markets: Regulatory proposals to dampen disruptions and deter distortions'. Paper presented at the Conference After Neoliberalism – Economic Policies That Work for the Poor: The New Rules for Global Finance Coalition, May 23–24, Washington, D.C.
Griffith-Jones, S. (2001) 'New financial architecture as a global public good'. Paper prepared for the United Nations Development Programme.
Griffith-Jones S. and J. Leape (2002) *Capital Flows to Developing Countries: Does The Emperor Have Clothes?* Brighton: Institute of Development Studies.
Griffith-Jones, S. and J. Ocampo (2003) 'What progress on international financial reform? Why so limited?' EGDI Study No. 1. Stockholm: Expert Group on Development Issues.
Griffith-Jones, S., S. Spratt and M. Segoviano (2003) 'Basel II and Developing Countries: Diversification and Portfolio Effects'. Brighton, Institute of Development Studies (In Mimeo).
Goodhart, C. (2002) *The Inter-Temporal Nature of Risk*. London: Financial Markets Group, London School of Economics.
IMF (2001) *World Economic Outlook*. Washington, D.C.: International Monetary Fund.
IMF (2003) 'IMF concludes discussion on access policy in the context of capital account crises; and review of access policies in the credit tranches and the extended fund facility'. Public Information Notice (PIN) No. 03/37. Washington, D.C.: International Monetary Fund.
Milne, A. (2003) *Basel Lite: Recommendations for the European Implementation of the Basel Accord*. London: Center for the Study of Financial Innovation.
Ocampo, J. and M. Chiappe (2003) 'Counter-cyclical prudential and capital account regulations in developing countries'. EGDI Study No. 1. Stockholm: Expert Group on Development Issues.
Persaud, A. (2003) 'The folly of value-at-risk: How modern risk management practices are creating risk'. Internet: www.gresham.ac.uk/commerce/2002-2003/Lect021202.htm. Accessed November.
Poveda, R. (2001) *La Reforma del Sistema de Provisiones de Insolvencia*. Banco de España, Madrid, January.
Skidelsky, R. (2001) *John Maynard Keynes. Vol.3, Fighting for Britain 1937–1946*. London: Macmillan.
United Nations (2002) 'Monterrey Consensus: International conference on financing for development', March. New York: United Nations.
United Nations (2003) *World Economic and Social Survey*. New York: UN Department of Economic and Social Affairs.

CHAPTER 7

THE DEBT CRISIS AND THE SOUTH IN AN ERA OF GLOBALISATION

KUNIBERT RAFFER

1. INTRODUCTION

The debt crisis is both an engine and a result of globalisation. International credit markets were an early example of globalisation after 1945. Long before the expression "globalisation" was coined, the Euromarket had established itself beyond national regulatory interventions and began to lend to the South in the late 1960s. After 1973, the recycling of "petrodollars" was widely acclaimed as proof of the success of the free and unregulated global market. Already the essential argument was that allowing market forces to operate – "globalisation", "liberalisation" and "deregulation" in modern lingo – would ultimately benefit everyone. Developing countries enjoying unprecedented access to capital were the proof. Their savings and foreign exchange gaps were efficiently closed. However, the international debacle of globalisation in 1982 and the bail-out of commercial banks using taxpayers' money put an end to that claim.

The international debt crisis was not the only crisis of neoliberal policies in the early 1980s. The liberalising and opening of the Chilean economy also produced a financial crash. Chile's military dictatorship provided an ideal context in which to realise neoliberal ideas. Price controls were eliminated, public expenditure reduced, state enterprises privatised and banking supervision cut down. The economy opened up, with moral backing from the Bretton Woods institutions. Voluntary financial transactions between private agents were seen as the business of only those directly involved, and presumably Pareto-optimal. The International Monetary Fund's director of the western hemisphere, E. Walter Robichek, even assured Chileans that private borrowers – as opposed to governments – were unlikely to over-borrow, even with official guarantees (Diaz-Alejandro 1985: 9). This view is sometimes called the Robichek doctrine. The bulk of booming credit expansion went into speculation and the trade deficit in 1982 reached 70 per cent of export revenues. After bailing out the

Max Spoor (ed.), Globalisation, Poverty and Conflict, 97–114.

Banco Osorno, the Chilean authorities realised that 'practically no inspection or supervision of bank portfolios existed' (ibid.: 8). Reserve requirements had been steadily reduced, and 'apparently little effort was spent on investigating the banking credentials of new entrants' when banks were privatised (ibid.). The Chilean miracle of 1981 turned into a catastrophe in 1982, when GDP fell by more than 14 per cent. The government was forced to socialise private losses. The similarities with Mexico in 1994–95 and with Asia in 1997 are striking.

The US Savings & Loans crisis of the 1980s is another case in which the public had to pick up the bill for neoliberal policies. Inflation-adjusted losses due to the debacle were 'several times larger than the losses experienced in the Great Depression' (Stiglitz 1998: 16). Yet, in relation to GDP this crisis is dwarfed by crises in the South and formerly communist countries: 'This debacle would not make the list of the top 25 international banking crises since the early 1980s' (ibid.). This means that more than one major crisis unfolded per year on average – effects of globalisation, liberalisation and deregulation that are simply ignored in official discourse. Recalling Mexico in 1994–95 and the volatility of global financial markets, the OECD (1996a: 57) demanded 'provision of a much larger officially provided safety net'. Less diplomatic language might simply speak of officially subsidised speculation, with losses covered by governments, possibly even from aid budgets. In contrast to problems of official aid, catastrophic private failures are interpreted as the manifest need for even more taxpayers' money. Debtor countries have picked up the bill for bail-outs plus interest on the money lent to finance them.

Unlike the less-developed countries in the South and especially the poor in debtor countries, creditors and speculators have enjoyed protection. Both Bretton Woods twins have gained dramatically in importance. Acting as judge, jury, expert and bailiff on their own behalf, public creditors have forced debtors to accumulate further unpayable debts and to "open" their economies much beyond what the creditor governments themselves would accept.

Rodrik (1996: 17) interpreted the debt crisis as an opportunity seized by orthodox economists for a 'wholesale reform of prevailing policies' offering the chance 'to wipe the slate clean and mount a frontal attack on the entire range of policies in use'. A crisis brought about by over-spending and over-lending in globalised credit markets and a sudden change of northern economic policy that sent interest rates skyrocketting – as the OECD (1996b: 18) admits – was simply declared to stem from too little globalisation: import substitution and "inward looking" policies. No distinction was made between "bad" and "appropriate" import substitution, even though the Asian tigers used these discredited policies to good effect before they embarked on neoliberal globalisation. The crash of the globalised credit market provided leverage for further globalisation in the South.

The debt crisis displays the typical pattern of neoliberal crises: euphoric eulogies on the efficiency and benevolence of shedding national boundaries and regulatory constraints right up to the crash. Once the crisis breaks, attitudes towards public intervention totally change, and socialising private losses created by deregulated markets is welcomed and demanded. Southern governments have repeatedly had to socialise private debts. Taxpayers, especially vulnerable and politically powerless groups, are left to bear the brunt. In contrast to domestic markets, mechanisms to protect the poor do not exist in the international sphere. Globalisation has not replicated the structures of domestic civilised legal and political systems – considered useless at best to creditors and speculators. No crisis resolution procedure with debtor protection yet exists for a sovereign debt overhang.

This chapter briefly analyses the global credit market. It identifies four waves of debt-creating flows which changed debt structures substantially while increasing the total amounts of poor countries' debts. Money from new sources has masked the problem of over-indebtedness instead of solving it. Meanwhile no further group of creditors is available. Therefore, in November 2001 the IMF started the discussion on the sharing of unavoidable losses. Globalising features of insolvency – an old idea fought bitterly by the IMF for nearly two decades – was taken up by Anne Krueger (2001a). Though she is to be commended for breaking the taboo of mentioning insolvency, her model aims at protecting the IMF and is unlikely to provide a solution. It was rejected during the 2003 spring meeting of the Bretton Woods institutions. Understandably, strong opposition also came from emerging markets – the very countries in whose favour the IMF claimed its model to be. After showing the model's faults, this chapter presents a more appropriate model based on globalising economic efficiency, the rule of law and respect for human rights.

2. WAVES OF RESOURCE FLOWS AND CHANGES IN DEBT STRUCTURES

Structural disequilibria were identified early on as the base of the debt problem by the Pearson report (Pearson et al. 1969: 153ff), which was prepared on request of the president of the International Bank for Reconstruction and Development (World Bank) at the very time when large-scale lending in the Euromarket took off. The report strongly recommended debt relief as necessary for less-developed countries. It warned of 'many serious difficulties' that could result from very large-scale lending:

> The accumulation of excessive debts is usually the combined result of errors of borrower governments and their foreign creditors. Failures on the part of the debtors will be obvious. The responsibility of foreign creditors is rarely mentioned (ibid.: 156).

This sounds as modern as the report's finding that debt management has emphasised spending cuts and credit restrictions while neglecting the need to sustain sound development outlays. The early debt problem resulted from the first wave of lending by official creditors after 1945. In spite of these warnings southern debts exploded during the 1970s, covering up the problem.

This borrowing and lending spree was fuelled by several factors. There was a nearly universal convergence of interests of borrowers, lenders and OECD governments. A period of negative real interest rates and relatively high commodity prices encouraged less-developed countries to borrow. Spreads fell dramatically. Differences of interest rates between North and South, supposedly reflecting differences in risk, were perceptibly reduced. Furthermore, less-developed countries saw commercial loans as an opportunity to diversify their sources of finance, to improve their position vis-à-vis official, especially multilateral lenders, by reducing creditor power to impose conditions. Declining official development assistance (ODA), the practice of tying aid and the frequent use of ODA by "donors" for political purposes or to obtain economic advantages (cf., Raffer and Singer 1996), made commercial loans attractive. Apart from very few exceptions, commercial banks have not tried to dictate their clients' policies.

Problematic facts, such as the first "adjustment measures" some thirty years ago, and warning voices raised throughout the 1970s were simply ignored (cf., Raffer and Singer 2001: 163ff). Even after August 1982 the Bretton Woods twins thought for a while that there was no crisis, even stating that the money market functioned well. They saw no signs of liquidity bottlenecks, nor of restrictions in the capital base for private bank lending to less-developed countries, which was supposed to continue on a large scale. But relatively soon they skilfully used this crisis to dramatically increase their importance. After the demise of the Bretton Woods system, economic reason would have suggested dissolving its supporting mechanism, the IMF, too. Yet instead the debt crisis provided a unique chance to find a new justification for its existence, as debt manager.

Conveniently, a scapegoat was easily identified: OPEC surpluses and increased import costs of crude oil. This view, which exonerated the Euromarket and globalisation by blaming southern commodity exporters, became textbook wisdom. Todaro and Smith (2002: 610) explained the whole debt build-up by OPEC surpluses. Yet quantitative evidence fails to corroborate this view (Raffer and Singer 2001: 136ff). Surpluses of some OPEC members were important but only part of global liquidity. The group's net surplus was considerably lower, as many OPEC members were themselves large borrowers. The "first" crisis country in 1982, Mexico, is even an important oil exporter.

During the 1970s commercial banks knew that their claims would be protected "against the market" by northern governments. Disregarding the most elementary rules of prudent banking, the assumption was made that sovereigns might get illiquid, but never insolvent. Massive misallocations of resources resulted, both directly within borrowing countries and indirectly by preventing otherwise viable investments elsewhere. If doubtful loans had not been urged upon the South, excess supply of international money would have driven down interest rates further, making investments viable that were unprofitable under actual interest rates. These imperfections and interventions which kept the market mechanism from functioning cannot be blamed on OPEC.

Debt management after 1982 was based on the so-called illiquidity theory: the assumption that there was no fundamental crisis, only a temporary inability to pay. Public funds poured in – the second wave of inflows – allowing commercial banks to receive higher (re)payments than would otherwise have been possible and producing a remarkable change in debt structures. The World Bank (1988: xxix) complained that concerted lending by banks since 1982 had been just sufficient to refinance around a quarter of interest payments by the highly indebted countries (HICs), 'making the World Bank the principal net lender to HICs' (ibid.). A substantial bail-out of private banks by public multilateral institutions and a major shift of risk to these public entities occurred. This hardened conditions for debtors since multilaterals, in marked contrast to private banks, refused to reschedule or reduce their claims until quite recently, on top of dictating economic policies. Financial merry-go-rounds kept up the pretence of multilateral debts being serviced on time. Funds from, say, the World Bank repaid the IMF, which lent again, enabling the debtor to service the World Bank's loan "on time". Often OECD governments participated as intermediary financiers. Debtors had to pick up the whole bill.

Apart from shifting risk to official international financial institutions, this also shifted risk from the United States to other countries. US banks were much more heavily exposed in HICs than those elsewhere. Unlike European banks they had provided virtually no safeguards against the risks of their exposure. Their percentage of total bank debts was larger than US shares in international financial institutions. Seen as a country, the United States could thus reduce its risk, being bailed out by other countries. One might call this a globalisation of risk.

In the South debt management caused great hardship, as recalled by the expression the "lost decade". Meanwhile it became officially recognised that structural adjustment had been unable to solve the problem. Poor countries got two Highly Indebted Poor Country (HIPC) initiatives – and HIPC II is already faltering. The Zedillo report, written at the request of the secretary general of the United Nations, stated that HIPC II 'in most cases' (Zedillo et al. 2001: 21) did not go far enough to reach sustainable debt levels, suggesting a 're-enhanced' HIPC III (ibid.: 54). By proposing its Sovereign Debt Restructuring Mechanism (SDRM) for relatively richer countries, the IMF itself acknowledged that its own and the World Bank's debt management efforts could not solve the problem. There is therefore no need to go through the large literature on whether structural adjustment worked (which could not prove it did). By adopting new mechanisms, IMF managers agreed that structural adjustment did not work. It suffices to recall that some thirty years of structural adjustment in sub-Saharan Africa produced not a single case of sustainable recovery.

The 1990s began with officially heralded hope and recovery, if not for all debtors so at least for those with "prudent" economic policies; that is, those implementing the advice of the World Bank and IMF. Once again the "market" worked. Global investors started to pour money into less-developed countries – the fourth and last wave of debt-creating flows. Latin America was used – with an occasional, cautious caveat regarding sustainability – as a practical vindication of adjustment policies based on the so-called Washington Consensus emphasising globalisation and deregulation.

Interestingly, the World Bank's own data available in the summer of 1994 showed no reason for this euphoria, nor did the data justify the vindication of the success of adjustment policies (cf., Raffer 1996). Latin America's debt overhang persisted. Conventional indicators such as the debt service ratio (DSR) did indeed fall because arrears increased steeply. Sub-Saharan Africa's conventional debt indicators were even dramatically lower than Latin America's at that time (Raffer and Singer 2001: 181), also due to arrears. However, the Bretton Woods twins did not interpret the data from Africa as optimistically as Latin America's, a difference that might be helpful in properly evaluating their optimism on Latin America's recovery.

An index proposed by Raffer (1996) portrays a true picture of the situation, reflecting the debt overhang properly by incorporating arrears:

$$0 \leq DSR/DSR_d{}^* \leq 1 \tag{1}$$

This index simply measures which percentages of payments contractually due are actually paid. DSR is the World Bank's debt service ratio based on actual payments (on a so-called "cash base"). The subscript d denotes payments contractually due. The denominator is actual debt service plus all payments due but not effected, such as arrears or capitalised interest (for details see Raffer 1996).

During 1990–92 Latin America honoured less than half its contractual obligations. Brazil and Argentina paid between 26 and 38 per cent (ibid.: 32) depending on the year, but paying particularly low percentages in 1992, the last year for which the World Bank published data before the summer of 1994. The perceived end of Latin America's debt crisis was due to the toleration of extremely large non-payments, or breaches of contract. If creditors had accepted much lower, let alone similar arrears in 1982 there would have been no debt crisis.

Improved developing country access to international capital markets was not the result of regained confidence of "old" creditors. Rather, it was due mainly to a new type of investor starting to lend: the bondholders. The last shift in the structure of creditors was thus effected. Bonds (the usual instrument of sovereign debts until World War II), foreign direct as well as short-term portfolio investment, poured into some less-developed countries, allowing voluntary repayments to commercial banks and easy servicing of multilateral debts. Direct investment was substantially fuelled by privatisation, which in most less-developed countries means selling public assets to foreigners. Commercial banks themselves knew better than to put substantial sums of their own money into these countries again. Mutual funds, pension funds and retirement accounts – or the public at large in the form of private bondholders – replaced lending by commercial banks and international financial institutions to a considerable extent. Regulatory changes, and a trend toward explicitly rating southern sovereign borrowers at least partially triggered by them, allowed these new investors in. It should be recalled that the Basel Capital Accord (known as Basel I) explicitly demanded lower capital weights for short-term exposure in the South, thus encouraging shifts to shorter maturities. Northern regulatory decisions thus contributed to the Asian crash in 1997.

It is difficult to see how the official euphoria by governments and international financial institutions in the early 1990s could not have encouraged these new, private flows. Similar to the shift from bank loans to international financial institutions during the 1980s another shift in lending to private bondholders took place in the 1990s. Without it, arrears would no doubt have been even larger.

Soon after the "Tequila Crisis" official euphoria found a new target: East Asia. The same cycle repeated itself. East Asian countries were hailed as model economies, successful because they had so fully embraced globalisation and the teachings of economic orthodoxy – like Mexico, the model pupil of the Bretton Woods twins right up to 1994–95. More recently, Argentina was still being praised for its economic policies while it was crashing.

The World Bank (1999: 2) acknowledged meanwhile having known 'the relevant institutional lessons' since the early 1990s, drawing attention to the Southern Cone crisis of 1982 and Mexico's in 1994–95. An audit report by its Operations Evaluation Department (OED) on Chile's structural adjustment loans highlighted 'the lack of prudential supervision of financial institutions in increasing the economy's vulnerability to the point of collapse' (sic!, ibid.). The OECD's 'key lesson' was that 'prudential rules and surveillance are necessary safeguards for the operation of domestic financial markets, rather than unnecessary restrictions' (ibid.). Yet this did not keep the World Bank or IMF from encouraging the same policies that led to Chile's crash to Asian countries in order to allow these countries too to reap the benefits of globalised financial markets. According to the World Bank, it was 'guided' by 'the lessons of the general debt crisis' (whatever that might mean), not by the 'more relevant' cases of Chile in the 1980s and Mexico in 1994–95. The neglect of proper sequencing and institution building 'featured prominently in the Chile and Mexico crises' (ibid.). Briefly, the problem was known years before the crash, and the unfolding of the Asian crisis could be watched like a movie whose script is known. The Argentine crisis of 1995 goes unmentioned, though it was of a variety similar to Asia, namely triggered by private-sector debts.

One has indeed to ask why policymakers and international financial institutions failed to give these weaknesses appropriate weight (World Bank 1999: 2). Why did neither the World Bank nor the IMF (both not normally known for their restraint in giving advice) warn Asian countries to proceed more slowly with cautious sequencing – as they do meanwhile – pointing to already available evidence, instead of supporting quick liberalisation and applauding inflows of volatile capital, as in Mexico before 1994? No credible answer is at hand. Instead, we can state that decision makers overlooked the Asian countries' failure to comply with the basic tenets of the much-abused Washington Consensus. Summarising that Mexico fulfilled most of the consensus conditions but East Asia did not, the World Bank (ibid.) reached the following conclusion: 'Washington consensus policies were neither the cause of high growth, nor the cause of the crisis.' This begs the question why adopt such policies under structural adjustment, and why did the World Bank and IMF propagate them so forcefully. One also wonders about the rapid U-turn in the evaluation of a region whose "Asian miracle" orthodoxy the Bretton Woods institutions had claimed to result from applying their policies (cf., Raffer and Singer 2001: 53, 138ff).

Twenty years on from 1982, debt management of the international financial institution style reached the end of its flagpole. After shifting lending from bilateral creditors to commercial banks, to public international financial institutions and then to bondholders, no further shift is possible. There is little left to sell to foreign investors either. The game of debt musical chairs is over. Reality must now be faced – debt reduction to sustainable levels can no longer be avoided.

One cannot rule out that international financial institutions themselves might still be prepared to finance one further shift towards more multilateral lending if they get enough resources and remain protected from losses. During the Asian crisis the IMF's first deputy managing director argued, using Thailand's crisis of 1997 and Mexico's of 1994–95 as supporting evidence, that the prospect of larger crises caused by capital account liberalisation would call for more resources for the IMF to cope with the very crises the Fund's proposal would create in the future (Fischer 1997). From the narrow viewpoint of institutional self-interest – which one, of course, hopes to be irrelevant – such crises that are better than using the membership right to capital controls guaranteed by the IMF Articles of Agreement, which would not require increased IMF resources. But, even though the huge bail-out of Brazil in August 2002 may be interpreted as a different signal, solvent member countries are unlikely to go on financing increasingly larger bail-outs.

A look at the time series of total southern debts shows that the losses necessary to regain sustainability are meanwhile much larger than they were in 1982. Traditional debt management was basically a Ponzi scheme, new loans repaying old ones and financing interest service. The expression "financing debt relief" – logically rather absurd – characterises this system. The share of unpayable debts, so-called "phantom debts" (cf., Raffer 1998), has grown. These are debts merely existing on paper because accounting practices refuse to recognise losses already suffered. The important point is that phantom debts can never be recouped by creditors. They are unreal, as they have no economic base. There is some justice in that because they owe their existence to creditor mismanagement anyway. "Forgiving" them (a term uniquely reserved for less-developed countries, by the way) does not really mean losing money as official creditors often claim. Money one cannot get cannot be lost. Reducing phantom debts is simply an acknowledgement of facts. Insufficient debt reductions have occasionally been granted, reducing the growth of debts somewhat. Total debts, however, including phantom debts, have grown.

Creditors as a group will have to accept larger losses, but not all creditors will necessarily be worse off. The structure of creditors has changed dramatically. While bondholders were practically non-existent in 1982, they are now an important class of creditors in quite a few cases, while banks were able to reduce their exposure. Argentina is a good illustration: the share of bonds in public and publicly guaranteed long-term debt rose from 8.2 per cent to 74.2 per cent, and the share of commercial banks declined from 59.6 per cent to 8.3 per cent between 1980 and 1999. Quite noteworthy distributional effects exist, exacerbated by the fact that the international financial institutions were able to secure a privileged status of *de facto* preferred creditors. Therefore others have to accept larger losses. This is not only at odds with economic sense, creating massive moral hazard, but in the case of multilateral development banks it is also a breach of their statutes. The articles of agreement of the World Bank and all regional development banks contain detailed rules on how to reduce claims if necessary. The European Bank for Reconstruction and Development (EBRD) recognises losses, proving that development banking and market discipline can be combined.

The record since 1982 recalls Mexico's default in 1914, which is a prime illustration of the economic reason behind an orderly and fair process. After many years during which creditors practically ran the country, they received less than 10 per cent of face values of amounts outstanding. This result – if not a better one – could have been obtained more easily and quickly by emulating insolvency, as suggested by the then US ambassador. A quick, fair and reasonable solution to Mexico's crisis of 1914 would also have been in the best interest of creditors, as debt management which tries to make insolvent debtors pay increases phantom debts. A speedy solution would have limited losses for creditors as a group.

3. THE WAY OUT: GLOBALISING INSOLVENCY PROCEDURES

Suggested soon after 1982 as the solution to the sovereign debt overhang, the idea of sovereign insolvency was shunned by official creditors, especially the IMF. This changed dramatically in November 2001, when IMF First Deputy Managing Director Anne Krueger (2001a) suddenly proposed an analogy to domestic insolvency laws. Opponents of sovereign insolvency, especially IMF staff, seemed to lose any recollection of their own grave reservations. As if touched by a magic wand, the "arguments" that had been used to assert that the principles of insolvency could not be applied to sovereign debtors disappeared.

Despite the "new" appearing in the titles of Krueger's first papers, and the fact that she did not quote any prior publications, the idea of adapting insolvency rules to sovereigns was by no means novel. Adam Smith – meanwhile quoted by Krueger – advocated it, arguably before her. Sovereign insolvency has had a variety of advocates since 1982 (cf., Raffer 2001, Rogoff and Zettelmeyer 2002). Krueger's first speech on the subject followed several statements in favour of sovereign insolvency by G-7 ministers and central bankers.

In contrast to earlier proponents, Krueger used the idea of sovereign insolvency – the IMF's SDRM – as a means to increase the IMF's influence and possibly also to improve its reputation after the Asian crisis and the Meltzer report. In her model and in all the variations that were to follow (cf., Raffer 2004) the IMF dominated absolutely, determining the outcome in several ways. The IMF Executive Board would determine sustainability and the "adequacy" of the debtors' economic policies. By determining sustainability the IMF automatically determined the amount of debt reduction. Most important – and most contentious – decisions would be taken by the IMF, both a creditor in its own right and dominated by a creditor majority. Multilateral debts, especially the IMF's own claims, remained exempt. The present *de facto* preferential status would thus be legalised. Although IMF staff churned out many different versions after November 2001, these important elements remained.

Krueger suggested initially that the IMF be given the right to endorse the stay triggered by the debtor's demand for insolvency relief. Later this was changed. In an attempt to please the private sector, the IMF eventually adjusted its wording: '*Activation would not automatically trigger any suspension of creditor rights. There would be no generalized stay on enforcement and no suspension of contractual provisions, (including provisions relating to the accrual of interest)*' (2002b: 9, italics original). This is not really a fundamental change. While any creditor would formally retain the right to litigation – a concern expressed by many Wall Street investors – the so-called "hotchpot" rule and the possibility of enjoining specific enforcement actions would render this right useless (cf., IMF 2002b, Raffer 2003, 2004). The hotchpot rule means that any amount recovered due to litigation would be deducted from the sum this creditor would finally be entitled to receive. Litigation thus becomes pointless. But since this rule would be practically inapplicable if litigating creditors had already received more than they would get under the SDRM, the IMF proposed that specific enforcement actions could be enjoined upon the debtor's request and subject to the approval of creditors if they threaten to seriously undermine the restructuring process. Assuming normal economic behaviour, other creditors would favour stopping litigants if and when litigants were likely to get more by litigation than what could be expected to be their share under the SDRM. Put frankly – this proposal is therefore a practical joke.

Using the problem of so-called "vulture funds", Krueger (2001a: 7) alleged that the IMF was necessary to implement sovereign insolvency. Arguing that laws barring "vultures" from interfering with the mechanism 'must have the force of law universally', she concluded that amending the IMF Articles of Agreement was needed. All member states would have to change their domestic laws to allow overruling the opposition of creditor minorities. So-called "super majorities", a quorum of all creditors demanding to reduce their claims, would then be able to bind dissenting creditors. Not all countries and territories are IMF members though. As both changing the Fund's statutes and the process of ratification would take time, her proposal would have 'no implications for our current negotiations with member countries – Argentina and Turkey, for example' (ibid.: 2). This solution, meanwhile called the "statutory approach", would firmly and officially install the IMF as the overlord of sovereign debt relief. The results are likely to match those of HIPC I, HIPC II, structural adjustment, and the Miyazawa/Brady deals, all characterised by a leading role of the IMF.

In contrast to Krueger's assertions it would be sufficient to change sovereign immunity laws in the very few jurisdictions stipulated by loan agreements. Inserting a clause voiding or suspending waivers of immunity during sovereign insolvency proceedings would solve the problem of litigation by vultures (cf., Raffer 2002a).

Finally, Krueger proposes a dispute resolution forum with various aims and characteristics:

> limited but exclusive powers for the orderly conduct of the restructuring process, including the resolution of disputes between a sovereign debtor and its creditors ... and amongst creditors (Krueger 2002: 1).

A dispute resolution forum would verify claims, oversee voting and certify key decisions by the debtor and "super majorities" of creditors. Their operations were described as follows:

> [independent] not only of the Executive Board, but also of the governors, management and staff of the IMF. The flipside of this independence is that the role of the dispute resolution forum should be strictly limited (ibid.: 4).

Apparently, there is a trade-off between independence and authority. A dependent forum would qualify for a larger role.

In spite of its limited role, Krueger (2002) proposed a complicated, clumsy and unnecessary five-stage process, again dominated by the IMF, for setting up dispute resolution. All member countries were to nominate one person. These nominees would then be vetted by a "neutral" committee established by the IMF's Executive Board, to reduce the list from 183 names to 21. External advice was suggested. The vetted and reduced list would be passed for approval to the governors, who would vote only on the entire list as a package. As voting is weighted at the IMF, creditor governments could overrule any opposition by debtors. They could thus establish a dispute resolution forum without any person nominated by the South. If approved, the managing director would appoint the members for a renewable term (four to five years). Those appointed would elect a president, who – when needed – would "impanel" three members for an actual case.

Krueger repeatedly claimed that this process would 'empower the creditor and debtor' (e.g. Krueger 2002: 4) or the super-majority of creditors. The outcome of the process 'will remain where it should be – in the hands of the debtor and creditors' (Krueger 2001a: 5). However, the IMF determines the amounts of reductions necessary and the policies of the debtor – in US legal parlance one may say that the IMF, not the debtor, submits the plan. There thus remains little for other creditors to actually vote on. They can refuse to accept what would essentially be the IMF's offer, but no more. The idea that creditor majorities be granted the right to vote on the extension of the stay beyond a maximum period during which the IMF alone 'endorses' it (Krueger 2001a: 7) would not really change the picture.

On the other hand, the IMF (2002a: 10) suggested creating many creditor classes with classification rules part of the amendment of the IMF Articles of Agreement, and 'each class vested with effective veto power over the terms offered to other classes'. This would likely result in at least one of the many classes blocking the procedure. It would be surprising if the IMF were unwilling to provide good services to help overcome such deadlocks. Considering the number of cases in need of a solution, critical minds might see this proposal as a substantial employment programme for the IMF.

The last variant proposed before the 2003 Bretton Woods spring meeting (IMF 2002b: 58ff) has a selection panel formed either by the Executive Board asking seven to eleven representative member countries to forward the name of one judge each or by the IMF managing director appointing seven to eleven independent and qualified persons based on external advice from respected organisations. Outside advice – certainly a positive element – was possible pursuant to the first proposal as well. The latter variant is preferred by IMF staff, who believe that appointment by the managing director would be perceived as less political than appointment by the Executive Board and that it would avoid the distortions that could result from weighted voting. This belief is odd, since the very appointment of the present managing director proved beyond doubt the influence of weighted voting rights. Is it likely that any managing director would be unaware of this when themselves appointing panels?

Understandably, Krueger's "statutory approach" has met substantial reservations. Even quite a few IMF executive directors have voiced concerns about a significantly extended role of the IMF. The secretariat of the G-24 distributed a paper critically commenting on the IMF's proposal to all executive directors representing less-developed countries (Raffer 2002b). US Treasury Under-Secretary for International Affairs John Taylor (2002) opposed an increase in the IMF's importance and proposed a package of new collective-action clauses. Wrongly discussed as an alternative to sovereign insolvency, these clauses combine perfectly with it. The representation clause for bondholders, for example, would indeed improve insolvency procedures substantially. Taylor stated the need to limit official sector support, seeing sovereign debt restructuring reform as a way to do so. He does not object to arbitration as a means to handle inconsistencies between different types of issues or jurisdictions One may thus conclude that Taylor is opposed to a specific model, but not against any form of insolvency. Obviously Taylor is not against arbitral awards as such, but against centralising the process with the IMF. Discriminated private creditors have also voiced strong and understandable reservations against a leading role of the IMF in determining debt reductions. It seems easy to guess what private creditors think about phrases such as "involving" or "bailing-in" the private sector, considering that the private sector has repeatedly granted – sometimes substantial – debt reductions, while the IMF has been protected.

4. SHORTCOMINGS OF THE IMF MODEL

All of the many variants of the IMF's model violate the fundamental principle of the rule of law that stipulates that one must not be judge in one's own cause. The IMF, a creditor, would decide on debt reductions. Yet all civilised legal systems require a neutral entity to preside over insolvency cases, indeed any legal procedure, and to decide if and when necessary.

Unconcerned about the rule of law, the IMF's proposal also fails to take the essential function of any insolvency into account: the resolution of a conflict between two fundamental legal principles. In a situation of over-indebtedness the right of creditors to interest and repayments collides with the principle recognised generally (not only in the case of loans) by all civilised legal systems that no one must be forced to fulfil contracts that lead to inhumane distress, endanger life or health or violate human dignity. Briefly put, debtors – unless they happen to be less-developed countries – cannot be forced to starve themselves or their children to pay. Although claims are recognised as legitimate, insolvency exempts resources from being seized by bona fide creditors. Debtors' human rights and dignity are given priority over unconditional repayment. It is important to emphasise that insolvency deals only with claims based on a solid and proper legal foundation. Debts without such firm base need no insolvency procedures.

The proposed SDRM is unfair to non-multilateral creditors, especially the private sector, and to the debtor. Regarding its own major shareholders – the members of the Paris Club – the IMF remained undecided, suggesting both that the Paris Club could participate as a separate class of creditors and that these bilateral official creditors might not participate at all. Apparently, the choice of whether to remain exempt from the SDRM would be left to the IMF's major shareholders. Should they decide to opt out, the private sector alone would have to accept reductions of their claims. The following passage (IMF 2002a: 21) sheds clear light on the Fund's repeated assertion of wanting to empower private creditors:

> [Paris Club members] would presumably continue their current policy of requiring that the debtor seek comparable treatment from private creditors when private claims on the

sovereign are judged to be material, and would also continue to assess whether the
agreement of private creditors meets this requirement.

This is logical since projections and advice by the Bretton Woods institutions
usually form the basis of the Paris Club's debt decisions – contradictions between the
sustainability assessed within the SDRM and Paris Club decisions would thus be
unlikely. The assessment of private creditors by the Club would put them under
additional control. In any case, all discriminated creditors would lose more money than
would be the case with equal treatment of all creditors. It is unfair to debtors forced
once again to accept IMF decisions, rather than having sustainability and the terms of
debt reduction determined by a fair, transparent and efficient procedure.

Krueger's (2001b: 4) defence of preferential treatment of the IMF is flawed. The
assertion that the IMF charges lower interest rates and has no commercial interest may
be right. But differences in interest rates are not a sufficient reason for exemption, and
the Fund certainly has an institutional interest. The IMF's record, especially but not
only during the Asian crisis, shows that it does not avoid disorderly adjustment, nor
does it 'discourage countries from adopting policies that would do unnecessary harm
to themselves' (ibid.). But most important, the IMF takes the economic decisions,
often forcing its views on its clients. There is therefore a significant difference between
lenders: commercial banks did lend aggressively but have seldom interfered with their
clients' economic policy, nor did bondholders. International financial institutions have
strongly influenced the use of loans, dictating economic policies. Rather than lending
only when others are reluctant to do so (ibid.), the IMF has routinely taken economic
decisions.

Like all international financial institutions, the IMF has refused to participate in the
risks involved. First, it insists on full repayment; the borrower even has to pay if
damages are negligently caused by IMF staff. This leads to the economically perverse
situation where high rates of failures engineered by the international financial
institutions might render "adjustment programmes" administered by those same
institutions necessary. Similarly, failed programmes are likely to give rise to new
programmes, as long as unconditional repayment to the international financial
institutions is upheld. Errors and negligent damages increase the IMF's importance.
Logically, one may speak of 'IFI-flops securing IFI-jobs' (Raffer 1993a: 158). The
IMF – like other international financial institutions – gains from its own errors and
negligence. Because of its involvement in the debtor's economic policies, all
international financial institutions must carry their fair share of risk (Raffer 1993a).
Bringing the market to the IMF is one way of achieving this.

Unlike the international development banks which are supposed to decide on
projects, the IMF initially was to be an emergency lender. Conditionality was not
foreseen at Bretton Woods. This justified preference, and loan loss provisions were
unnecessary. When conditionality was introduced, no appropriate changes were made
regarding IMF financial accountability. This became particularly problematic once the
IMF started massive debt management operations.

The present perverted incentive system is totally at odds with any market economy
– with unsurprising results. Like any other creditor, international financial institutions
must carry the risk of losses. Like consultants they must be financially responsible for

their advice. Connecting decisions and risks is the most basic condition for the functioning of the market mechanism. If this link is severed market efficiency is disturbed, as proven by former Communist economies. Economic efficiency, but also fairness to debtors and other creditors, demands that international financial institutions no longer be preferentially treated. HIPC I already broke the taboo of multilateral debt reductions, pointing in the right direction.

None of Krueger's papers nor any IMF document on the SDRM contains the smallest hint of any kind of debtor protection. Neither did the affected populations or vulnerable groups have any possibility to voice their views. Participation and protection of the poor are, meanwhile – at least officially –, part and parcel of HIPC II. Civil society is to participate in designing anti-poverty strategies. Krueger's SDRM falls behind this standard, reflecting the failed ideas of HIPC I and structural adjustment. It lacks fairness. It does not deal with the problem of sovereignty. Any protection of sovereignty and governmental powers would, however, limit the IMF's absolute dominance. Fairness would mean financial losses for the IMF.

5. A FAIR AND TRANSPARENT PROCESS

A mechanism that is 'demonstrably fair to all parties' (Krueger 2002: 4) must not be dominated absolutely by one creditor. Moreover, it must take the special problem of public/sovereign debtors into account. The need to deal satisfactorily with sovereignty was a powerful argument against the first generation of proposals advocating the application of corporate insolvency – as Krueger does now. My proposal, called the Fair and Transparent Arbitration Process (FTAP) by NGO campaigns, was developed to answer this legally correct point: Chapter 11, Title 11 of the US Code (corporate bankruptcy) cannot be applied to sovereigns. Therefore I proposed in 1987 the US Chapter 9, Title 11 – municipal insolvency – as the model. Since it solves the problem of governmental powers, this model is adaptable to sovereigns (Raffer 1989). Naturally, only the basic principles of domestic Chapter 9 should form the basis of arbitral proceedings. Some important and necessary details of domestic Chapter 9 are unnecessary and inapplicable internationally. Eligibility and authorisation to be a Chapter 9 debtor, fundamental and useful as they are within the United States for constitutional reasons, is but one example.

Raffer (1990, 2001) described in detail Chapter 9-based debt arbitration. The discussion below highlights its main differences with the IMF's model, including problems IMF staff have failed to tackle.

5.1 Fair Arbitration

Instead of relying on an IMF body, ad hoc arbitration panels must be formed by the parties in the traditional way: each side – creditors and the debtor – nominate one or two persons who then elect one further member to reach an uneven number. Sovereigns could "file" for insolvency protection by depositing their demand, for example, at the United Nations. This would automatically trigger a stay. The panel must endorse the standstill immediately on being formed. It would reject the debtor's demand if unfounded, denying the debtor any advantage from starting the procedure. It would also verify claims, as is routine in any domestic case. This proposal (Raffer 1990: 309) did in fact become part of Krueger's "new approach". Meanwhile the IMF (2002b: 68) demanded specific checks regarding 'for example, the authority of an official to borrow on behalf of the debtor', echoing what I had demanded some ten years before (Raffer 1993b: 68), nearly in my own words. In many discussions in which I participated representatives of official creditors, especially international financial institutions, had declared this idea to be utopian and impossible. Meanwhile the IMF has embraced it – though not to the extent of quoting me.

Arbitrators would have the task of mediating between debtor and creditors, chairing and supporting negotiations with advice, providing adequate possibilities to be heard for those affected by the plan and – if necessary – making decisions. As facts would be presented by both parties and by representatives of the population in a transparent procedure, decisions are unlikely to affect substantial sums of money but would rather solve deadlocks. Agreements between debtor and creditors would need the panel's confirmation, in analogy to Section 943 of Chapter 9. Panels would have to take particular care to safeguard fairness and a minimum of human dignity of the poor – in analogy to the protection enjoyed by a municipality's inhabitants.

5.2 Protecting Sovereignty

None of the IMF's papers proposes measures to protect the governmental powers of debtors. In the United States the court's jurisdiction depends on the municipality's volition, beyond which it cannot be extended, similar to the jurisdiction of international arbitrators. Municipalities cannot go into receivership. This makes Chapter 9 especially suited for sovereign cases.

5.3 Right to Be Heard

A municipal debtor's population has a right to be heard during the proceedings. This would have to be exercised by representation in sovereign cases (Raffer 1990). In US domestic cases, §943(b)(6) even explicitly demands electoral approval of any provisions of the plan regarding matters where such is required by non-bankruptcy law.

5.4 Protecting the Poor

Chapter 9 guarantees an appropriate form of debtor protection – a human right presently granted to anyone but the globe's poorest. US municipalities are allowed to maintain basic social services essential to the health, safety and welfare of their inhabitants. It is mandatory that schemes to protect humane minimum standards be part and parcel of every international composition plan. Subsidies and transfers necessary to guarantee these standards to the poor must be maintained. Funds necessary for sustainable economic recovery ("fresh start") must be set aside. The principle of debtor protection demands that – in analogy to the protection granted to the inhabitants of indebted US municipalities – the money to service a country's debts must not be raised by destroying the debtor's economic future.

5.5 Equal Treatment of Creditors

Generally, all creditors of one sovereign debtor – including domestic creditors – should be treated strictly equally. The reasons for the equal treatment of international financial institutions in an insolvency, as an easy and necessary way to make them financially accountable, were presented above.

One possible exception may be other less-developed countries. It would not make sense to bankrupt country A by relieving country B. In such cases preferential and differential treatment, strictly according to objective criteria, such as claims affected, creditors' per capita GDP or export income, seems worth considering. Regarding some types of domestic debts exceptions must be discussed, for instance, pension funds forced by law to buy government bonds.

Fresh money lent during the Fair and Transparent Arbitration Process must, of course, have unconditional seniority, as is usual practice in all insolvency procedures. Should the IMF provide these resources without conditionality – as initially stipulated at Bretton Woods – this money must enjoy the same status.

5.6 Quick Implementation

Ad hoc panels could start work immediately on the basis of the main principles of Chapter 9 if important creditors agree. There is no need to lose time by negotiating a statutory approach and ratifications.

5.7 Regulatory Changes

Although not a necessary part of an international Chapter 9, it could be used as an opportunity to introduce stabilising regulatory changes into the international financial architecture. One example is encouraging appropriate loan loss provisioning, as do continental European tax laws, by allowing tax deductible provisioning at negligible – if any – real economic costs to the budget (cf., Raffer 2001).

6. CONCLUDING REMARKS

The Euromarket's lending spree, an early example of globalisation, masked structural debt problems until crisis violently erupted in 1982. Denying southern debtors the obvious and humane solution, new creditor groups were brought into the game, prolonging unnecessary suffering by debtors and increasing total debts in a huge Ponzi scheme. Now, reality has to be faced: 'At the moment too many countries with insurmountable debt problems wait too long, imposing unnecessary costs on themselves, and on the international community that has to help pick up the pieces.' Krueger (2001a: 8) fails to mention that creditors, most notably the IMF, have blocked this sensible solution, forcing debtors to 'wait too long'. It is to be hoped that this missing link in the international financial architecture will finally be established, not by the SDRM but by a fair mechanism respecting the rule of law, human rights and economic efficiency, also when it comes to the poor in the South.

The rejection of the SDRM during the 2003 spring meeting of the Bretton Woods institutions by the United States and emerging market economies precluded the introduction of an unfair, self-serving and inefficient system. Yet it does not mean the end of the search for a viable solution. On 17 May 2003 the G-7 finance ministers issued a statement at Deauville reaffirming their commitment to 'greater transparency and more orderly, timely and predictable workouts of unsustainable debts'. The Paris Club was encouraged to improve its methods, for instance, by adjusting its "cut-off date" after which debts incurred are not eligible for relief by the Paris Club. This limits the share of debts that qualify for relief. It has transmitted errors, producing insufficient debt relief initially and compromising meaningful solutions later on. Too little relief at the first try spawns new problems and becomes difficult if not impossible to remedy later (Raffer 1989). The present inflexible cut-off date has therefore been criticised by NGOs and academics. According to the G-7 finance ministers, the Paris Club should also tailor its response to the specific financial situation of each country rather than defining standard terms, as done up till now, under this new approach. All this is good news for efficient debt management.

In any case, the problem will not go away. The discussion on a viable solution to the debt overhang is likely to erupt again with the next big crisis, especially if and when the sums needed for further bail-outs are large. Unfortunately, political actors go on delaying, imposing unnecessary costs on debtors and the international community, which Krueger rightly decried. But it is to be hoped that the globalisation of human rights and the rule of law for all human beings will eventually catch up with globalised financial markets, making them more humane and more efficient.

REFERENCES

Diaz-Alejandro, C. (1985) 'Good-bye financial repression, hello financial crash', *Journal of Development Economics* 19 (1&2): 1–25.

Fischer, S. (1997) 'Capital account liberalization and the role of the IMF'. Paper presented at the IMF Seminar *Asia and the IMF*, Hong Kong, (19 September).
    <http://www.imf.org/external/np/apd/asia/fischer.htm>

World Bank (1998) *World Debt Tables 1988–89*. Vol. 1, Washington, D.C.: World Bank.

World Bank (1999) *1998 Annual Review of Development Effectiveness*, OED, Task Manager Robert Buckley. Washington, D.C.: World Bank.

IMF (2002a) 'Sovereign Debt Restructuring Mechanism: Further considerations'. Prepared by the International Capital Markets, Legal, and Policy Development and Review Departments in consultation with other Departments (14 August).

IMF (2002b) 'The design of the Sovereign Debt Restructuring Mechanism: Further considerations'. Paper prepared by the Legal and Policy Development and Review Departments in consultation with the International Capital Markets and Research Departments (November 27).

Krueger, A. (2001a) 'International financial architecture for 2002: A new approach to sovereign debt restructuring', (26 November).< http://www.imf.org/external/np/speeches/2001/112601.htm>

Krueger, A. (2001b) 'A new approach to sovereign debt restructuring', (20 December).
    <http://www.imf.org/external/np/speeches/2001/1122001.htm>

Krueger, A. (2002) 'Sovereign debt restructuring and dispute resolution', (6 June).
    <http://www.imf.org/external/np/speeches/2002/060602.htm>

OECD (1996a) *Development Co-operation, Efforts and Policies of the Members of the Development Assistance Committee, 1995 Report*. Paris: OECD.

OECD (1996b) *Shaping the 21st Century: The Contribution of Development Co-operation*. Paris: OECD.

Pearson, L. B. et al. (1969) *Partners in Development: Report of the Commission on International Development*. New York: Praeger.

Raffer, K. (1989) 'International debts: A crisis for whom?', in H.W. Singer and S. Sharma (Eds) *Economic Development and World Debt* [Selected papers of a Conference at Zagreb University in 1987], pp. 51–63. Basingstoke and London: Macmillan Press; New York: St Martin's Press.

Raffer, K. (1990) 'Applying Chapter 9 insolvency to international debts: An economically efficient solution with a human face', *World Development* 18 (2): 301–313.

Raffer, K. (1993a) 'International financial institutions and accountability: The need for drastic change', in S. M. Murshed and K. Raffer (Eds) *Trade, Transfers and Development, Problems and Prospects for the Twenty-First Century*, pp. 151–166. Aldershot: Elgar. <http://mailbox.univie.ac.at/~rafferk5>

Raffer, K. (1993b) 'What's good for the United States must be good for the world: Advocating an international Chapter 9 insolvency', in Bruno Kreisky Forum for International Dialogue (Ed.) *From Cancún to Vienna. International Development in a New World*, pp. 64–74. Vienna: Bruno Kreisky Forum for International Dialogue.

Raffer, K. (1996) 'Is the debt crisis largely over? A critical look at the data of international financial institutions', in R. Auty and J. Toye (Eds) *Challenging the Orthodoxies*, pp. 23–39 [Paper presented at the Development Studies Association Conference, Lancaster, 7–9 September 1994]. Basingstoke and London: Macmillan Press; New York: St Martin's Press.

Raffer, K. (1998) 'The necessity of international Chapter 9 insolvency procedures', in Eurodad (Ed.) *Taking Stock of Debt, Creditor Policy in the Face of Debtor Poverty*, pp. 25–32. Brussels: Eurodad.

Raffer, K. (2001) 'Solving sovereign debt overhang by internationalising Chapter 9 procedures', *Arbeitspapier 35*, ÖIIP (Österreichisches Institut für Internationale Politik/ Austrian Institute for International Affairs), Vienna (June). Slightly updated version available from the Internet. <http://mailbox.univie.ac.at/~rafferk5>

Raffer, K. (2002a) 'Shopping for jurisdictions: A problem for international Chapter 9 insolvency?', (24 January). <http://www.jubilee.org/raffer.htm>

Raffer, K. (2002b) 'The final demise of unfair debtor discrimination? Comments on Ms Krueger's speeches'. Paper prepared for the G-24 Liaison Office to be distributed to the IMF's Executive Directors representing Developing Countries (31 January). <http://mailbox.univie.ac.at/~rafferk5>

Raffer, K. (2003) 'To stay or not to stay: A short remark on differing versions of the SDRM'. <http://www.jubileeplus.org/latest/raffer310103.htm>

Raffer, K (2004) 'The IMF's SDRM: Another form of simply disastrous rescheduling management?' in C. Jochnick and F. Preston (Eds) *Sovereign Debt at the Crossroads*. Oxford: Oxford University Press .

Raffer, K. and H.W. Singer (1996) *The Foreign Aid Business: Economic Assistance and Development Co-operation*. Cheltenham (UK) and Brookfield (US): Edward Elgar.

Raffer, K. and H.W. Singer (2001) *The Economic North-South Divide: Six Decades of Unequal Development*. Cheltenham (UK) and Northampton (US): Edward Elgar.

Rodrik, D. (1996) 'Understanding policy reform', *Journal of Economic Literature* 34 (1): 9–42.

Rogoff, K. and J. Zettelmeyer (2002) 'Bankruptcy procedures for sovereigns: A history of ideas, 1976–2001', *IMF Staff Papers* 49 (39): 470–507.

Stiglitz, J. (1998) 'More instruments and broader goals: Moving toward the "Post-Washington Consensus"'. WIDER Annual Lecture. <http://www.wider.unu.edu/stiglitz.htm>

Taylor, J. B. (2002) 'Sovereign debt restructuring: A US perspective' (2 April). <http://www.iie.com/papers/taylor0402.htm>

Todaro, M. P. and S. C. Smith (2002) *Economic Development*, 8th Edition. Boston: Addison Wesley.

Zedillo, E. et al. (2001) 'Recommendations of the high-level panel on financing for development', UN, General Assembly, 26 June (A/55/1000).

PART II

GOVERNANCE, CIVIL SOCIETY AND POVERTY

CHAPTER 8

DISEMPOWERING NEW DEMOCRACIES AND THE PERSISTENCE OF POVERTY

THANDIKA MKANDAWIRE

A country does not have to be deemed fit *for* democracy; rather, it has to become fit *through* democracy. This is indeed a momentous change, extending the potential reach of democracy to cover billions of people, with their varying histories and cultures and disparate levels of affluence (Sen 1999).

1. INTRODUCTION

This chapter considers two simultaneous processes taking place in developing countries: (i) the adoption of orthodox economic policies during a period of growing awareness of the pervasiveness and persistence of poverty and (ii) the growing political empowerment of populations throughout the world through processes of democratisation. Over the last decade, international conferences, pronouncements by international organisations and bilateral donors, NGO campaigns and declarations made by national governments have brought the issue of poverty back onto international and national agendas. This follows decades in which poverty was displaced by excessive focus on structural adjustment and stabilisation. At the same time, significant steps have been made towards democracy in many countries. This wave of democratisation has highlighted the blight of poverty, because of the greater transparency in political and economic affairs, the political empowerment of the poor themselves and growing recognition that poverty impinges on democracy's own prospects.

Until recently, it was assumed either that democracy was a luxury that poor countries could ill afford or that socio-economic conditions in these countries were not auspicious for the implantation of democracy. The emergence of democracies in social and economic conditions that had been ruled out by theories that insisted on a number of economic preconditions for the emergence of democratic government has led to new optimism about the prospects for democracy under widely divergent economic and social conditions. Unfortunately, however, it has also led to a view on democratic

117

consolidation that assumes an extremely voluntaristic character, overemphasising the role of political leadership, strategic choices about basic institutional arrangements or economic policy and other contingent process variables. The focus on political crafting of democracies has bred complacency about the possibility of consolidating democracies in unfavourable structural contexts. This chapter argues that both ideational and many structural impediments must be borne in mind when studying the consolidation of democracy in the developing countries. One such constraint is the predominance of economic policies that hamper democracies in their addressing issues of equity and poverty. The focus here is on the fact that new democracies have tended to be more orthodox than older democracies.

2. THE CENTRALITY OF GROWTH AND EQUITY

Much of the recent discontent with the new democracies and the consolidation process has been with respect to their institutional weaknesses, such as presidentialism, lack of horizontal accountability and persistence of "authoritarian enclaves" that at times hold democracies at ransom. This has led to a flurry of epithets such as "low-intensity democracies", "exclusionary democracies", "*démocracie tropicalisé*", "delegative democracy" and "low-intensity citizenship". The problems that these epithets highlight are often essentially procedural in nature. Yet they also point to discontent on the substantive issues of equity and material well-being. Amartya Sen observed that there has never been a famine in a democracy. This observation points to the ability of democracies to respond to extreme conditions. It does not, however, tell us much about the persistence of the everyday forms of poverty in many democracies. Among contemporary developing countries, democracies can be found among both the good and the bad performers in terms of poverty reduction (Table 1). This fact is addressed in the *Human Development Report 2002* on deepening democracy in a fragmented world:

> Now, 10 to 20 years later, democracy has not produced dividends in the lives of ordinary people in too many countries. Income inequality and poverty have risen sharply in Eastern Europe and the former Soviet Union, sometimes at unprecedented rates.... Poverty has continued to increase in a more democratic Africa. And many newly democratic regimes in Latin America seem no better equipped to tackle the region's high poverty and inequality than their authoritarian predecessors (UNDP 2002: 63).

Democracy *per se* does not eliminate poverty. It is rather the strategies of development that do, with the result that some of the best performers in the eradication of poverty have been authoritarian countries pursuing developmentalist and socially inclusive policies, while some democracies have been among the worst performers (such as India, Botswana, the Philippines and Venezuela). The best performers have been Sri Lanka and Jamaica. However, even the best do not compare with South Korea, Taiwan and Singapore, where the percentage of population below the poverty line is zero. Many of the new democracies are pursuing policies that are unlikely to address the problems of poverty. For developing countries where poverty is acute, the legitimacy of democracy cannot rest only on their procedures, but must rest on their performance as well. How the fight against poverty is pursued has enormous

Table 1. *Poverty-Reduction Performance of Counties According to Political Regime*

Performance	Authoritarian Regimes	New Democracies	Old Democracies	Grand Total
High	9	7	9	25
Low	34	10	5	49
Medium	15	15	11	41
Grand total	58	32	25	115

Note: This table uses the index developed by Moore et al. (1999) recalculated (using 2002 data) to measure the efficiency with which national income is converted into longevity, literacy and education. Data for both the degree and endurance of democracy is from the Freedom House database. New democracies are countries that made the shift in 1982 and have remained democratic.

implications for democracy. Or as stated by Gordon White, 'It is our thesis that the capacity of democratic regimes to secure sustained and equitable socio-economic development depends heavily on the extent to which they can construct effective developmental states' (White 1998: 29). From time to time democracies will have to respond to the challenge of the "full belly thesis", which claims that democracy is a luxury the poor cannot afford and gives precedence to the "right to development" over all other rights.

Political regimes can affect poverty through two major channels: economic growth and redistribution. We know that equity without growth may have one-off benefits for the poor, but the dynamics of demographics and depreciation of physical and social infrastructure will eventually lead to the impoverishment of everyone. Numerous countries have made laudable achievements in poverty reduction with limited resources (see e.g. the studies in Ghai 2000). However, in many cases the experiments have foundered for lack of sustained growth and balanced development. We know too that a process of growth in the context of unchanged income distribution can improve the incomes of all, including the poor. But growth can also be "immiserising" as incomes of the poor decline due to the anti-poor bias of the growth process. In any case, equitable growth does better than a distribution-neutral or anti-poor growth pattern.

In recent years the international community has set a number of goals for the reduction of poverty (e.g. the United Nations Millennium Development Goals). As Table 2 shows, fairly high levels of growth in per capita incomes are required to meet some of the goals set by the international community to halve poverty by 2015, especially if one assumes no significant changes in income distribution. Obviously economic growth that engenders greater equity will benefit the poor more than growth that is equity-neutral. With greater equity, the levels of growth required to halve poverty by 2015 are quite feasible (Table 3). For Latin America and the Caribbean a rate of only 0.6 per cent instead of 7.0 per cent would be required, and for sub-Saharan Africa 2.4 per cent growth would be required rather than 5.9 per cent. Note, however, that even this lower rate for Africa is higher than the forecast by the World Bank for 2001–10.

Significantly, although it is now agreed that equity measures would improve the efficacy of growth in addressing problems of poverty, the orthodox policy regime

Table 2. Growth Rates Required to Halve Poverty by 2015 and Income Shares

	Per Capita Growth Rates			Target Minus			
	To Meet Targets 2001–15	Actual 1965–2001	Actual 1990–2001	Actual 1965–2001	Actual 1990–2001	World Bank Projections 2001–10	Income Share Top 20%
EAP	3.50	4.68	5.83	−1.18	−2.33	5.1	44.0
EE & CA	3.80	2.97	−0.68	0.83	4.48	3.3	44.0
LAC	7.00	0.98	1.43	6.02	5.57	2.1	53.0
ME & NA	2.80	0.63	0.93	2.17	1.87	1.4	n.a.
South Asia	3.90	2.67	3.17	1.23	0.73	3.8	40.0
Sub-Saharan Africa	5.90	0.12	−0.13	5.78	6.03	1.3	52.0

Source: Calculated from World Bank CD-Rom online, Dagdeviren et al. 2002 and World Bank (2002).

Table 3. Required Rates of Growth Under Different Policy Regimes (%)

	Forecast Growth (2001–10)	Growth Required to Halve Poverty by 2015	
		Broader Based	No Change
Sub-Saharan Africa	1.3	2.4	5.9
High inequality		3.5	10.4
Low inequality		2.1	4.6
Latin America and Caribbean	2.1	0.6	7.0
High inequality		0.5	7.0
Low inequality		2.1	4.5

Sources: Hamner and Naschold 2001, World Bank 2001.

basically rules out explicit redistributive policies for several reasons. One is the basic faith in the efficacy of the market to bring about the desired results and another is the fear that redistributive measures will scare away private investors.

The point here is that growth is important for the alleviation of poverty, and different patterns of growth have different effects on poverty, depending on initial levels of inequality and contemporaneous patterns of redistribution of the additional resources generated by growth. A further point is that the challenge for new democracies is to pursue policies that are both growth-enhancing and equitable.

3. AN ELECTIVE AFFINITY?

In earlier literature on structural adjustment and democracy a commonly held assumption (by both advocates and opponents of adjustment) was that democracy would hinder adjustment. The arguments advanced were similar to those put forward in the earlier debate on compatibility, which posited a trade-off between democracy and growth. Essentially it was said that democracy would push policy towards short-term gratification of myopic voters by either increasing public consumption or pressing for redistributive policies that could produce disincentives among potential investors. This would reduce savings and investment and, hence, growth. In the longer term, moreover, democracies were perceived as more likely to develop powerful entrenched interest groups that would block flexible adaptation to changing technology or international trends (Nelson 1989). A similar argument was taken up in the 1980s but given a new twist by the "new political economy" in ascendance at the time. Democracies, it was said, would capitulate to the pent-up demands of the newly mobilised social forces that had borne them to power. They would thus be unable to pursue the tough austerity measures necessary for structural adjustment and would also tend to resort to "macroeconomic populist" strategies, inducing fiscal laxity (running high deficits and subsidies), insisting on price controls (e.g. food subsidies, minimum wages) and promoting nationalisation measures (Dornbusch and Edwards 1992).

In contrast, authoritarian governments were said to be more likely to adopt and enforce unpopular economic stabilisation and adjustment measures because such policies required, in the words of Deepak Lal, 'a courageous, ruthless and perhaps undemocratic government to ride roughshod over newly-created special interests groups' in developing countries (Lal 1983). Authoritarian regimes, moreover, are assumed to be better placed to make long-term plans, less influenced by popular pressures and better able to both forestall and suppress protest. Thus, Ronald Findlay suggested that while authoritarian rule may not be sufficient, it is likely necessary for implementing orthodox policies: 'It is very difficult to imagine a genuinely democratic regime that can insulate itself from domestic pressures to the extent necessary, even if the outward-looking strategy is to everyone's best interest in the long run' (Findlay 1988: 93). Stephen Haggard states the case as follows:

> Since authoritarian political arrangements give political elites autonomy from distribution pressures, they increase the government's ability to extract resources, provide public goods, and impose the short term costs associated with efficient economic adjustment. Weak legislatures that limit the representative role of parties, the corporatist organization of interest groups, and recourse to coercion in the face of resistance should all expand government's freedom to manoeuvre on economic policy (Haggard and Webb 1993: 262).

By the mid-1980s, the adoption of orthodox stabilisation and adjustment programmes by such democracies as India and virtually all the new democracies undermined that view. Empirical evidence seemed to suggest that, contrary to the earlier belief that only authoritarian regimes would implement these policies, democracies could do just as well, if not better. First, it became clear that the association between economic populism and democratic rule was not historically accurate. Karen Remmer's study of ten South American countries and Mexico showed that democracy did not reduce government's capacity to manage debt crises. Specifically, new democracies outperformed their authoritarian counterparts 'in promoting growth, containing the growth of fiscal deficits, and limiting the growth of the debt burden' (Remmer 1990: 327). Drawing on the experiences of Latin America and Southern Europe, Bresser Pereira (1993) went further and turned the tables arguing, 'Populism is an endogenous product of technocratic policy styles.'[1] Populist pressures to pursue immediate particularistic interests can be attenuated by the strength of democratic institutions (through representative organisations and institutions participating actively in the formulation and implementation of economic policy) and not the exhortations of technocrats. Maravall (1994) noted that for the authoritarian state the only source of legitimacy is usually high economic growth while the legitimacy of democracy is not as dependent on economic performance. Consequently, authoritarian regimes are more likely than democracies to engage in macroeconomic populism.

All this raises questions about the "elective affinity" between authoritarianism and the effective implementation of neoliberal policies. Indeed, some observers suggest that if there is any affinity at all it is between democracy and market liberalism. If democracies are able to consistently pursue the principles of liberalisation it is because they are, at least, not incompatible with their political agenda. For Duquette (1999: 221), 'democracies are the only true bearers of a genuine process of structural change'. Diamond and colleagues state the case thus:

> Increasingly, it appears that the conditions conducive to successful economic reform are not incompatible with democratic governance. These conditions include political leadership strongly committed to basic structural reform and possessing the political skill necessary to mobilize and craft supporting coalitions [and] a 'relatively strong consensus' among elites on certain fundamental policy principles (Diamond et al. 1988b).

Economic reform is more likely to be sustainable, and to effect a fundamental economic restructuring over time, if the governments imposing the transitory pain of adjustment are viewed as legitimate by society, if they consult major social and interest groups and involve them in the design of policies and if they (along with independent media and policy centres) educate the public about the need for reform. Democracies are advantaged in all of these respects.

4. SOME EVIDENCE AND ILLUSTRATIVE EXAMPLES

The first thing to do is to provide empirical evidence that new democracies are implementing orthodox policies and, more pointedly, that new democracies have tended to be even more orthodox in their policies than the more consolidated democracies. Ideally our analysis would be facilitated by a set of standard policy indicators for all countries. For Latin America there is, fortunately, the set prepared by Morley, Machado and Pettinato (1999).[2] They constructed an index that measures government efforts to implement a reform package. The index was calculated for each of the following indicators: tax reform, domestic financial reform, international financial liberalisation, trade reform and privatisation. Using this index the policies of the old and new democracies can be compared. Old democracies are defined as countries that were democratic before 1983. For the new democracies, the average of the index is calculated for the five years before and after democratisation. For the old democracies, the value is calculated for the five years before 1983 and the five years after. This gives us Table 4, which shows that (i) the new democracies are indeed more "orthodox" and (ii) in many cases, the reform process was led by the regimes in place before the democratic transition. Thus, the greater orthodoxy of the new democracies cannot be attributed to their efforts to "catch up" by effecting reforms that their predecessor may have failed to implement.

Table 4. *Policy Reform Index for Old and New Democracies*

	Average Five Years Before	Average Five Years After	Transition Year
Old Democracies			
Venezuela	0.5382	0.4302	<1983
Dominican Republic	0.5983	0.3876	<1983
Jamaica	0.6405	0.4346	<1983
Ecuador	0.6618	0.5396	<1983
Brazil	0.6720	0.4970	<1983
Honduras	0.6779	0.6434	<1983
Colombia	0.6881	0.6072	<1983
Cost Rica	0.7708	0.5272	<1983
Honduras	0.6779	0.6434	<1983
Bolivia	0.7340	0.5430	<1983
Average	*0.6660*	*0.5253*	
New Democracies			
Paraguay	0.7056	0.6410	1992
Argentina	0.6844	0.6666	1983
El Salvador	0.5754	0.5276	1985
Uruguay	0.8288	0.7856	1985
Chile	0.8320	0.7380	1990
Guatemala			1996
Mexico			1997
Peru			2000
Average	*0.8488*	*0.7252*	

Source: Calculated from Morley et al. (1999).

4.1 Old Latin American Democracies

Costa Rica is the oldest democracy in Latin America, belonging to what Leftwich (1998) referred to as "party-alternation non-development democratic states". Party alternance is fairly well established, with two dominant, multi-class parties, the National Liberation Party and the Social Christian Unity Party, having alternated in power since 1949. These two parties have reached broad consensus about development policy and both are committed to a mixed economy.

Costa Rica faced a critical economic situation in 1980. It turned to the IMF and a letter of intent between the two was considered in late 1981. It was not signed, however, and the government soon after suspended its international debt obligations.

The new government of Alberto Monge took office in May 1982, with substantial economic support from the United States. This government instituted a "100-day" stabilisation plan which included an appreciation of the colon (the currency), together with controls on the outflow of capital, unification of the official and free-market exchange rates, increases in income, sales and consumption taxes, decreases in subsidies, increases in the prices of some public utilities, a public-sector wage freeze and a credit restraint/contraction of the money supply. The stabilisation plan lowered the inflation rate, consequently leading to a rise in real minimum wages (which act as a guide in the setting of private-sector wages), which were effectively indexed to the inflation rate of the recent past. Average real salaries increased by over 40 per cent between 1982 and 1985.

From 1987 Costa Rica adopted a structural adjustment programme that continued the standard features with respect to trade, fiscal and monetary policies. The key instruments were positive interest rates, subsidies to exports, reduction of barriers to imports, promotion of investments, especially in duty-free zones, and institutional reforms. However, the policies were highly contested even during this period. Thus, José Maria Figueres Olsen of the National Liberation Party became president in 1994. Olsen opposed economic suggestions made by the IMF, instead favouring greater government intervention in the economy. The World Bank subsequently withheld US $100 million of financing. In 1998, Miguel Angel Rodríguez of the Social Christian Unity Party became president, pledging economic reforms such as privatisation. However, even he did not go far along the orthodox route, implementing little privatisation of state-owned utilities and services. Today, the large government-owned service enterprises continue as public enterprises, including the largest banks, health care and government monopolies in insurance and utilities.

Costa Rica differs from many other democracies – both old and new – in its social policies. In contrast with Chile, for instance, where the neoliberal transformation of social policy led to privatisation, reductions of universalistic benefits, more means-tested programmes and a decline in social expenditures by more than a quarter, Costa Rica rejected similar individualisation and privatisation. Instead it strengthened the universalistic character of its pension and health policies and sought to put them on a sounder financial basis. In addition, during adjustment Costa Rica maintained its complex and generally well-enforced system of legal minimum wages.

Jamaica is another of the region's oldest democracies. Under different regimes, the country has had 'a long history of highly selective implementation of reforms determined by political calculation' (Killick 1998). Even when it eventually made adjustments, the World Bank and other donors were obliged to adopt a "soft belly" approach which concentrated on backing measures that the government supported (Killick 1998). Jamaica had a long and often troubled history of involvement with the IMF prior to 1990. By the early 1990s, however, it had met its outstanding commitments to the Fund and decided to avoid further engagement with it. Jamaica, therefore, provides a particularly interesting (and atypical) case of a developing country that managed the crisis in its financial sector during the 1990s without IMF assistance or involvement (Kirkpatrick and Tennant 2001).

4.2 New Latin American Democracies

Some of the region's new democracies tried expansionist heterodox policies. Two well-known examples are Argentina and Brazil, the former with its *Plan Real* of Alfonsin and the latter with its *Plan Cruzada*. Yet these plans unravelled and were immediately abandoned, further reinforcing the view that there was, indeed, no alternative to orthodoxy. In the case of Argentina, by the time Carlos Menem took over in May 1989, inflation was running at 200 per cent per month; external debt was 338 per cent of total exports that year. The hardship caused by the ensuing hyperinflation forced Menem to distance himself from the corporatist entities that had backed him and from the populist platform on which he had campaigned. His administration adopted orthodox policies. The failure of "heterodox policies" was partly the result of polarised and fragmented systems that impeded efforts in negotiated agreements. Successful income policies, for example, often involve corporatist arrangements that require highly organised labour, tacit agreement of business and technical coherence of state policies. Attempts at *concertación* were doomed to failure in the Argentinean context of that time.

In the case of Chile a new democracy emerged after one of the most ideologically orthodox regimes and opted, in the name of *continuismo*, to pursue the same orthodoxy as its abhorred predecessor. Chile also belongs to the category of countries that Haggard and Kaufman (1995) classified as "non-crisis democracies", where regime change was not the result of macroeconomic crisis. In such democracies the new regime is likely to emphasise social policy and equity while maintaining the main features of the predecessor's economic policy.

4.3 Old African Democracies

The two oldest democracies in Africa are Botswana and Mauritius. These states have the distinction of being among the few countries often cited as "democratic developmental states" (Leftwich 1998, Meisenhelder 1997). Although their economic policies are generally touted as evidence of the benefits of relying on the market, the actual policies pursued by these economies are not orthodox.

Botswana, a democracy, has enjoyed the highest economic growth rate in the developing world – averaging 7.3 per cent between 1970 and 1995. Though it is at times cited by the Bretton Woods institutions as an economy managed according to its precepts, this is patently misleading. A more accurate description of Botswana's fiscal policy is as "conservative Keynes-inspired expenditure policy" in the sense that taxation is low and expenditure is distributive or, at least, demand-stimulating (Weimar 1989). This is an economy with a large public sector, whose parastatal Botswana Development Corporation has 114 holdings and which has exhibited lacklustre interest in privatisation because there is no compelling financial reason for it with the budget in healthy surplus. Moreover, because of the soundness of its fiscal condition there has been no effective donor pressure on the country to accelerate the pace of privatisation.

Mauritius is also considered one of the most successful economies in Africa, with a growth rate of 5.4 per cent between 1980 and 1999. It too is cited as an example where openness has paid off and is presented as a country in which adjustment has worked. However, much of this description rests on tendentious classification and the desire to claim paternity to what has been an obviously successful development experience. The evidence shows that Mauritius has pursued rather heterodox adjustment policies, including a industry protected from the import-substitution era, an export promotion zone, price controls and tight monetary policy (Bräutigam 1994). Indeed the IMF gave Mauritius its highest (i.e. "worst") score on its policy restrictiveness index in the early 1990s and reckoned that the country remained one of the world's most protected economies even into the early 1990s (Subramanian 2001, Subramanian and Roy 2001).

A recent World Bank study (Hinkle, Herrou-Aragon and Kubota 2003) reinforces this view. The study calculated an index of anti-export bias B.[3] If the B index is equal to one, then on average commercial policies are neutral between import-competing and exporting. Where B turns out to be less than one, the trade regime is partial to exporting rather than to import-competing activities. As is clear from the last column of Table 5, because Mauritius had the highest nominal protection tax both on domestically produced manufactured goods and on imported inputs, the B index for its domestic industries, at 1.9, was higher than that of the eight new democracies included in the table. For our purposes, the significant point is that even in this limited case we see that the new democracies such as Benin, Malawi and Mali are more "orthodox" in their trade policies than Mauritius.

Mauritius's policies are non-orthodox not only with respect to trade, but also in relation to social policies. Its development strategies demanded reconciling the obvious need for such a small island economy to adopt an export-oriented strategy with the political requisite to shield industries that had emerged under the import-substitution phase and the social pacts that had been crafted during that period. The solution was to allow the setting up of export-free zones while protecting inward-looking industry. It also included the management of potentially explosive ethnic and racial relations by maintaining a fairly sophisticated welfare state. Subramanian and Roy (2001) pointed out that Mauritius has sustained its economic performance by OECD-type social protection that has taken several forms: a large and active presence of trade unions with centralised wage bargaining; price controls, especially on a number of socially sensitive items; and generous social security, particularly for the elderly and civil servants. Unlike in the OECD, however, generous social protection has not yet necessitated high taxes, reflecting both strong growth and a favourable demographic structure with a high proportion of the population being of working age (Bräutigam 1994).

4.4 New African Democracies

The 1990s witnessed a dramatic wave of democratisation in Africa. Unlike Latin America, none of the new democracies in Africa have experimented with heterodox macroeconomic policies. They went straight to orthodox policies.

Table 5. The B Index of Overall Anti-Export Bias Measured in Terms of Output Prices

	Year	E_m/E_x	Unweighted Average NPTR on Domestically Produced Manuf. Goods	Effect of NTBs on Average Price of Import-Competing Manuf. Sector Output (b)	Average Tax on Export Industry Output (estimate) (c)	Taxes and Duties on Tradable Inputs to Exports (d)	B Index
South Africa	1996	1.03	32.9	0.0	0.0	8.5	1.5
Ghana	1996	1.01	30.3	0.0	9.0	7.3	1.6
Mali	1997	1.00	31.0	2.0	8.0	8.9	1.6
Senegal	1996	1.00	50.9	0.0	0.0	6.9	1.6
Malawi	1995	1.04	43.3	0.0	0.0	9.4	1.6
Tanzania	1996	1.03	44.6	0.0	0.0	12.1	1.7
Benin	1996	1.00	18.1	4.0	23.0	8.6	1.8
Zimbabwe	1997	1.06	53.4	1.0	0.0	10.4	1.8
Mauritius	1996	1.05	71.3	0.0	0.0	6.1	1.9
Mean for Africa	*1.02*	*40.4*	*1.5*	*5.0*	*9.0*	*1.7*	
Median for Africa	*1.01*	*42.3*	*0.0*	*0.0*	*8.9*	*1.6*	

Note: NTB = Non-tariff trade barriers; NPTR = Normal permanent trade relations.
Source: Hinkle, Herrou-Aragon and Kubota (2003).

Few of the new democracies embarked on orthodox policies with as much fanfare and conviction as Zambia (Abrahamsen 2000: 118, Kayizzi-Mugerwa 2001). President Frederick Chiluba was not only a "born-again Christian" but also a born-again adherent of neoliberalism who could declare, 'We will privatise everything from a toothbrush to a car assembly plant' (cited in Abrahamsen 2000). Yet in no other sub-Saharan African country was the organisational base of the unions stronger and ideological aversion to adjustment better articulated. Zambia, the most urbanised country in sub-Saharan Africa, had had its share of "IMF riots". Its powerful labour unions had played an important role in giving the Movement for Multi-Party Democracy (MMD) a popular appeal and ensuring the party's victory. Indeed, President Chiluba himself came from this movement. In its election manifesto released in 1991, the MMD declared that it would control inflation in collaboration with international donors, but that this would not be achieved through 'an unbalanced suppression' of workers' earnings (Bratton 1994). Yet on assuming power, the MMD, which blamed past poor performance on failed socialist policies embarked on "tough policies" under what it called the New Economic Recovery Programme. The financial sector was rapidly liberalised and the central bank made independent. Food subsidies were eliminated. Trade reform led to sharp falls in tariffs – from 100 per cent to 40 per cent. The government 'embarked on one of the fastest rates of privatization in Africa', according to Kayizzi-Mugerwa (2001: 139).[4]

A year after the elections, the deficit was reduced from 7.7 to 2.4 per cent of GDP. It shot up sharply in 1993, but government showed a surplus in 1996 and 1997. Reforms received initial support from donors, who increased their balance of payments support and rescheduled about US $2.5 billion of Zambia's debt, with some debt cancelled altogether. The social consequences of these policies were summarised by Mcculloch, Baulch and Cherel-Robson:

> Our study finds a dramatic increase in poverty and inequality in urban areas between 1991 and 1996 due to stabilization, the removal of maize meal subsidies, and job losses resulting from trade liberalization and the privatization programme. Between 1996 and 1998, despite economic recovery at the national level, the reduction in urban poverty and inequality has been small. In rural areas, drought devastated rural livelihoods in the early 1990s, while maize marketing reforms principally benefited those near the major urban centres, and hurt more remote rural farmers. Consequently there was little change in the overall poverty headcount for rural areas between 1991 and 1996 although there was a substantial reduction in rural inequality during this period. The rural sector experienced strong growth between 1996 and 1998 and this translated into a substantial reduction in poverty in rural areas between the two years. However, differential access to inputs, transport and marketing services has led to an increase in rural inequality (Mcculloch, Baulch and Cherel-Robson 2000: 1).

Tanzania is another of Africa's new democracies. Benjamin Mkapa laid to rest Tanzania's *Ujamaa*, one of the many variants of African socialism, when he became president in 1995. This change in policy approach much improved the country's image among donors, as observed by Danileson and Skoog:

> From being the ugly duckling that implemented reforms reluctantly and only when pressed, the country is now being lauded by the entire donor community for meeting ESAF [the Enhanced Structural Adjustment Facility of the IMF] benchmarks, for rapid and consistent implementation, and for showing a willingness to reform and a thorough understanding of the need for drastic change of economic policies. Now Tanzania is no longer looked at as the ugly duckling. Now she has transformed into a beautiful swan, a keen reformer who actively and enthusiastically participates in reform negotiation, suggesting even more drastic measures than donor organizations do (Danielson and Skoog 2001: 148–149).

The Mkapa government took on the adjustment programme with an alacrity that surprised donors, often outdoing the international institutions themselves as a reformer. In its new system of fiscal discipline, to be ensured by a cash budget, debt servicing ranked first on the list of priorities, followed by payment of salaries and then the rest. Policy improvements were symbolised by reductions in the current account and budget deficits. The current account deficit fell from 22 per cent of exports to just 12 per cent. The budget moved from a deficit of 2.7 per cent of GDP in 1994 to a surplus of 1.1 per cent in 1998. Ominously though, with these improvements came declining investments and savings. The improvement in the current account was itself a reflection of the decline in imports of producer goods, while imports of consumer goods surged following foreign exchange liberalisation. In addition, funds for maintenance, textbooks and medications fell sharply, eroding both the physical and human capital of the country.

South Africa is probably the more dramatic illustration of a new African democracy adopting orthodoxy. In no new democracy was a radical shift in macroeconomic policies as widely expected as in South Africa. The new government was confronted with high unemployment and one of the world's most unequal distributions of income. Its political base, its close ties with the trade union movements and its historical ties with the South African Communist Party suggested that the African National Congress (ANC) would adopt a radical nationalist programme. The view was that the combination of high inequality, pent-up expectations for social change, ideological predisposition and high levels of labour militancy and urbanisation would lead to a more heterodox policy agenda. In the words of Herbst, 'a future South African government will face a much more demanding population that is more concentrated, easier to organize, and better armed than was the case in the rest of the continent' (Herbst 1994: 37–38). Although South Africa entered the period of transition in the early 1990s 'with only an impressionistic economic vision' (Habib and Padyachee 2000), the initial programme had a state-led, developmentalist thrust directed at alleviating the legacy of poverty and inequality. Eventually, however the actual policy (Growth, Employment and Redistribution or "GEAR") was, in the words of Habib, a 'fairly orthodox neoliberal one'.

After some years of orthodoxy, a number of these new democracies have fallen foul with the Bretton Woods institutions. In the more formulaic accounts of African politics, the problems faced by the new democracies flow simply from the assumption that African politics is driven by neopatrimonialism. Democracy has made no difference to the internal factors that have accounted for Africa's poor performance: neopatrimonial institutions have led to fiscal crisis through neoclientelism, patronage, rent-seeking and corruption (Van de Walle 2001). A fairer assessment would be that they, like other African countries, have been subject to the same deflationary policies. In most cases, the deficits that emerged were largely due to increased debt servicing (as a result of high interest rates) rather than increases in the primary deficit. And although they all voiced discontent with the adjustment programmes,[5] none has shifted towards heterodox policies.

4.5 Old Asian Democracies

Because of the high rates of growth enjoyed by India in recent years, there have been attempts to include India among the list of "strong adjusters" or "globalisers" pursuing neoliberal policies (see e.g. Dollar and Kraay 2001). For much of the 1980s, India resisted pressures from the Bretton Woods institutions to change its economic policies in the direction of the Washington Consensus. However, in 1991 India was hit by a serious financial crisis.[6] The current account deficit doubled from an annual average of 1.3 per cent to 2.2 per cent of GDP during the second half of the 1980s. In 1990–91, the gross fiscal deficit of the government (centre and states) reached 10 per cent of GDP and the annual inflation rate peaked at nearly 17 per cent by August 1991. An unprecedented balance of payments crisis emerged. 'For the first time in modern history, India was faced with the prospect of defaulting on external commitments since the foreign currency reserves had fallen to a mere US $1 billion by mid-1991' (Bajpai 2002).

However, even after the reforms, India's overall economic policy diverged significantly from orthodoxy. The share of public expenditure in GDP remained high (33 per cent). The state continued to pursue a fairly active industrial policy. India's continued protection of its industries, with tariff rates averaging 27 per cent, vastly exceeds the average tariff rates of other economies in the region. Government policy also reserves certain items for production in the small-scale sector. While the government did relax the law on entry of new firms in various activities, it maintains constraints on exit by requiring government permission before businesses can close. India also continues to maintain high barriers to foreign direct investment, in contrast to most of the fast-growing Asian economies. The state continues to play a key role in finance, infrastructure, port facilities and road building. In addition, the agricultural sector was largely excluded from trade liberalisation measures. No wonder Rodrik lists India among the 'countries that have marched to their own drummers and that are hardly poster children for neoliberalism', and that have 'violated virtually all the rules in the neoliberal guidebook even while moving in a more market-oriented direction' (Rodrik 2002). India, like other East Asian counties and China, while espousing trade and investment liberalisation has done so in an unorthodox manner – gradually, sequentially and only after an initial period of high growth – and as part of a broader package with many unconventional features (Rodrik 2001).

4.6 New Asian Democracies

The developmental states of South Korea and Taiwan provide the rare cases where success in economic development have led to political pressures for democratisation. This evolution is predicted by the modernisation school, which argues that economic development produces middle classes that eventually clamour for political rights. In such cases, the developmental model itself is not in economic crisis, but rather, it suffers an erosion of political legitimacy. Nonetheless, as noted in the case of Chile, in such "non-crisis" democracies there has been greater pressure for social equity and welfare than for economic reform. This is not to rule out pressures for orthodox policies. The demand for such policies increased during the financial crisis of 1996–97 when the new democracies were entreated to abandon the polices that had undermined the "Korean miracle". Furthermore, some political actors associate a number of development institutions, such as the huge industrial conglomerates (the *chaebols*) and the universal banking system, with authoritarian rule and are pushing for more market-oriented policies so as to weaken these perceived threats to democracy. Even some of the businesses that were nurtured by the developmental state now want to cut the umbilical cord and support market-oriented policies.

In cases such as Indonesia, where the economic development model seems to have run out of steam, democratisation may be the outcome of economic crisis. Since the government drew its political legitimacy largely from strong economic performance, the collapse of the economy led to a clamour for democratisation. The tendency of the new regime is then towards more orthodox economic policies, partly because of domestic pressures but also due to vulnerabilities to external pressure. Democratisation and gains in political rights may in such a case be accompanied by dramatic reversals in social rights, as some of the development institutions associated with the *ancien régime* are dismantled indiscriminately either because they are inherently incompatible with the new political dispensations or because they are seen as guilty by association.

5. THE PROBLEMATIQUE

Much of the literature on democracy and policymaking starts with the assumption that "good" policies are those encompassed by the Washington Consensus. The question is then posed, "Can democracies implement such good policies?" This type of analysis, however, fails to recognise the problem of reconciling the inflation of demands that comes along with democratisation with the fiscal deflation that was *de rigour* under adjustment (Hutchful 1995). It also fails to problematise the adherence of democracies to orthodox policies, indeed their rather paradoxical proclivity towards these policies, by asking whether emergent democracies should implement orthodox policies, especially given the considerable evidence of their negative social effects.

The evidence that democratic states are not necessarily "soft states" and can take tough policy measures when necessary suggests some level of political and institutional capacity on the part of democracies and, therefore, is a good sign for the prospects of development. However, the use of such capacity to adopt orthodox policies is not. While orthodox economic policies have been successful in the stabilisation of economies, they have usually done poorly in achieving high economic growth and more equitable distribution of income – two other economic aspects that help new democracies to endure. There is a fairly widely accepted view that structural adjustment is not pro-poor because it is not particularly pro-growth and also because it tends to worsen income distribution (Cornia 2000).

While IMF programmes have reduced deficits and improved countries' balance of payments, their effects on economic growth have been "mixed". Many studies find negative effects of IMF programmes on growth (Easterly 2000, Goldstein and Montiel 1986, Khan 1990, Przeworski and Vreeland 2000). An IMF background study for the evaluation of its Enhanced Structural Adjustment Facility (ESAF) reported the effects of structural adjustment policies on growth as 'barely discernible when full account is taken of macroeconomic policies, human capital accumulation, initial conditions and exogenous shocks' (Kochnar et al. 1999). Conway (1994) found that initial negative effects of adjustment on growth were offset by subsequent growth. Przeworski and Vreeland (2000), using a broadly similar approach, found significantly negative and persistent effects on growth. A very recent study by Barro and Lee (2002), which used a different (instrumental variable) approach to take account of the endogeneity problem, concluded that while adjustment programmes do not have a significant contemporaneous effect on growth, they do have a lagged effect that is negative.

As for poverty, the United Nations Conference on Trade and Development (UNCTAD) states 'it is clear that even when well implemented, past adjustment programmes have not delivered sustainable growth to make a significant dent in poverty in most LDCs' (UNCTAD 2002). Even where growth has been achieved, IMF and World Bank involvement lowers the growth elasticity of poverty (i.e. the responsiveness of poverty rates to a given amount of growth) (Easterly 2000). The implication is that under structural adjustment the poor benefit less from economic expansion, but they also suffer less from economic contraction. Yet the fact that the poor are hurt less during the downturn can be no consolation since one of the objectives of lending by the World Bank and IMF is ultimately to restore growth. So, while growth and equity are good for the poor, orthodox policies are neither pro-growth nor pro-poor. As noted above, inequality has increased in the era of adjustment.

There is a growing literature on "pro-poor macroeconomics" (see e.g. Cornia 2003, Lustig 2000) which suggests that the current orthodoxy is not pro-poor in nature. Even the Bretton Woods institutions have jumped on the pro-poor policies bandwagon, as they rather begrudgingly concede that their policies have at best not been pro-poor and, all too often, have actually been anti-poor.[7] On the equity front, orthodox policies have rarely bothered to explicitly address the issue, often deductively deriving the outcomes of policies through an axiomatic account of the effects of neoliberal economic policies on poverty and equity. This has often obviated the need for social policy. It is simply stated that "getting prices right" will lead to high growth and greater equity through improved competitiveness in labour-intensive goods. The resultant increase in demand for labour would lead to higher wages, thus generating higher incomes for the poor in both rural and urban areas. In addition, competition would reduce the monopoly rents accruing to the well-off rent-seeking elites and free agriculture from the many indirect taxes imposed on it through protected domestic markets for industrial goods and overvalued foreign exchange.

The mechanisms that lead to low growth and greater inequality are fairly well known. The excessive and dogma-driven focus by orthodox programmes on sharp demand compression in order to reduce inflation to single digits leads to sharp falls in output and employment. Too rapid deficit reduction is also often a source of deflation. And reduction of the fiscal deficit has at times been achieved through cuts of pro-poor expenditure rather than through higher taxation (Cornia and Court 2001).

If there was any immediate "elective affinity" between neoliberal policies and democracies, it was not obvious to those who militated for democracy or voted in new governments. In Africa, Yusuf Bangura noted that demonstrations for democracy have been organised by opposition groups and parties with traditional sympathies for the aspirations of the poor:

> Contrary to the neoliberal formulations, democratisation is seen by the majority of dissident groups as an instrument for obstructing structural adjustment and protecting some of the gains in public welfare and living standards threatened by the reforms (Bangura 1992).

In fact, opposition to the negative economic and social effects of structural adjustment and the orthodox stabilisation process has driven movements for democracy, especially in Africa. These bread-and-butter issues were at the heart of the wave of protest that swept the African continent in the late 1980s and early 1990s. What began as a protest by urban groups against growing pauperisation increasingly led to the linking of economic demands to more explicitly political demands for constitutional change.

So a number of questions arise: Why are democracies pursuing economic policies that are known to be deflationary, seemingly backed by their supporters? How does one explain the 'neoliberalism by surprise' (Stokes 2001) which has led to adoption of orthodox policies by movements that were catapulted into power by opposition to these very types of policies? Why are new democracies more orthodox in this respect than older ones? If income distribution remains constant, then any growth will benefit everyone. This argument has been used to buttress the view that we really need not worry about equity when thinking about poverty. But, if as we now know both growth and equity are good for the poor and if there is no trade-off between growth and equity and we take the elimination of poverty as a matter of extreme urgency, why can't we think of strategies that are pro-poor in their bias? And why have new democracies not pushed for more egalitarian policies in order to reduce poverty? Why have the poor not used their voting power to push for policies that are pro-poor and lead to sustainable improvements in the lives of the poor? Or, as Putterman, Roemer and Silvestre (1998: 90) express it,

> if equalization of the distribution of wealth is possible through the electoral process, and if it is in the interest of the large majority of people (as would appear to be the case since median wealth is far below mean wealth in all capitalist democracies) why is it not implemented through political action by rational citizens?

One obvious explanation is that in many countries movements for democracy have been dominated by elites for whom equity and poverty alleviation are not high on the policy agenda. Such elite-driven democracies may have proven unwilling to transgress the narrow confines imposed on them by both the domestic and foreign elites that perhaps led or supported the democratisation process. This said, we must consider the fact that a number of elite-dominated democracies, such as that in Costa Rica, have accommodated an egalitarian ethos that has allowed the state to pursue policies that advance the interests of the poor. Even in the African context there are significant differences between, for example, Botswana and Mauritius. This suggests that politics matters.

Thus what needs to be explained is how in all too many cases there was a 'metamorphosis from heterodox candidate to orthodox candidate', to use Teivainen's (2002) apt phrase. The next section ventures to provide some possible answers to the paradox. Indeed, in a number of cases the political unrest provoked by adjustment in the form of so-called "IMF riots" has accelerated the demise of authoritarian regimes.

6. SOME EXPLANATIONS

6.1 Ideological Shifts

One of the remarkable transformations of the latter part of the twentieth century was the ideological shifts in the major industrial countries and the international institutions over which they exercised considerable influence and the collapse of the "actually existing socialism". New democracies were said to be simply partaking in the new global *zeitgeist* (Diamond, Linz and Lipset 1988a) or "liberal moment" that not only sanctioned individual and human rights but also claimed a close affinity of these rights to markets. A feature of this period was a more positive attitude by the United States towards democracy in the developing countries than was the case during the Cold War.[9] The "third wave" of democratisation was a global systemic and normative integration, that has since led to an understanding that political and economic liberalisation are produced in tandem: they are two sides of the same coin (Huntington 1991, Simensen 1999). For Fukuyama (1992) this was not simply a passing "liberal moment", but the final triumph of liberalism against other ideologies, the battles against which had constituted history.

Significantly, egalitarian ideologies were on the defensive during the period and neoliberalism triumphed. First, there was the crisis of the welfare state and the eventual triumph of conservative political movements symbolised by Ronald Reagan and Margaret Thatcher. These ideological shifts were imposed on the Bretton Woods institutions, bringing to an end the earlier and short-lived engagement by the World Bank in "growth with equity" strategies. Neoliberal policies became the Washington Consensus. In Africa[10] in the immediate post-colonial period ideologies of nation-building, developmentalism and assorted idiosyncratic "socialisms" pushed for policies with considerable pro-poor bias: free education and health services, pan-territorial pricing and food subsidies. In many cases, these social policies were associated with economic policies that ran into deep trouble in the late 1970s. Guilty by association, in the era of adjustment, these "welfare policies" were dismissed as fiscally irresponsible or as "market distortions". Moral and ideological premises of such social policies were deliberately associated with rent-seeking, urban bias and clientelism. In addition, there was a shift in political leadership. Most populist nationalist elites have by now been replaced by an elite which, while spawned by state policies, has a much more pro-market orientation. This is described by Mafeje in a paper presented in 1992:

> All evidence points to the fact [that] in the so-called 'wave of democratisation' sweeping through Africa a new class of compradors will gain ascendancy. They will be largely technocrats who will try their best to ingratiate themselves with the World Bank and to give structural adjustment programmes in Africa a longer lease on life. Unlike their predecessors, they will be less nationalistic, more pro-West and will espouse some naïve and anachronistic ideas about liberal democracy. In the hope of achieving the long awaited democracy since independence, the people will vote for them. But disillusionment will come first (Mafeje 1995: 25).

Another argument has to do with the domestic ideological interpretation of the recent past and the conflation of interventionism with authoritarianism. In much of Africa, authoritarian rule has been linked to an interventionist state. This historical experience is adduced to conclude that democracies must be non-interventionist.[11] Matters are made worse by the personalisation of policies in the past so that where there have been regime shifts, there is the tendency to reverse all past policies. Projects and programmes initiated by a dictator are summarily abandoned regardless of their economic merit. Dismantling the state is part of laying to rest the demons of authoritarianism; and the zeal with which this is done often confounds economic sense. At times this dismantling fits well with the international financial institutions' own demolition job, which may have been stalled by the resistance of the authoritarian regime.

Such turnarounds are not exclusively African. In Latin America, past experience and interpretation of the devastating impact of macroeconomic populism also informs how democracies respond. In some cases it has been argued that social movements learned that linking democracies to substantial demands only leads to macroeconomic populism, which eventually leads to military *coups d'état*. In the Latin American literature, there is the widespread view that substantive demands made on democratisation in the past have rendered societies ungovernable (crisis of governability) and "overloaded" the system, leading to macroeconomic populism which invited military intervention. This time around, so the argument goes, the focus should be on the formal aspects of democracy. Montecinos noted that through the 1970s and 1980s, the evaluation that parties and political analysts made of Chile's democratic breakdown led to "self-criticism": 'Intellectuals recognized that in the past their dogmatic quest for ideological purity had been a main factor in the polarization that preceded the military coup' (Montecinos 1993). Indeed, the new political leadership in democratic movements must avoid raising the expectations of the followers to economically untenable and destabilising levels.[12]

The view emerged that nothing should be done to upset the elite and the military (see e.g. Di Palma 1990, O'Donnell and Schmitter 1986). A more manipulative formulation of the new arguments even suggested that some screening of participants was necessary: 'parties of the Right-Centre must be "helped" to do well. And parties of the Left-Centre should not win by an overwhelming majority' (O'Donnell and Schmitter 1986: 62). Di Palma goes even further, giving this view a classical conservative twist by insisting that 'in the interests of democratization, the corporate demands of business and the state may have to take precedence over those of labour'. According to this view, democracies must demonstrate their capacity to pursue orthodox policies and disavow their past populist tendencies or electoral promises.

In some cases, such as Argentina, Bolivia and Brazil, the hyperinflation that followed their experimentation with heterodox policies taught the new democracies to choose a more cautious approach.[13] This choice may be reinforced by the significant success of the policies of the *ancien régime* with respect to a number of macroeconomic variables including economic growth and economic stability. This shift in the ideologies and composition of the elites needs to be better understood in thinking about possible coalitions for more egalitarian and pro-poor policies.[14]

All these factors – the conditionalities and impositions by the Washington Consensus and Bretton Woods institutions, the collapse of the "real existing socialisms" in the East, the tribulations of the particularistic and often idiosyncratic socialisms in the "Third World", the weakening of post-World War II social pacts, the delegitimation and shredding of the state by corrupt elites – may suggest that Margaret Thatcher was correct when she pronounced, "There is no alternative!" Significantly, this state of mind leaves the new democratic movements with no clearly articulated transformative and socially inclusive model on which to build.

6.2 Absence of Political Coalitions

In democracies numbers matter. A frequently raised question is therefore, "What prevents popular majorities from exercising their numerical strength to influence policy in their favour?" Or, as Kenneth Roberts asked with respect to Latin America, 'Why do institutions that are supposed to embody popular sovereignty, produce elitist and exclusivistic outcomes when subaltern sectors constitute a large majority of the population?' (Roberts 1998: 2). An answer is that in democracies organisational capacities and politics determine the alignment and weight of the numbers.

The newness of democratic government often means a lack of organisational capacity and therefore an inability to regulate interest groups by organic representation of new political interests and coalitions. Ironically, precisely because of the lack of a culture of coalition-building, there are often no coalitions that can create a sustainable budgetary formula, and austerity packages generally unravel due to lack of such coalitions. Older democracies enjoy neo-corporatist institutional arrangements consisting of "social pacts" designed to complement traditional mechanisms requiring compromise and greater cooperation and burden sharing – the prerequisites for successful stabilisation under a democratic regime. Such arrangements may indeed have been the source of their stability, having allowed the emergence of political coalitions whose own stability will, in turn, be the result of the political culture that evolved through coalition building. Under such conditions primacy is more on political coherence than the pursuit of technically coherent economic policies.

An important point that emerges from the resilience of the welfare regimes in the older democracies is that national configurations of democracy directly and indirectly shape the capacity of domestic institutions to resist or deflect external pressures (see e.g. Swank 2001). Democracy allows groups opposed to a certain set of policies to emerge and, over time, be part of the "social pacts" that sustain the democratic order. Consequently, their views receive attention. Negotiation of binding agreements implies that all actors must be internally cohesive and their representative institutions must speak authoritatively for them and guarantee their compliance – a characteristic that is only achieved with passage of time (Haggard 1997). New democracies, almost by definition, lack such a culture of coalition building and the institutions that go along with it. This is illustrated by the case of Tanzania, where no significant level of compromise and negotiation among domestic actors has yet been achieved in policymaking. As Therkildsen noted,

> policy decisions… do not necessarily reflect collectively binding political compromises nor genuine political support for the reform package as a whole. Rather such decisions are often influenced by larger political aims (which may not be relevant to the reform per se) or by accommodation to perceived or real donor pressures, or to individual ministries' resource-mobilizing strategies vis-à-vis the donors (Therkildsen 1999).

One central feature of new democracies undergoing market-oriented reforms is the strengthening of private capital, which wields tremendous veto power over macroeconomic policies – and the consequent weakening of the state's capacity to regulate the economy and to mediate class and sectoral conflicts. In a surprisingly large number of new democracies, businessmen and businesswomen have assumed leading roles. In addition, some of the social groups opposed to orthodox policies will have been weakened by retrenchment and a general decline of well-being. The new political dispensation of democracy has opened space for new deliberative mechanisms between the state and business. However, the same cannot be said about the poor due to their lack of institutionalised channels that service their needs:

> The skewed and exclusionary nature of policymaking that continues to prevail in much of Africa gives reason to believe that policy outcomes will reflect the interests and concerns of business with privileged access to the policy process, which was previously identified as a source of blockage of effective reform (Robinson 1998).

In many cases the new democratic leaders have lacked political-organisational skills and the ability to forge coherent multi-party coalitions that would enable them to pursue equity and welfare-enhancing policies within an essentially free-market neoliberal economic context. Comparing India and Chile, Sharma argued that the ability to introduce a number of progressive social measures may be partially explained by the political capacity of institutionalised parties to maintain a coalition and to articulate fairly coherent policies and strong links to autonomous civil organisations (Sharma 1999, Weyland 1999). Understandably, new political movements emerging after years of subterranean existence will not yet have had time to acquire these skills and may lack coherent policy positions on a number of critical issues. Chile seems to be a rare case in which, as Montecinos (1993) reported, even before the assumption of power, the opposition movement had developed a fairly coherent model of policy options, partly because of the significant influence of the technocracy upon which it could draw. She further argued that this, combined with a deep-rooted political organisational culture, may have accounted for the measured additions to an essentially neoliberal model.

The nature of civil society and the capacities and preoccupation of its key organisations also matter in shaping the post-transition agenda. Weyland (1996) blamed the virtual absence of redistributive reform in Brazil on the inability of actors in civil society to put together an effective pro-redistribution coalition due to social fragmentation, corporativism, clientelism and weak parties. More specifically, there was an absence of organised and coordinated social movements and universalistic, programmatic political parties. Weyland argued that if civil society has expanded, it has been more in the form of networks of small associations, not national peak organisations. This could well describe the African situation, where the literature suggests that ethnic politics, fluidity of class identities and clientelism have played a significant role in preventing the emergence of cohesive political movements with clear transformative projects.

Note here the role of NGOs within civil society. The combination of neoliberal ideology, the weakening of the state and the new public management theories has resulted in a dramatic increase in funding for NGOs. Loss of state legitimacy and capacity necessitated the widespread faith in the capacity of NGOs. This bred the myth that NGOs could substitute for the state in combating poverty, and both the Bretton Woods institutions and NGOs found themselves on the anti-statist side of the debate on policy. Part of this trend was due to conflation of NGOs and civil society so that supporting NGOs was seen as creating a vibrant civil society and, therefore, as a contribution to democracy. Old social movements were seen as either irrelevant or a spent force. However, as it turns out, in many countries the old movements – labour unions, student movements, professional associations (especially lawyers) and churches – have been central to the struggles for democratisation. These movements are membership movements and so tend to carry more political weight nationally than most NGOs. In addition, many of the "new social movements" have eschewed linking the struggle for democracy to substantive overarching macro-issues through which the fundamental decisions affecting the poor are framed. Their demands have been particularistic (e.g. ethnic claims, gender, intellectual freedom) or confined to what Tendler refers to as 'projectizing and micro-izing' (Tendler 2000). Yet although poverty is lived at the micro level, its causes are largely macro. With their focus on service delivery at the micro level, NGOs, as such, are unlikely to constitute a major political force in combating poverty. Recall also that while poverty reduction may suggest empowerment at the micro level through "participation", it often entails disempowerment at the macro level.

Probably the greatest obstacles to the mobilisation of democratic institutions for poverty alleviation are the organisational weaknesses of the poor themselves. The role played by the poor, their capacity for self-organisation, alliance building and articulation of interests are often important factors in placing poverty on the national agenda. This is recognised in the new rhetoric about the "empowerment" of the poor. The organisational capacity of the poor is often undermined by a number of factors: problems of collective action, especially for dispersed rural populations and informal labour; the cross-cutting nature of rural identities and interests; and the capacity of elites to manipulate these identities in a manner that rarely advances the interests of the poor.[15] It is sometimes argued that the policies that might benefit the poor may be counterintuitive and therefore unlikely to win political support, and also whatever gains they promise are long-term in nature. 'Long-run and indirect links do not work well in democratic and mass politics: the effect has to be simple, intuitively graspable, clearly visible, and capable of arousing mass action' (Varsheny 1998: 17).

This time discrepancy allows the losers to organise against adjustment while the potential gainers are still not sufficiently organised or even aware that they would benefit in the long term, given the rather counterintuitive nature of the case in their favour. The irony of this view is that it proposes non-democratic solutions in which a benevolent technocracy can pursue the counterintuitive policies on behalf of the benighted poor. Consequently, early literature on adjustment emphasised the need to find ways that would insulate policymakers from popular pressure while they pursued the social good. Where democracy was conceded as a solution, the strategies proposed to circumvent democratic politics included "shock treatments", "insulation" of key policy instruments and "external agents of restraint". This view is premised on a neoliberal populism that assumes that market-friendly policies undermine special interests and rent-seekers.

Most arguments assume that all the immediately popular, pro-poor policies lead only to consumption, which reduces long-term growth. And whatever gains the poor get from state policies are deemed as short-sighted and, at worst, a kind of "macroeconomic populism" and therefore likely to self-destruct. However, today there is a rediscovery of Myrdal's insistence that consumption by the poor is investment. A whole range of pro-poor policies enhance long-term growth through "human capital" effects, such as better education and health and political stability. Pro-poor polices such as land reform or targeted credit may enhance the performance of markets and thus produce both equity and efficiency, which are good for growth.

6.3 The "New Broom" Argument

The "new broom" argument is that new regimes (democratic or otherwise) may enjoy a "honeymoon" period during which they may be able to press forward harsh austerity measures associated with orthodox economic policies (Williamson and Haggard 1994). In many cases, those who have just ascended to power through democracy may be aware that the policies they will introduce are unpopular and unlikely to be implemented under other political conditions. The opportunistic view then is to immediately introduce the policies since one never knows how long the window will stay open. The crisis will then have provided the window of opportunity for the executive to rely on support of the technocracy to carry out reforms, unencumbered by political contestation. Those who hold this view also tend to argue that new democracies should impose "shock treatment" before the enemies of reform can organise and long before the next election, by which time the fruits of reform will be visible.

One should also point out that new democracies are likely to be beholden to the technocracy. The reliance on technocracy may not be merely the result of imposition from outside, but a reflex reaction to the bad governance of the past. For instance, where patrimonialism and clientelism are identified with authoritarian rule and crisis, the new democracies may be inclined to rely on technocracy. In such situations, technocracies inclined towards orthodox economic policies may be in a stronger position in the new democracies. Another factor is major donors' instrumentalisation of democracy in pursuit of orthodox policies. Among donors who had feared that democracy might scuttle adjustment programmes, it was both a relief and politically correct to view democracy as the most efficacious instrument for creating the political framework best able to manage orthodox economic policies. In this new understanding, 'liberal democracy, social pluralism and market orientation are now the three pillars of African reform' (Sandbrook 1996: 2). Democracy became one aspect of "getting the politics right" for adjustment and orthodox economic policy.

The "new broom" has its downside, however. First, it can be captured by groups not particularly interested in anti-poverty policies, as seems to have been the case in many new democracies. Second, for effective anti-poverty policies, the state must have a capacity to process popular demands, manage conflicts over policy and implement policies systematically. New brooms normally lack such capacity.

7. GLOBALISATION, STRUCTURAL ADJUSTMENT AND "CHOICELESS DEMOCRACIES"

Perhaps one remarkable feature of the current wave of democratisation is the strong convergence in economic policies and institutional reforms, despite very different initial conditions and paths traversed. This suggests the overwhelming assertion of strong conditioning factors that have made themselves felt in all of these countries. The most obvious one is globalisation, both in terms of the preponderance of certain ideological predilections and the impositions and binding nature of certain economic constraints.

The welfare state and many of the post-independence "national development plans" were based on socially or politically "embedded" domestic markets, government responsibility for aggregate demand growth and state control over cross-border economic activity. Policymaking was built on the assumption that state policies were "national", not only with respect to objectives but also with respect to instruments. The post-war Bretton Woods international architecture itself was based on a "liberal embeddedness" that combined trade openness with domestic compensation mechanisms to mitigate the social costs of the volatility of trade (Ruggie 1983). This same order permitted certain latitude for the emergence of "developmental states", by allowing developing countries to adopt flexible exchange rates, capital controls and politically controlled central banks "designed" to serve the nationalists' domestic goals of rapid industrial development and nation-building (Helleiner 2003). Pre-globalisation, developing countries could thus potentially control a wide range of policy instruments, enabling them to pursue national development objectives. To be sure there was "dependency" and "neo-colonialism", but the possibility of "national policies" was never excluded (whether this meant the "new international economic order", renegotiating one's integration into the global system, "delinking" or something else).

With globalisation, state capacity has been severely eroded; nations must comply with the exigencies of global market forces or be marginalised. This loss of sovereignty is supposed to be compensated for by higher levels of growth. Failure to achieve high rates of growth in the era of neoliberalism is perceived to be evidence of failure of internal economic policies. Thus the marginalisation of whole continents and the persistence of such national problems as unemployment are blamed on the failure of policymakers to remove domestic market distortions and rigidities. In this way, a whole range of policies that states have pursued in the name of social welfare, national cohesion or development is associated with "distortions" and rigidities.

Most democracies emerged during the era of structural adjustment and thus may be hampered in their policies by the state of the economy they inherited. Many new democracies are largely products of the crisis of the interventionist model, and the collapsing dictatorships will have bequeathed serious economic problems to their successors. Poor economic performance is likely to have been one the causes of the collapse of the old regime and the outgoing authoritarian government may have engaged in fiscal profligacy to gain political support, leaving the state bankrupt or highly indebted.[16] The new governments may feign ignorance of the state of the economy before assuming power and may justify their subsequent sharp policy turns away from electoral promises by claiming that the state coffers were in worse shape than they could ever have imagined.[17] The new regime is thus compelled to seek assistance from the Washington institutions and is forced to adopt their standard programmes. The conditionalities that come with such aid severely limit the choices of new democracies, tending to push them towards a standard set of policies and producing, in effect, 'societies which can vote but cannot choose' (Przeworski, Stokes and Manin 1999: 84) or what is elsewhere herein labelled "choiceless democracies" (Mkandawire 1999a).[18]

7.1 Institutional and Pre-emptive Policy Lockup

Another determinant of policies in new democracies is the constraints imposed on them by new constitutions and institutional arrangements. Teivainen (2002) referred to this aspect as the "constitutional politics of economism" which establishes "reserved domains" by the insulation of specific concerns of government authority and substantive policymaking from elected bodies. Di Palma calls it "pre-empted democracy", designed to freeze or pre-commit the government and government initiatives (Di Palma 1997). Chile is often cited as the case where the resulting government 'lacks de jure and de facto power to determine policy in many significant areas because the executive, the legislative and judicial powers are still decisively constrained by an interlocking set of "reserve domains", military "prerogatives" or "authoritarian enclaves"' (Linz and Stepan 1998: 48).

The furtive search for "insulated" institutions is often intensified on the eve of democratic governance, since in authoritarian regimes no institution can really be independent of the authoritarian leadership. As countries move from authoritarian to democratic rule there has ensued a spate of activities seeking to isolate key policy instruments from democratic oversight. This is seen most clearly in the significant increase of central bank independence in the 1990s, the decade of democratisation. This is not by mere coincidence. The issue of central bank independence only arises under democratic governments, since again, no institution can really be independent of an authoritarian leadership.[19] These institutional arrangements – what Maravall (1994) referred to as "authoritarian enclaves" – mushroomed in the 1990s in tandem with the emergence of new democracies. Such constraints undermine the ideal of deliberative democracy and tend to strengthen the hand of selected groups.[20]

The "institutional deficits" created by this new order are well illustrated by the programmes of poverty eradication to which a large number of developing countries, including new democracies, adhere. For all the talk about participation and consultation in Poverty Reduction Strategy Papers (PRSPs), there are neither institutional arrangements for, nor political understanding of, the role of democratic institutions. Conventional economic wisdom argues that the general public, including elected political leaders, cannot understand the counterintuitive nature of good macroeconomic advice. Therefore, there has been a systematic attempt to circumvent elected bodies in the consultative process of drawing up PRSPs. In many cases, the NGOs selected as partners and proxy representatives for the poor in the process have lacked the legitimacy enjoyed by such membership associations as trade unions and professional associations. When the poor have been consulted it has been over residual spaces left for them or their putative spokes-institutions. The new model seeks empowerment for the poor at the micro level while disempowering them at the macro level. It sets out to address poverty, even as the macroeconomic model maintains its deflationary and non-developmental characteristics. PRSPs look uncannily similar everywhere. They are essentially linked to the disbursement of debt relief and pay no attention to political sustainability and, even less, to the productive capacity and surplus generation of the poor themselves. The "dialogues" that take place leave untouched the core adjustment model. They simply add some "soft" ingredients to the hard macroeconomic model, which has remained essentially the same despite the evidence that it is a failed development strategy. And so those aspects of stabilisation that have contributed to increase poverty through their deflationary effects on labour markets and reduced state expenditures on social services and infrastructure remain untouched.[21] Thus, although poverty eradication is premised on high economic growth rates, it is tethered to a macroeconomic framework whose principle focus is still stabilisation and which has so far produced miserly growth rates even among the "success stories".

7.2 "Signalling" Capital

The orthodox model of adjustment places great weight on attracting foreign capital. Yet the process of democratisation often causes uncertainty among investors, inducing among them a wait-and-see attitude.[22] This behaviour of the private sector may cause a heightened need to "signal" private capital that a new government is stable and favourable to foreign capital. As Dailmani argued, the higher the degree of democracy, the greater the need to balance the threat of capital flight, which is more likely with the opening of capital markets, with political demands including the need for political incentives to increase government intervention in cushioning market dislocations (Dailmani 2000). To attract foreign capital, new democracies must take great pains to conceal any populist inclinations they may have harboured. Democracies must demonstrate their capacity to pursue orthodox policies and disavow their past populist tendencies, the need to do this being higher among political parties that may in the past have identified themselves with radical ideologies.

South Africa is a poignant case in point. Its new policies rested 'on the assumption that restrictive fiscal policies will send such positive signals to investors that growth will leap forward on a wave of confidence-driven investment' (Nattrass and Seekings 1998: 32). On monetary policy the government appointed Tito Mboweni, a "leftist" to the post of governor of the reserve bank. He understudied the outgoing governor for a year, during which time considerable resources were invested in a public relations campaign to convince the "market" that Mboweni had shed his leftist ideological baggage. Currently, in Brazil, Lula has been under enormous pressure to show that he can pursue "responsible" fiscal policies. One should add here that the measures of "good governance" that go with the current wave of reforms and appeal to private investors could be at odds with poverty eradication.[23]

7.3 Limited Social Policy Instruments[24]

In the developing countries, the first victim of globalisation has been the state's power to intervene in the economy to ensure certain social outcomes, such as equity and poverty alleviation. The rudimentary "welfare states" that post-colonial regimes had instituted became a target of both ideological and fiscal attack. Social expenditures were seen as straining the fiscal budget and as a source of financial instability. On the ideological plane, whatever gains had accrued to the workers in the formal sector were now seen as "distortions" in labour markets brought about by the activities of rent-seeking urban coalitions. In cases of dislocation, social safety nets might be recommended as a temporary measure but, in such a scheme of things, there was no need for any comprehensive social policy specifically aimed at addressing issues of poverty and equity.

Together with the disappearance of poverty from the policy agenda came the disappearance of "social development" as something that state policies deliberately pursued (beyond simply overseeing the spontaneous market processes). Earlier "developmentalist" arguments for social policy as a key instrument of development simply disappeared. Macroeconomics, with all its attention firmly fixed on stabilisation and debt servicing, had a jaundiced view of all public expenditures, including social expenditure. This trend in macroeconomic policy has important implications for social policy and therefore outcomes in terms of equity and poverty eradication. In general, under orthodox policies new democracies have tended to opt for targeted policies to address specific pockets of poverty while avoiding redistributive social policies. The two key instruments proposed by the IMF and World Bank have been (i) social safety nets, which were introduced to address the adverse effects of structural adjustment programmes, and (ii) "targeting the poor". Initially, these measures were viewed as temporary, since their need would be diminished by the high employment elasticity of growth associated with structural adjustment programmes.

Almost since the inception of structural adjustment programmes critics have pointed to their negative effects in terms of poverty. As noted earlier, discontent against these policies has contributed to the mobilisation for democratisation. With evidence growing that structural adjustment programmes were adversely affecting large numbers of people, the Bretton Woods institutions were compelled to shift positions. Poverty was brought back into the adjustment agenda. In the late 1980s and early 1990s, bilateral and multilateral donors set aside significant amounts of funds aimed at "mitigating" the "social dimensions of adjustment". Such programmes were to act as palliatives to minimise the more glaring inequalities that their policies had perpetuated. Funds were made available to provide a so-called "safety net" of social services for the "vulnerable" – but this time not by the state (which had after all been forced to "retrench" away from the social sector) but by the ever-willing NGO sector.

Under structural adjustment, social policy has been limited and targeted, the argument being that, given finite resources, social policy must be aimed at the needy poor to prevent funds being captured by the well off, who can meet their needs by drawing on the private sector. This preference for targeting in the social policy arena is rather paradoxical in light of the World Bank's aversion to targeting in many economic activities, such as selective industrial policies or credit rationing in the financial sector. Arguments deployed against targeting in the economic field revolve around possible distortions it might generate: information distortions, incentive distortions, moral hazards and administrative costs, invasive loss and corruption. The assertion is that governments do not have the knowledge to pick winners or to monitor the performance of selected institutions. The solution is "universal" policies; that is, policies that create a level playing field, applying equally to all entrepreneurs. Lump sum transfers or uniform tariffs that apply to all are strongly recommended. Paradoxically, when it comes to social policy, such "universalism" is rejected on both equity and fiscal grounds. Instead, selectivity and rationing are recommended – apparently in total oblivion of the many arguments against selectivity raised with respect to economic policy. Suddenly, governments lambasted elsewhere for their ineptitude and clientelism are expected to put in place well-crafted institutions and be able to monitor their performance.

Yet there is nothing to exclude the possibility that "targeting" in the social sector may be as complex and amenable to "capture" as "targeting" with respect to economic policy. It is definitely the case that the criteria for selection are at least as complicated, as controversial and as ambiguous as those for economic policy. Social indicators are extremely difficult to construct and poverty itself is multidimensional. Sen (1999) raised exactly the same arguments against targeting in the social sphere. Asymmetry of information and the attendant moral hazard would always pose the danger of including the non-needy among the needy, or of not including some very needy. Targeting makes difficult demands on the administrative capacities of most developing countries and can easily lead to inefficiencies and corruption, especially where the majority of the population is in fact the poor.

Furthermore, it is necessary to consider the kind of political coalitions that would be expected to make such policies politically sustainable. The World Bank's approach concentrates on the problem of optimally disbursing *given* external resources (aid), not on generating and disbursing domestic resources. Not surprisingly, such an approach fails to deal with the relationship between targeting and the political economy of domestic resource mobilisation. The experience in developed and middle-income countries is that universal access is one of the most effective ways to ensure middle-class support of taxes to finance welfare programmes.

The attraction of targeting presumably is that it not only allows for prioritisation in the context of budget cuts and dwindling aid, but it also allows earmarking, and thus severely limits the discretionary expenditures of the state. The preference for targeting is probably based on recognition that there is very little room for redistributive measures in a policy package in which the state is reduced to the night watchman. Targeting allows the state (or rather the donors) to franchise their responsibilities to NGOs. The question that immediately arises is that posed by Wood:

> To what extent do citizens lose basic political rights if the delivery of universal services and entitlements is entrusted to non-state bodies which would at best only be accountable to the state rather than directly to those who service entitlements? (Wood 1997: 81)

In the process of the "franchising", the state loses control over policy and therefore loses responsibility for upholding the rights of all its citizens, producing what Wood refers to as a "franchise state" whose creation dilutes those dimensions of responsibility and accountability associated with the much-trumpeted "good governance".

Finally, there is the danger that the inherent selective and discriminatory nature of targeting may actually polarise societies by accentuating differences. This resultant fragmentation within the state and society can foster clientelism and segmentation over universalism. Universalism has always been associated with notions of citizenship. The new approach poses the risk of hollowing out citizenship by severely limiting citizens' rights. Current thinking about poverty heralds a shift in the orientation of development from the promotion of equality as part of the development agenda to the promotion of social order against a backdrop of increased inequality and insecurity.

8. CONCLUSION

Eradication of poverty requires high economic growth rates, structural change and redistribution. The great challenge, then, remains democratically devising strategies that simultaneously ensure high and sustained rates of growth, equitable distribution and rapid reduction of poverty within a highly competitive global environment. Success stories suggest that key components include carefully orchestrated policies on trade, investment and technology and social policies to promote health, education and social cohesion in a context of political stability. While most of the well-known success stories have been authoritarian, there is a moral imperative for such strategies to be democratically anchored. There is now compelling evidence that nothing prevents democracies from performing well in these tasks. However, so far democracies, especially the new ones, have been compelled by both ideas and structural factors to pursue polices that are not developmental under the Washington Consensus and "second generation" variants of good governance. These policies are definitely not socially inclusive, and their relationship to democracy has been problematic to say the least. This chapter reviewed some of the reasons why democracies choose these policies.

A useful distinction was made between democratic institutions and democratic politics. The former is concerned with methods and procedures for legitimising rules and assuring that political contestation is free and fair, while the latter emphasises participation, equality and emancipation. We have learned that concern for democratic politics without due respect for institutions can lead to populist authoritarian regimes. But we also know that democracy is not simply a question of rules and institutions, but also of the content and purpose of these institutions and rules and that the failure by democratic institutions to foster democratic politics has produced lifeless institutions that have done little to address serious issues of poverty and inequality. They have produced instead "democracy with tears", which in many cases has rebounded on itself. The hollowing of the democratic process makes it rather pointless to use democratic spaces to compete over state resources. Instead, it encourages extra-parliamentary struggles, including personalism, factionalism and use of other means inimical to a democratic order. The current danger is that the emerging political order, while liberal and democratic, may preside over societies that will be strongly elitist and socially quite regressive.

This chapter has argued that the asserted "elective affinity" between democracy and orthodox neoliberal policies overlooks serious problems that new democracies face in consolidating themselves under the prevailing national and global economic regimes. One implication is the need to give fledgling democracies more instruments and room for manoeuvre, not only to go beyond the Washington Consensus, but also to make democracy a meaningful institutional arrangement for dealing with the serious problems of poverty and inequality, as well as divergent interests.

NOTES

1. Maravall similarly observes that dramatic fiscal crisis and inflation in Latin America are more attributable to development efforts of dictatorships than to democratic populism (Maravall 1994).

2. The policy indicator for variable I for county j in any given year is given by

$$I_{ij} = \left| \frac{\underset{j}{Max} X_{ij} - X_{ij}}{\underset{j}{Max} X_{ij} - \underset{j}{Min} X_{ij}} \right| \quad (1)$$

    where
    Xij is the actual value of variable i for county j,
    Iij is the index value of variable i for country j,
    Max Xij is the maximum value of variable i for variable j for all countries,
    Min Xij is the minimum value of variable i for variable j for all countries.

    The policy index is then the average of the n indices.

$$I_j = \frac{1}{n} \sum_{i=1}^{j} I_{ij} \quad (2)$$

    The higher the index the more orthodox the policies.

3. The paper measures B which is the measure of anti-export bias and its formula is as follows:

$$B = \frac{E_m (1 + t + n + PR)}{E_x (1 + s - t_l + r)} \quad (3)$$

    where E_m and E_x are nominal exchange rates applied on imports (m) or exports (x); t is the average import duty, n is any additional differential domestic taxation of imports, PR is the differential between the domestic and border prices of importable commodities subject to quantitative restrictions or import monopolies, s is any export subsidy ($s>0$) or export tax ($s<0$), t_l is the taxes and duties on inputs used in production of exportable goods (that is, the tax rate on inputs multiplied by the share of that input in total production costs), and r is any import duty. If B is higher than one, as is usually the case, the index indicates the degree to which commercial policies favour import-substitution relative to exporting.

4. To be fair to MMD it did in fact promise a liberal economic policy. In its manifesto released in February 1991, MMD explicitly stated that the state would not be a 'central participant' in the economy and that, instead, it would encourage a 'wide spectrum of entrepreneurship' (cited in Bratton 1994). As Kayizzi-Mugerwa (2001) noted, 'While in other Africa countries privatisation was undertaken as part of the conditionality attached to economic reforms, Zambia was one of the few countries where a party with an election Manifesto, that included privatisation was elected'. It also promised to collaborate with international financial institutions with whom Kenneth Kaunda had open quarrels. In addition, Chiluba warned his followers of hard times ahead. However, within the MMD a streak of populist rhetoric was allowed to flourish.

5. Chilumba's remarks on privatisation are typical of the disillusionment: 'We were blind when we sold some parastatals, and made mistakes. How can you have a parastatal buying off another parastatal, and call it privatisation? Government was asleep when it sold Chilanga Cement and Zambia Sugar to CDC which is a British parastatal. We are wondering why some countries are advocating the dismantling of parastatals here while on the other hand keeping their parastatals back home' (cited in Africa Business 2001).

6. This account draws mainly on Bajpai (2002).

7. Over the years the Bretton Woods institutions have taken great pains (i) to argue that structural adjustment programmes do not hurt the poor and (ii) to insist on the search for a post-Washington Consensus and pro-poor and pro-growth macroeconomic policy. Indeed the whole idea of Poverty Reduction Strategy Papers is an indictment of the set of policies. A Google search for "pro-poor" and "macroeconomics" yielded almost 1,800 hits! Many are documents of the Bretton Woods institutions and the UN system.

9. To the extent that this matters, the turn by the United States towards security issues during its anti-terrorism campaign is likely to erode the commitment to democracy not only at home, as has been suggested by many observers, but also abroad. Already a number of authoritarian regimes have jumped onto the anti-terrorism bandwagon thus easing pressures on them to democratise.

10. I discuss these "shifting commitments" in Mkandawire (1996b).

11. This is a non-sequitor. Democracies such as India, the Western European welfare and liberal states and "developmental states" such as Mauritius and Botswana have been interventionist without thereby undermining their democratic credentials.

12. In a sense this view takes us back to Huntington (1991) whose dread of "revolution of rising expectations" seems to have persuaded him that stable democracy depends on "disillusionment and lower expectations" on the part of the general population: Democracies become consolidated when people learn that democracy is a solution to the problem of tyranny but not necessarily to anything else". However, Huntington himself recognises the need for new democracies to be effective in addressing substantive problems of society when he states that new democracies are faced with a serious dilemma because 'lacking legitimacy, they cannot become effective, lacking effectiveness, they cannot develop legitimacy' (ibid.: 258)

13. Weyland attributes the caution to risk aversion: 'Conscious political learning from these dramatic failures provided an important motivation for the Alwyin administration to pre-empt or limit demands of its supporters and followers so as not to endanger an economic stability' (1999: 69). This not to say the new government ignored the "social debt" left behind by the outgoing fascist regime. Rather the choice of the government was to address a number of social issues within the fiscal parameters of the inherited policy regime.

14. In a number of cases, the intellectual leaders of the democratisation movements have had to make stunning intellectual somersaults. The case of Ferdiando Cardoso of Brazil is probably the most spectacular. One of the key figures in the Latin American critique of *dependencia*, 'once Cardoso was in power, the question of dependency and development was turned on its head. As President, Cardoso sought explicitly to make the Brazilian economy as dependent as possible on the multinationals and financial institutions of the core in order to develop the Country.' One should add here that this change is not simply a result of ideological evolution but can partly be explained by the immobilism produced by the violence of authoritarian regimes on the body politic. For an interesting set of studies on the post-Banda politics in Malawi see Englund (2002).

15. Jenkins (2000) argues that in India obfuscating tactics have been used to defuse political resistance to policy shifts. He argues that in India informal institutions have driven economic elites towards negotiation, while allowing governing elites to divide the opponents of reform through a range of political tactics.

16. This seems to have been the case in countries such as Malawi where a hitherto fiscally conservative regime went on a spending spree just before the 1994 elections, dramatically increasing the deficit and causing the national currency to fall.

17. Fujimori, who adopted the "shock treatment" policies that had been pushed by his opponent, used this argument to justify his policy switch although the real reason may actually have been the pressures from the international financial institutions.

18. Thomas Friedman has stated this constraint most graphically and with some exaggeration. 'Two things tend to happen: your economy grows and your politics shrinks…The Golden Straightjacket narrows the political and economic choices of those in power to relatively tight parameters. That is why it is increasingly difficult these days to find any real differences between ruling and opposition parties in those countries that have put on the Golden Straightjacket. Once your country puts on the Golden Straightjacket, its political choices get reduced to Pepsi or Coke – to slight nuances of policy, slight alterations in design to account for local traditions, some loosening here or there, but never any major deviation from the core golden cites' (cited in Economist 2001: 22).

19. This statement is contradicted by the evidence from Cukierman which suggests that in non-OECD countries central bank independence has been highest among authoritarian regimes. However these results depend on a rather poor measure of such independence (governor turnover relative to change in government leadership) which in a sense endogenises the length and stability of terms of office. In Banda's Malawi, the head of the Reserve Bank was a close relative of the mistress of the state. He served for a long time. It would be perverse to consider this as evidence of central bank independence. Such independence means little if laws are highly personalised or not respected.

20. In Latin America the new arrangements have often been part of the "pacts" for democratisation. Chile illustrates the case where the outgoing authoritarian regime imposes "authoritarian enclaves" that deliberately exclude key elements of policymaking from parliamentary oversight. Writing on Mexico, Boylan (2001) argued that the domestic threat of policy change is the primary motivation driving authoritarian elites to insulate their preferences in autonomous agencies. Boylan's argument is that when authoritarian elites fear the populism that may come along with new democracy and expect a regime shift, they may be tempted to create autonomous central banks to lock in a commitment to orthodox policies.

21. This is openly acknowledged by the Bretton Woods institutions themselves. Thus the IMF/World Bank's review of PRSP notes, 'The macroeconomic policy and structural reform agenda – for example, trade liberalization and privatization – are, however, sometimes not even on the table for discussion. Even countries like Uganda that have a rich history of macro-level participation do not indicate that civic inputs have substantially shaped the direction of ongoing fiscal and agricultural reform' (IMF/World Bank 2001 cited in Craig 2003:58).

22. This is an aspect of the "political business cycles" phenomenon which, as Block and Vaaler (2001) pointed out may have implications not only for incumbent governments and their electorates but also for foreign actors involved in allocating credit and pricing it. They find that agency sovereign risk ratings decrease and bond spreads increase for developing countries during election periods because both agencies and bondholders appear to view elections in developing countries negatively, and impose additional credit costs.

23. It is now widely assumed that "good governance" is essential for attracting private investment. A number of measures of such good governance have been developed to rank countries. A widely used one in econometric studies is the International Country Risk Guide (ICRG), which consists of measures based on "expert judgement" of countries. Moore et al. (1999) have regressed this to an index (the relative income conversion efficiency or RICE) that measures a country's capacity to translate national material resources into human development (i.e. life expectancy and education levels). They found a negative correlation between ICRG and RICE. In other words the higher the government institutions are scored from the perspective of international investors and lenders, the worse the governments perform

in converting national income into human development. Moore concluded, 'This is strong evidence that 'governance' factors that matter to international investors and lenders are significantly different from those that relate to poverty.'

24. This section draws heavily from Mkandawire and Rodrigues (2000).

REFERENCES

Abrahamsen, Rita (2000) *Disciplining Democracy: Development Discourse and Good Governance in Africa.* London: Zed Books.

Bajpai, Nirupam (2002) 'A decade of economic reforms in India: The unfinished agenda'. Working Paper 89. Cambridge, Mass.: Center for International Development at Harvard University.

Bangura, Yusuf (1992) 'Authoritarian rule and democracy in Africa: A theoretical discourse', in Peter Gibbon, Y. Bangura and Arve Ofstad (Eds) *Authoritarianism, Democracy and Adjustment: The Politics of Economic Reform in Africa*, pp. 39-82. Uppsala: The Scandinavian Institute of African Studies.

Barro, Robert and Jong-Hwa Lee (2002) 'Who is chosen and what are the effects?' NBER Working Paper. 8951. Cambridge, Mass.: National Bureau of Economic Research.

Block, Steven and Paul Vaaler (2001) 'The Price of Democracy: Sovereign Risk Ratings, Bond Spreads and Political Business Cycles in Developing Countries'. CID Working Paper. 82. Boston, Mass.: Center for International Development at Harvard University.

Boylan, Delia M. (2001) *Defusing Democracy: Central Bank Autonomy and the Transition from Authoritarian Rule.* Ann Arbor: University of Michigan Press.

Bratton, Michael (1994) 'Economic crisis and political realignment in Zambia', in Jennifer Widner (Ed.) *Economic Change and Political Liberalisation in Sub-Saharan Africa*, pp. 101–128. Baltimore, Mass.: John Hopkins University Press.

Bräutigam, Deborah (1994) 'Institutions, economic reform, and democratic consolidation in Mauritius'. *Comparative Politics* 30 (1): 45–62.

Bresser Pereira, Luiz Carlos (1993) 'Introduction', in Luiz Carlos Bresser Pereira, Jose Maria Maravall and Adam Przeworski (Eds) *Economic Reforms in New Democracies: A Social Democratic Approach.* Cambridge: Cambridge University Press.

Conway, Patrick (1994) 'An atheoretic evaluation of success in structural adjustment', *Economic Development and Structural Change* 42: 267–292.

Cornia, Giovanni Andrea (2000) 'Inequality and poverty in the era of liberalisation and globalisation'. Paper presented at meeting of the G-24 held at Lima, Peru 29 February to 2 March.

Cornia, Giovanni Andrea (2003) 'Pro-poor macroeconomics'. Paper presented at Ford Foundation conference held in Antigua, Guatemala.

Cornia, Giovanni Andrea and Julius Court (2001) 'Inequality, growth and poverty in the era of liberalisation and globalisation'. Policy Brief No. 4. Helsinki. UNU WIDER.

Craig, David (2003) 'Poverty reduction strategy papers: A new convergence', *World Development* 31 (1): 53–69.

Cukierman, Alex, Pantelis Kalaitzidakis, Lawrence Summers and Steven Webb (1993) 'Central bank independence, investment, growth and real rates'. Paper presented at Carnegie-Rochester Conference on Public Policy held at Geneva.

Dagdeviren, Hulya, Rolf van der Hoeven and John Weeks (2002) "Poverty reduction with growth and redistribution'. *Development and Change* 33 (5): 383–414.

Dailmani, Mansoor (2000) *Financial Openness, Democracy, and Redistributive Policy.* Washington, D.C.: World Bank.

Danielson, Anders and Gun Eriksson Skoog (2001) 'From stagnation to growth in Tanzania: Breaking the vicious circle of high aid and bad governance', in Mats Lundahl (Ed.) *From Crisis to Growth in Africa?* pp. 147–168. London: Routledge.

Di Palma, G. (1990) *To Craft Democracies: An Essay in Democratic Transition.* Berkeley: University of California Press.

Di Palma, Guiseppe (1997) 'Markets, state, and citizenship in new democracies', in Manus Midlarsky (Ed.) *Inequality, Democracy and Economic Development*, pp. 290–317. Cambridge: Cambridge University Press.

Diamond, Larry Jay, Juan J. Linz and Seymour Martin Lipset (1988a) *Democracy in Developing Countries*. Boulder, Colorado: Lynne Rienner.

Diamond, Larry Jay, Juan J. Linz and Seymour Martin Lipset (1988b) 'Introduction: What makes for democracy?' in Larry Jay Diamond, Juan J. Linz and Seymour Martin Lipset (Eds) *Democracy in Developing Countries*, pp. 1–66. Boulder, Colorado: Lynne Rienner.

Dollar, David and Aart Kraay (2001) *Growth Is Good for the Poor*. Washington, D.C. World Bank.(Draft).

Dornbursch and S. Edwards (Eds) (1992) The macroeconomics of populism', in Dornbursch and Edwards (Eds) *The Macroeconomics of Populism in Latin America*. Chicago: University of Chicago Press.

Duquette, Michel (1999) *Building New Democracies: Economic and Social Reform in Brazil, Chile and Mexico*. Toronto, Buffalo: University of Toronto Press.

Easterly, William (2000) 'The lost decades: Developing countries stagnation in spite of policy reform, 1980–1998', *Journal of Economic Growth* 6: 135–157.

Easterly, William (2001). 'The effect of IMF and World Bank programs on poverty'. Paper presented at WIDER Conference on Economic Growth and Poverty reduction held at WIDER, Helsinki, 25–26 May.

Economist (2001) 'Globalisation and its Critics: A Survey'. September 29th to October 5.

Englund, Harri (2002) *A Democracy of Chameleons: Politics and Culture in the new Malawi*. Uppsala: Nordiska Afrikainstitutet.

Findlay, R. (1988) 'Trade, Development and the State', in T. Ranis. and G. Schultz (Ed.) *The State and Development Economics: Progress and Perspectives*, pp. 78–99. London: Basil Blackwell.

Fukuyama (1992) *The End of History and the Last Man*. London: Hamish Hamilton.

Ghai, Dharam (Ed.) (2000) *Social Development and Public Policy*. Geneva: UNRISD/Macmillan.

Goldstein, Morris and Peter J. Montiel (1986) 'Fund stabilization programs with multicountry data: Some methodological pitfalls', *IMF Staff Papers* 33 (June): 304–344.

Habib, A. and V. Padayachee (2000) 'Economic policy and power relations in South Africa's transition to democracy', *World Development* 28 (2): 245–263.

Haggard, Stephen and Steven Webb (1993) 'What do we know about the political economy of economic policy reform?' *World Bank Research Observer* 8 (2): 143–168.

Haggard, Stephen and Robert R. Kaufman (1995) *The Political Economy of Democratic Transitions*. Princeton, New Jersey: Princeton University Press.

Haggard, Stephen. (1997) 'Democratic institutions, economic policy and development', in Christopher Clague (Ed.) *Institutions and Economic Development: Growth and Governance in Less-Developed and Post-Socialist Countries,* pp. 121–149. Baltimore: John Hopkins University Press.

Hamner, Lucia and Felix Naschold (2001) 'Attaining the international development targets: Will growth be enough.' Paper presented at the Conference on Growth and Poverty, Helsinki.

Helleiner, Eric (2003) 'The southern side of embedded liberalism: The politics of postwar monetary policy in the Third World', in Jonathan Kirshner (Ed.) *Monetary Orders: Ambiguous Economics, Ubiquitous Politics*. Ithaca: Cornell University Press.

Herbst, Jeffrey (1994) 'South Africa: economic crises and distributional imperatives', in Stephen Steedman (Ed.) *South Africa; The Political Economy of Transformation*. Boulder: Lynne Rienner Publishers.

Hinkle, Lawrence, Alberto Herrou-Aragon and Keiko Kubota (2003) 'How far did Africa's first generation trade reforms go? An intermediate methodology for comparative analysis of trade policies'. Africa Region Working Paper Series. 58a (Volume I). Washington, D.C.: World Bank.

Huntington, Samuel P. (1991) *The Third Wave: Democratization in the Late Twentieth Century*. Norman: University of Oklahoma Press.

Hutchful, Eboe (1995) 'Adjustments, regimes and politics in Africa', in Thandika Mkandawire and Adebayo Olukoshi (Eds) *Between Liberalisation and Repression: The Politics of Adjustment in Africa*, pp. 52–76. Dakar: CODESRIA.

IMF/World Bank (2001) *The Review of the Poverty Reduction Paper (PRSP) Approach; Early Experience with Interim PRSPs and full PRSPs*. Washington, D.C.: IMF and World Bank.

Jenkins, Rob (2000) *Democratic Politics and Economic Reform in India*. Cambridge: Cambridge University Press.

Kayizzi-Mugerwa, Steve (2001) 'Explaining Zambia's elusive growth: Credibility gap, external shocks or reluctant donors?' in Mats Lundahl (Ed.) *From Crisis to Growth in Africa?*, pp. 132–146. London: Routledge.

Khan, Mohsin S. (1990) 'Macroeconomic effects of fund-supported adjustment programs', *IMF Staff Papers*, 37 (June): 193–225.

Killick, Tony (1998) *Aid and the Political Economy of Policy Change*. London: Routledge.

Kirkpatrick, Colin and David Tennant (2001) 'Responding to financial crisis: Better off without the IMF? The case of Jamaica'. Discussion Paper Series Working Paper 38. Manchester: University of Manchester.

Kochnar, K., S. Coorey, H. Bredenkamp and S. Schadler (1999) *Economic Adjustment and Reform in Low Income Countries: Studies by the Staff of the International Monetary Fund*. Washington, D.C.: International Monetary Fund.

Lal, Deepak (1983) *The Poverty of Development Economics*. London: Institution of International Affairs.

Leftwich, Adrian (1998) 'Forms of democratic developmental state: Democratic practices and development capacity', in Mark Robinson and Gordon White (Eds) *The Democratic Developmental State: Political and Institutional Design*, pp. 52–83. Oxford: Oxford University Press.

Linz, Juan and Alfred Stepan (1998) 'Toward consolidated democracies', in Takashi Inoguchi, Edward Newman and John Kean (Eds) *The Changing Nature of Democracy*, pp. 48-67. Tokyo: United Nations University Press.

Lustig, Nora (2000) 'Crisis and the poor: Socially responsible macroeconomics'. Technical Papers Series.Washington, D.C.: Sustainable Development Department, Inter-American Development Bank.

Mafeje, Archie (1995) 'Theory of democracy and the African discourse: Breaking bread with my fellow-travellers', in Eshetu Chole and Jibrin Ibrahim (Eds) *Democratisation Processes in Africa: Problems and Prospects*, pp. 5–28. Dakar: CODESRIA.

Maravall, J. M. (1994) 'The myth of the authoritarian advantage, *Journal of Democracy*, 5 (October): 17–31.

Mcculloch, Neil A., Bob Baulch and Milasoa Cherel-Robson (2000) 'Poverty, inequality and growth in Zambia during the 1990s'. Working Paper 114. Brighton, Sussex:. IDS.

Meisenhelder, Tom (1997) 'The developmental state in Mauritius', *Journal of Modern African Studies* 35 (2): 279–297.

Mkandawire, Thandika (1999a) 'Crisis management and the making of "choiceless democracies" in Africa', in Richard Joseph (Ed.) *The State, Conflict and Democracy in Africa*, pp. 119–136. Boulder: Lynne Rienner.

Mkandawire, Thandika (1999b) 'Shifting commitments and national cohesion in African countries', in Lennart Wohlegemuth, Samantha Gibson, Stephan Klasen and Emma Rothchild (Eds) *Common Security and Civil Society in Africa*, pp. 14–41. Uppsala: Nordiska Afrikainstitutet.

Mkandawire, Thandika and Virginia Rodrigues (2000) 'Globalisation and social development after Copenhagen'. Occasional Paper 10. Geneva: UNRISD.

Montecinos, Veronica (1993) 'Economic policy elites and democratisation', *Studies in Comparative International Development* 28 (1): 25–53.

Moore, Mick, Jennifer Leavy, Peter Houtzager and Howard White (1999) *Polity Qualities: How Goverance Affects Poverty*. Washington, D.C.: World Bank (In Mimeo).

Morley, Samuel, Roberto Machado and Stefano Pettinato (1999) 'Index of structural reforms in Latin America'. Serie Reformas Economicas 12. Santiago: CEPAL.

Nattrass, Nicole and Jeremy Seekings (1998) 'Growth, democracy and expectations in South Africa', in Abel Abedian and Michael Biggs (Eds) *Economic Globalisation and Fiscal Policy*, pp. 25–53. Cape Town: Oxford University Press.

Nelson, Joan (1989) *Fragile Coalitions: The Politics of Economic Adjustment*. Washington, D.C.: Overseas Development Council.

O'Donnell, Guillermo and P. Schmitter (1986) *Transitions from Authoritarian Rule*. Baltimore, Maryland: John Hopkins University Press.

Parra, Bosco (1996) 'Governance and democratisation in post-Pinochet Chile', in Lars Rudebeck and Olle Törnquist (Eds) *Democratisation in the Third World: Concrete Cases in Comparative and Theoretical Perspective*. Uppsala: Uppsala University.

Przeworski, Adam, Susan Carol Stokes and Bernard Manin (1999) *Democracy, Accountability, and Representation*. Cambridge: Cambridge University Press.

Przeworski, Adam and James Vreeland (2000) 'The effects of IMF programs on economic growth," *Journal of Development Economics* 62: 385–421.

152   THANDIKA MKANDAWIRE

Putterman, Louis, John Roemer and Joaquim Silvestre (1998) 'Does egalitarianism have a future?', *Journal of Economic Literature* 36 (June): 861–902.

Remmer, Karen (1990) 'Democracy and economic crisis: The Latin American experience', *World Politics* 42: 315–335.

Roberts, Kenneth (1998) *Deepening Democracy: The Modern Left and Social Movements in Chile and Peru*. Stanford: Stanford University Press.

Robinson, Mark (1998) 'Democracy, participation and public policy: The politics of institutional design', in Mark Robinson and Gordon White (Eds) *The Democratic Developmental State: Political and Institutional Design*, pp. 150–186. Oxford: Oxford University Press.

Rocha, Geisa Maria (2002) 'Neo-dependency in Brazil," *New Left Review* 16 (July–August):. 5–33.

Rodrik, Dani (2001) 'The developing countries' hazardous obsession with global integration'. <http://ksghome.harvard.edu/~.drodrik.academic.ksg/obsession.pdf>

Rodrik, Dani (2002) 'After neoliberalism, what?' Paper presented at the Alternatives to Neoliberalism Conference, Washington, 23–24 May. <http://new-rules.org/docs/afterneolib/rodrik.pdf>

Ruggie, John Gerard (1983) *The Antinomies of Interdependence: National Welfare and the International Division of Labor*. New York: Columbia University Press.

Sandbrook, Richard (1996) 'Transitions without consolidation: Democratisation in six African cases', *Third World Quarterly* 17 (1): 69–87.

Sen, Amartya (1999a) *Development as Freedom*. Oxford: Oxford University Press.

Sen, Amartya (1999b) 'Democracy as a universal value', *Journal of Democracy* 10 (3): 3–17.

Sharma, Shalendra (1999) 'Democracy, neoliberalism and growth with equity: Lessons from India and Chile', *Contemporary South Asia* 8 (3): 347–371.

Simensen, Jarle (1999) 'Democracy and globalisation: Nineteen Eighty-Nine and the "Third Wave"', *Journal of World History* 10 (2): 391–411.

Stokes, Susan Carol (1997) 'Democratic accountability and policy change: Economic policy in Fujimori's Peru', *Comparative Politics* 29 (2): 209–227.

Stokes, Susan Carol (2001) *Mandates and Democracy: Neoliberalism by Surprise in Latin America*. Cambridge: Cambridge University Press.

Subramanian, Arvind (2001) 'Mauritius' trade and development strategy: What lessons does it offer?' Paper presented at IMF High-Level Seminar on Globalisation and Africa held in Washington, D.C., March.

Subramanian, Arvind and Devesh Roy (2001) 'Who can explain the Mauritian miracle? Meade, Romer, Sachs, or Rodrik?' Working Paper WP/01/116. Washington, D.C.: International Monetary Fund.

Swank, Duane (2001) 'Political institutions and welfare state restructuring: The impact of institutions on social policy change in developed democracies', in Paul Pierson (Ed.) *The New Politics of the Welfare State*. Oxford: Oxford University Press.

Teivainen, Teivo (2002) *Enter Economism, Exit Politics: Experts, Economic Policy and the Damage to Democracy*. London: Zed Books.

Tendler, Judith (2000) 'Thoughts on a research agenda for UNRISD'. Paper presented at UNRISD Conference on Social Policy in a Development Context held at Tammvik, Sweden, September.

Therkildsen, O. (1999) *Efficiency and Accountability: Public Sector Reform in East and Southern Africa*. Geneva: UNRISD.

UNCTAD (2002) *From Adjustment to Poverty Reduction: What is New*. Geneva: United Nations.

UNDP (2002) *Human Development Report*. New York: United Nations.

Van de Walle, Nicolas (2001) 'The impact of multi-party politics in sub-Saharan Africa', *Forum for Development Studies* 1 (June): 5–42.

Varsheny, Ashotosh (1998) 'Democracy and poverty'. Paper presented at the Conference on the World Development Report 2000 *The Responsiveness of Political Systems to Poverty reduction*, organised by DIFD and IDS held at Caste Downingtown, U.K., August 15–16.

Weimar, Bernhard (1989) "Botswana: African economic miracle or dependent South African quasi-homeland', in *African Development Perspectives Yearbook*. Hamburg: Lit Verlag.

Weyland, Kurt (1996) *Democracy without Equity: Failures of Reform in Brazil*. Pittsburgh: University of Pittsburgh.

Weyland, Kurt (1999). 'Economic policy in Chile's new democracy', *Journal of Interamerican studies and World Affairs* 41: 3.

White, Gordon (1998) 'Constructing a democratic developmental state', in Mark Robinson and Gordon White (Eds) *The Democratic Developmental State: Political and Institutional Design*, pp. 17–51. Oxford: Oxford University Press.

Williamson, John and Stephen Haggard (1994) 'The political conditions for economic reform', in John Williamson (Ed.) *The Political Economy of Policy Reform*, pp. 525–596. Washington, D.C.: Institute for International Economics.

Wood, Geoff (1997) 'States without citizens: The problem of the franchise state', in David Hulme and Michael Edwards (Eds) *NGOs, States and Donors*, pp. 79–92. London: Macmillan.

World Bank (2001) *Global Economic Prospects*. Washington, D.C.: World Bank.

World Bank (2002) *World Development Prospects*. Washington, D.C.: World Bank.

CHAPTER 9

LOCAL GOVERNANCE AND RURAL POVERTY IN AFRICA

PASCHAL B. MIHYO

1. THE COMMON IDENTITY OF SUB-SAHARAN AFRICA

African cities defy generalisation. A visit to Cape Town and Sandton in South Africa exposes one to architecture of a beauty not found even in many developed countries. Some of these cities have their counterparts in Southern Europe and affluent parts of South America. But the majority of African cities seem to be fatigued and on the decline. Most, such as Harare, Lusaka, Nairobi and Maputo, were once built for settlers, with the local people confined to reserves and allowed to enter only to work for the settlers and made to leave after their job was done. As these cities were taken over by local populations, their objective disappeared but their characteristics remained. Today they reflect both the present and the past. They are clean at the centre and in the areas where the new rulers live, but completely dilapidated in the areas where the masses live. Still, compared to the countryside the cities are prosperous environs, and their disadvantaged populations at least enjoy some basic services in terms of transport, access to markets and employment. However, education, health, water, sanitation and housing are highly differentiated, depending on the status of the occupants of the various neighbourhoods.

Though Africa's urban centres are widely diverse, its rural areas are not. In Tanzania, five kilometres from Dar-es-Salaam, and a similar distance from Addis Ababa in Ethiopia, Bujumbura in Burundi or Freetown in Sierra Leone, strikingly similar situations are encountered: deep pockets of poverty and insecurity, inadequate systems of justice, crowded schools with insufficient teachers and materials, poor transport facilities that exact heavy penalties on producers and traders, dispensaries that receive only one kit of essential drugs each month. Leaving Kinshasa in what was mistakenly referred to by the colonialists as Belgian Congo, one notices that the semblances of Belgium that used to embellish the city have crumbled due to decades of looting and neglect. As the rural areas begin, the once passable roads slowly transform

155

Max Spoor (ed.), Globalisation, Poverty and Conflict, 155–175.
© *2004 Kluwer Academic Publishers. Printed in the Netherlands.*

into dead ends. The railway system lies abandoned, deliberately uprooted by General Joseph Mobutu as a security measure to impede adversaries from mobilising troops against him (Callagy 1984: 165). The rural areas essentially remain dependent on canoe, donkey and human power for transport; and insecurity is so entrenched that it drastically limits the population's alternatives.

While there are some extremes, such as the two Congos, infrastructure is the same throughout most of Africa's rural areas.[1] Few countries outside the Southern Cone have reliable road systems. Even in the Southern Cone it was the security policies of Apartheid, requiring rapid deployment of troops, that mandated that road infrastructure be well-maintained. This includes the Malawi in which President Kamuzu Banda with the help of the Apartheid regime in South Africa kept the road system intact in order to confront the challenges of the liberation war. Elsewhere in Africa, most of the best roads either go to the areas where the top brass originate or serve the neighbourhoods inhabited by former settlers or dominated by multinational farming or mining companies. Extension services target commercial farming areas and rarely serve poor farmers. Education and training in agriculture and extension support are devoted mainly to cash crops, with subsistence crops and small livestock getting much less attention. Though farming for subsistence is the primary domain of poor rural communities, it features neither in the strategies of formal education nor in production statistics.[2]

Neglect of rural communities is systematic both in least developed and in middle-income countries in Africa. When it comes to poverty, Botswana, one of the continent's most prosperous countries, suffers the same fate as Tanzania, one of the poorest. In both countries, 40 per cent of the population lives in absolute poverty. Most of the poor inhabit rural areas. In Tanzania, malnutrition is most rampant in the north where there is also the highest concentration of cattle; in Zimbabwe the highest rates of malnutrition are found on commercial farms (ZCTU 1993).

The San people in Botswana and South Africa, like their counterparts the Hadzabe and Sandawe of Tanzania and the Batwa of the rainforest in Congo and Uganda, have never had positive relations with the state or the market. They live in the bush and fend for themselves without support from their governments. In their systems of survival they have no common structures for food security. Each individual fetches food for their immediate needs and only those who are too sick or too young to hunt are cared for.

These three communities are despised in their countries and constantly under attack from their neighbours. The only contacts they have with the state or the markets are based on their alienation from their surroundings. Recently the Batwa were smoked out of their natural habitat so that the forests could be sold off to logging companies, while the San are kept on the move because their land is being allocated to mining companies.[3] This fate is experienced by many of Africa's minorities, a number of whom live in the rainforests. According to Scarrit and McMillan (1995: 328), Africa has a higher preponderance of ethnic minorities than any other continent.

It is not only the minorities that seem to share a common fate. Pastoral communities are rich because they own and control immense livestock. But most lack the capacity to use their wealth to control their environment. Apart from the Herero of Namibia, the Tuareg of Mali and Mauritania, the Tutsi in Burundi and Rwanda, the Samburu and

Kalenjin of Kenya, the Neur of Sudan and the Somalis who have combined agriculture and pastoralism and acquired some skills to improve on both, most of the pastoral communities comprise the "wealthy poor" in Africa. The Maasai of Kenya and Tanzania live a nomadic life, although they have what they consider semi-permanent or permanent settlements. The Nandi, Karamajong and Pokots who live in Kenya and Uganda are also nomadic. The same is true of the Fulani in Nigeria, Cameroon and Niger.

The colonial powers found these communities strong and in control of their resources and environment. They were mainly warrior communities and put up great resistance to colonisation. The Tuareg, for example, fought the French long and hard before they lost and the Hereros fought the Germans and lost only after they were exterminated in large numbers.[4] The Maasai not only fought battles with the Germans in East Africa but also took the British government to its own courts over ownership of their lands. In order to control them, some nomadic groups, such as the Maasai and Samburu, were relegated to national reserves.[5] On these reserves they are not allowed to hunt or use their resources without government permission. In Tanzania the Maasai have struggled to convince the government to make the national reserves where they live Maasai reserves. But government has so far rejected the proposition (Hodgson and Schroeder 2002). Weaker pastoral communities such as the Fulanis and the Galla of Somalia were made landless and remain marginalised. Today some still live on the reserves; most of them are stateless, especially the Fulani in West Africa, some of whom do not even regard themselves as belonging to the countries they live in.[6]

Another common factor in much of rural Africa is state-managed illiteracy. In 1984 a project on universal primary education was rejected in Sierra Leone and the experts who had gone there to advise on it were sent away prematurely.[7] The government officials said categorically that universal primary education was not a priority. A few years later conflict began, and uneducated and unemployed youths fed the ranks of the feuding vicious armies. Ignorance and illiteracy are a resource for conflict, and Sierra Leone provides an illuminating example of how it can be manipulated by the elite to promote their own programmes of greed and plunder.

Sierra Leone is not alone in the perpetuation of illiteracy as a means of controlling populations, especially rural inhabitants. The high preponderance of illiterate rural populations has maintained many regimes in power, in Angola, Benin, Congo, Kenya, Gabon, Guinea, Malawi, Mauritania, Mozambique, South Africa and Swaziland to mention only a few. At the same time indoctrination and lack of exposure helped regimes retain power in Ethiopia, Eritrea, Tanzania and Uganda.[8] Ignorance among rural people is crucial if the African elite is to continue managing resources on behalf of rural communities. They control resources without sharing benefits, using cheap rural labour to produce the commodities essential to fuel and run the enterprise systems, grease the bureaucratic machineries and cushion their own consumption needs and aspirations.

The main yardstick used to measure poverty is income, and a dollar a day seems unbelievably low.[9] But for most inhabitants of Africa's vast countryside, a dollar a week is the norm, including their subsistence resources. The rural areas produce at least 80 per cent of the food and raw materials but receive less than 20 per cent of the proceeds on these products. Seventy per cent of Africa's population lives in rural areas, yet they get less than 20 per cent of health budgets. They carry the highest burden of births and deaths and yet spend the least on health. The rural people are exposed to environmental hazards such as vector-borne diseases that they contract due to agricultural, forestry and survival activities; exposure to venoms; environmental diseases such as malaria, bilharzia, diahorhea and diseases of poverty that arise from improper nutrition, poor ventilation, inadequate sanitation, unclean water and sharing their living environment, water and food with wild animals.

Rural poverty in Africa is mainly rooted in capacity poverty, which the UNDP *Human Development Report 1996* defined as the inability to live a healthy and well-nourished life, to have a healthy reproductive life and to acquire and use education and knowledge (UNDP 1996: 27). It is the lack of these capabilities across the board that gives rural sub-Saharan Africa its common identity.

2. THE SEARCH FOR A GOVERNANCE PARADIGM CONDUCIVE TO DEVELOPMENT IN AFRICA

Over the last three decades, international organisations and African governments have organised numerous meetings on the causes of bad governance and economic failure in Africa and how to rectify the situation. Initially, the problem was located in the failure of education to equip African people with capacity to use their resources to confront challenges that arise from their environment. Capital formation through education was held to be the key to Africa's transformation. Therefore until the early 1980s the budgets of African governments reflected a high proportion of investments in education. Salaries of education personnel took a substantial share of government expenditures and teaching was among the few respectable professions (Maclure 1997: 16). But even at that stage it was clear that the investment pattern was urban-biased. Not only were more investments made in higher and tertiary education than in primary and secondary education, but most of the institutions of tertiary and higher education were located in urban and peri-urban areas. The rural areas, where most people live and where the resources for transforming the economy were based, received less investment and attention.

Nearly parallel with the above, another paradigm was adopted: the basic needs approach. The focus was on jobs and basic social services. It was designed on the basis of the welfare ideology dominant at that time. The state was expected to provide jobs and services. The jobs were expected to offset the cost of services by increasing earnings and purchasing power and thus widening the taxation base and consumption. However for Africa, the basic needs approach was introduced during the period of state control and regulation of the economy, disruption of rural production through collectivisation programmes in some countries and land policies that were based on large-scale state farms and limited private-sector participation in agriculture.[10]

Moreover, the basic needs strategies were more concerned with the assumed rather than ascertained needs of the poor and unemployed. The focus was on how to manage rather than how to reduce poverty. Most of the efforts were, as expected, deployed in urban areas through public works, work-for-food programmes and public-sector employment including on state farms. For the local and international elite, the urban centres were more dangerous than the rural areas. The cities' unemployed and homeless had more contact with organised groups such as trade unions and religious bodies. They could easily riot and disrupt the peace that was required to service the national and global economies. The rural populations seemed less potentially explosive (except in the remaining colonies which were close to attaining independence). Rural people did not pose a threat and therefore got minimal attention. As a result, they lost most of their active and skilled labour, which drifted into towns to tap the dividends of public works and similar programmes.

In the 1980s most of the African countries began experiencing technological stagnation, steep rises in unemployment, decreasing productivity and overstaffed public service departments and enterprises. Communication and transport infrastructure deteriorated and primary commodity prices declined. This prompted deep reflection on the causes of economic decline and the way out. Stabilisation and structural adjustment programmes, supported by the International Monetary Fund (IMF) and the World Bank became the prime movers of economic reforms, starting in the early 1980s and continuing for more than two decades. The emphasis was on strengthening markets, the rule of law and private-sector participation in the provision of services. Civil service systems were reformed and public wage bills trimmed. Indeed, public deficits fell, but the state's capacity to steer the development process was also substantially weakened.

Finally, in the late 1980s environmental issues were added to the development paradigms. The United Nations Conference on Environment and Development (UNCED 1987) report *Our Common Future* apart from re-emphasising the imperative of meeting basic needs such as jobs, food, energy, water and sanitation, stressed poverty alleviation, reviving growth, ensuring sustainable population levels and a balance between economy and ecology. The civil society organisational infrastructure that mushroomed in response to this paradigm confined itself to urban areas. The problem was and still is that the African elite is trained to manage and not to solve development problems; it is less oriented towards understanding let alone to adequately addressing rural problems. As they wrapped themselves into the environmental agenda, they concentrated more on conservation and ecotourism, which again more addressed the consumption needs of the international elite and the revenue needs of the local elite. The elites were systematically socialised through the school system to fear and shun rural areas, so it was unsurprising that the programmes of the 1980s again failed to give the rural sectors the necessary focus.

Beginning in the early 1990s the development debate and accompanying efforts turned their focus towards governance in general. The UNDP (1993) defined "good governance" as constituting accountability, equity, transparency, participation, rule of law, responsiveness, effectiveness and consensus-based decision-making. The World Bank in its *World Development Report* of 1999/2000 later added decentralisation, devolution and accountability (World Bank 2000). UNDP (1994: 18–22) emphasised the universality of the right to life, whether people lived in poor or rich countries, families or classes. It emphasised equity and redistribution of wealth by improving the health, education and nutrition of the poor. As a strategy for poverty reduction it identified seven principles that could help: employment creation, basic social services, social safety nets, agrarian reform, decentralisation, economic growth and sustainable development.[11] Subsequent reports added gender equality, health, increased life expectancy, educational attainment, reduction of conflict, improvement of housing[12] and human security and reduction of both income and capability poverty (UNDP 1996: 18–25). Consumption patterns and their underlying inequalities on the one hand and conspicuous consumption and its environmental impacts on the other were added by UNDP (1998: 1–5), while the *Human Development Report 2000* addressed the link between human development and human rights, emphasising the availability and affordability of justice, equal access to it for all and effective mechanisms for its delivery (UNDP 2000: Chp. 5).

We may ask to what extent these development paradigms, including the current one emphasising good governance, have focused on and attempted to address issues of rural poverty. Three major arguments may be advanced. The first is that for rural people, past development theories did not change the relations between them and the ruling elites. The colonial system was based on a dialogue failure through which rulers constructed the needs of rural people and pushed these down their throats without discussion or modification. Rulers were assumed to be custodians of the natural resources and labour and charged to develop them in the interests of industry and commerce in the coloniser economies. The relations were those of slave and master, unequal and non-negotiable. These relations do not seem to have changed. Dialogue failure continues between rural people and their local and national elites.

The second argument is that the rural economies have always been managed by intrusion. The colonial powers imposed rules and regulations that opened rural economies up to exploitation by external markets and imposed controls on their capability to move, organise and transact. During the period after colonialism intrusion continued, characterised by new forms of "indirect rule" instituted through decentralisation policies. These have now been supplemented by deregulation and trade liberalisation, which though professed to have reduced the degree of intrusion have actually increased the intruders and made rural people more vulnerable.

The third argument is that the main membrane that blocks communication and dialogue between the rural communities and the elite is the regulatory system and how it shapes entitlement systems. Instead of empowering them and enabling them to solve their own problems and control their environment and resources, the regulatory mechanisms socialise and integrate populations into inequitable systems of power, production, exchange and distribution. Instead of widening their capability to tame their environments and overcome poverty, regulation exposes them to entitlement systems that make them more vulnerable. The conclusion is that the development decades that changed the fortunes of many rural people in Asia and Latin America completely bypassed Africa's rural people. If the dialogue failure continues, Africa's rural poor will be equally bypassed by the new, less porous development paradigms including the current one on good governance.

3. GOVERNANCE AND THE ADMINISTRATION OF RURAL POVERTY

The rural decline that has been systematic in many African countries has been aggravated by several factors. Rapid urbanisation is one of them. Although urbanisation started during the colonial period it has increased in recent times due to rural development failure. In Asia and Latin America urbanisation is mainly due to industrialisation, but in Africa, while the early trends were linked to industrialisation efforts, recent trends are due more to rural neglect, environmental degradation, decline of employment opportunities, collapse of primary commodity prices, corruption in the school systems, intra- and inter-community conflicts and oppressive and exclusive patterns of power and production based on patriarchy and matriarchy.[13] Aina (1995) listed unequal development, population pressure, low agricultural productivity and the attraction of a better life in urban areas as the major factors pushing migration to urban areas. A second and related factor is the decline in rural services. Declines in rural services have been persistent and have recently become worse. Public services generally come under stress during periods of economic reform.

A 1999 study looked at the impact of economic reforms on poverty in Ghana, Mali, Tanzania and Uganda. In Ghana, it found that while incomes among urban people had decreased due to retrenchment policies, the rural poor had become relatively poorer. Rural Savannah was categorised as the poorest area in Ghana, with half of its population living below the poverty line and one-third classified as very poor (Ministry of Foreign Affairs 1999: 106). In Uganda it found scanty evidence of changes in patterns of poverty, though household surveys had been undertaken for the first time in many years. It also noted that 92 per cent of Uganda's poor and 96 per cent of the poorest live in the rural areas (ibid.: 152). In the case of Mali it concluded that due to structural adjustment the poor had become poorer (ibid.: 179).

The comments on Tanzania focused more on how aid had supported recovery efforts. UNDP (2002) indicated that while a lot had been achieved in terms of social development, poverty had increased due to, among other things, increased death rates caused by HIV/AIDS, reduced nutrition levels especially among rural communities and environmental degradation. While there was a marginal reduction in food insecurity in Dar-es-Salaam, in other urban areas food insecurity had increased and in the rural areas it was also on the rise (ibid.: 8). In 2002 the Malawian government estimated the poor in Malawi to be 65 per cent of the population (Government of Malawi 2002: 5) and structural adjustment was said to have worsened poverty (ibid.: 12). The decline of services to rural communities accounts most for these communities' lack of capacity to utilise their resources to confront poverty.

Even in urban areas public services remain constrained by lack of personnel. The small staffs retained after civil service reforms have tended to be concentrated in cities or the districts from which government leaders hail. Public officials often avoid setting up their residence in the distressed rural areas. It is not unusual to find rural development officers living in the urban centre at the district headquarters or a district commissioner living outside his or her district.[14] Political leaders too shun rural areas. In most sub-Saharan African countries, with the exception of Ghana, Tanzania, Uganda and South Africa, members of parliament are not required to live in their constituencies. Most live in a capital city and make occasional visits to their rural constituencies. This deprives the rural areas of opportunities to consult with and engage their leaders in some form of dialogue that could reduce the communication gap. The rural areas are therefore essentially detached, not only from the leaders they elect but also from the bureaucrats appointed by the state to serve them.

The allocation of human resources also tends to be urban-biased. Rural areas are rarely allocated the best public servants. On the contrary, they are often used as a dumping grounds for inefficient and uncooperative workers. In other cases, elderly civil servants, no longer energetic and dynamic, ask to be posted to rural areas as part of their transition to retirement. The framework of public service provides little room for the rural areas to feature in the improvement of public services, including human resources. The Poverty Reduction Strategy Papers (PRSPs) of Burkina Faso, Malawi, Tanzania and Zambia, for example, make no mention of improving the quality of public servants or even rural public services.

These same PRSPs, however, discuss governance issues in the framework of instilling political will and changing the mindset of public officials, improving access to justice and reducing crime and improving the management of public resources.[15] While these factors are important aspects of governance, they have an obvious macro-level bias and fail to address rural problems directly. The crisis of governance in these areas centres on lack of facilities and the centre's neglect of rural leadership. Village leaders have no offices. Most perform their administrative duties in their own homes. They receive little support in terms of materials. Even writing pads and pens are rare. Although they are supposed to transmit policy decisions to the village population, few village leaders can afford a radio. They cannot follow parliamentary debates; they do not get newspapers and have no reliable sources of public information. Neither are they paid wages, and they therefore end up extracting rent from their fellow villagers or charging – sometimes ridiculous amounts – for their services.[16]

Procedures for the administration and management of public functions are often unreliable and have not been adjusted to global changes. There might be a long lag between the time when a decision is taken at the centre and when the information reaches the implementers or people at the grassroots level. An example is the decision to abolish poll and other taxes in Tanzania that was made in June 2003. By the end of August some people in the rural areas were still being charged those taxes.[17]

At the extreme are cases where land laws passed in 2000 were in 2003 still unknown to the local magistrates in the rural areas. Village Land Act No. 5 was passed by Tanzania in 2000. That act requires village committees to be involved in decisions on land allocation. It sets a minimum number of members for the land allocation committees and the proportion of these who are to come from the village communities, including how many of the latter should be women. Yet the land office in Bagamoyo district continued allocating land without regard for the new law. A seminar organised by an NGO called Social Watch made the villagers aware of the new law, after which the community went to court to seek an injunction to restrain the district authorities from acting outside the law. The local magistrate and district officials were completely unaware of the law, however, and had to use the brochure obtained by the villagers as their source of information about the law. Therefore the villagers were more knowledgeable than public officials on this law. The irony of the episode is that Bagamoyo is only 40 kilometres from Dar-es-Salaam. If the magistrates in such close proximity to the capital are unaware of the current law, it is highly probable that those in remote areas do not know even the old laws.

In addition to information bottlenecks there are serious transport infrastructure problems. Mozambique's road network, for example, totals only 26,000 kilometres and roads in most of the fertile rural areas are impassable year-round (Government of Mozambique 2001: 56–57). In Zambia, transportation accounts for 60–70 per cent of the total cost of production (Government of Zambia 2002: 102). Most rural areas are still unreachable by lorry. Though people have to travel long distances to get water and firewood, they lack accessible roads. As pointed out earlier, in Congo (DRC) rural communities depend on canoes, human and animal power and limited numbers of bicycles and carts to ferry their goods. In some cases bicycles are used to bring pregnant women to distant clinics or hospitals. In Nachingwea district in southern Tanzania, for example, the author saw women who had delivered on their way to the hospital because bicycles induce labour. These women deliver on roadsides with no support, neither from the formal medical system nor from traditional birth attendants.

Communication bottlenecks further complicate life in rural areas, particularly when combined with poor health services. For example, in many rural areas women who are lucky enough to reach hospitals must have their own lanterns, sanitary materials and other accessories necessary for delivery. In most cases, pregnant women prefer to deliver at home,[18] usually with the help of traditional birth attendants. The *Interim Poverty Reduction Strategy Paper* of Burkina Faso reported high rates of morbidity and mortality among women. The report noted further,

> In the country as a whole, only 38.4 percent of pregnant women receive prenatal care. Deliveries under poor sanitary conditions resulted in a prenatal mortality rate of 126 per thousand in 1995. In addition to ignorance and poverty, women's health is affected by the burden of household chores, harmful traditional practices and inadequate sanitation and water-supply facilities (Government of Burkina Faso 2000).

Rural literacy is also low, although figures in national statistics are not always disaggregated to reflect the rural-urban divide. However, in the poorest African countries literacy levels are generally below 50 per cent of the population. In Malawi, for example, literacy was put at 58 per cent nationally while literacy among women was 44 per cent. The PRSP for Malawi in 2001 noted that only 11 per cent of the population had attained class eight education, and only 6 per cent of them were women (Government of Malawi 2002: 7). In Burkina Faso the literacy rate dropped from 19 per cent in 1994 to 18 per cent in 1998. The figure for urban areas was 52 per cent in 1994 and had fallen to 51 per cent in 1998. For the rural areas the literacy rate was 12 per cent in 1994 and it had dropped to 11 per cent in 1998. Only 6 per cent of rural women were literate in 1994, though this figure seems to have risen to 7 per cent by 1998 (Government of Burkina Faso 2002: Table 4). These are very low rates by any standards. Such an ill-educated population has immense difficulties to surmount in learning and reading about new production methods or health and safety measures. In addition, high rates of illiteracy among women affect entire societies. Women are the primary educators of children and society as a whole. They are also the primary health care providers in the community and the main household producers. If they have limited capacity to understand symptoms of diseases or absorb technical skills, they cannot adequately perform those roles.

School enrolment trends are also worrying. They have increased for some regions and dropped for others. In Burkina Faso, as a whole enrolment in primary education rose from 35 per cent in 1994 to 41 per cent in 1998. But while urban enrolment increased by 28 per cent, enrolment in rural areas fell by 4 per cent within the same period. While enrolment for boys increased by 27 per cent in urban areas it only rose by 3 per cent in rural areas. For the girls, in urban areas primary school enrolment rose by 29 per cent between 1994 and 1998 while it dropped by 5 per cent for girls in rural areas (ibid.). At the other extreme, in Malawi overall enrolment in primary schools rose explosively after the introduction of free primary education. But the pupil-to-teacher ratio was 114:1 in 2000. The quality of education has also deteriorated, with dropout rates increasing tremendously, along with an increase in pupils having to repeat a grade. That last, for example, reached 45 per cent for grade one in 2002. In the rural areas, the repetition rate was reported as 47 per cent (Government of Malawi 2002: 7–8). The same seems to have happened in Uganda where reintroduction of free education was not accompanied by expansion of facilities to match the rise in enrolment. As a result, overcrowding in classes and unmanageable class sizes have reduced the quality of education, increasing dropout rates.

In the face of these findings, we may question the extent to which the education system in Africa is oriented towards rural development. The school system seems to train people for outward migration. Primary schools are normally located in the villages and are accessible to the local people. Footpaths lead to the school and the teachers live in the village. Some teachers belong to these villages and perform other roles in village management. There is almost no division between the school and the community. After primary school the situation changes. Most secondary schools are in urban centres or removed from both urban and village communities. They are fenced and have gates. There are no footpaths leading to the school and the neighbouring communities can visit them only on parents' day or the annual "open day" when they are shown what the pupils have done during the year.

What the children are taught within these confines bears little relation with local systems of production. By the time the children finish secondary education they are ready to take off to the urban centres for which they have been adequately socialised. Those who chance to go to university find the campuses equally removed from communities. Most African campuses are built like airports, far from the cities. The students live on these campuses without mingling with the surrounding communities. Occasionally they come out to riot or demonstrate, demanding more privileges for themselves or supporting some political or social cause. By the time they graduate, they have been completely uprooted from their origins.

Even at university level the curriculum of African educational institutions has little connection with local, and particularly rural, systems of existence and survival. It is unsurprising therefore that despite the research ongoing in many universities, Africa is rapidly losing its traditional food crops and indigenous knowledge. Some 300 such crops have disappeared over a period of 50 years. Curricula for agricultural and veterinary sciences emphasise cash crops and livestock – the domains of large-scale commercial farms. They pay scant if any attention to the food crops and small stocks which are the mainstay of smallholder farmers and peasants. This process of education has reinforced the urban bias in Africa's development profiles.

4. DECENTRALISATION OF CONFLICT AND POVERTY

In theory decentralisation policies aim to stimulate local and regional development. They are means of enabling communities to participate in most spheres of decision-making, to increase their political, social and economic citizenship and ensure them opportunities to enjoy their social, political and economic rights as subjects and not objects of governance and development. For people everywhere, decision-making is more meaningful if it enables them to expand their scope of knowledge and information and provides them with the means to establish and maintain a stable, secure and peaceful environment. It is also more meaningful if it strengthens their institutions of power and production and enhances their ability to interact and transact equitably with other communities. Decentralisation, therefore, should aim at creating dynamic and participatory systems that add value to systems of governance at the national level.

According to Olowu (2001), decentralisation policies should reflect a recognition by those in central government that the people at the grassroots level and in local communities are capable of managing their own affairs. Their systems should therefore enable local people to do so democratically and to resolve conflicts.[19] However, Olowu's study of decentralisation programmes in Kenya, Sudan, Tanzania and Zambia indicated that central authorities did not make efforts to help grassroots-level governance take off. In some cases, the ministries in charge were not given adequate resources and no officials were appointed to implement the programmes. Hence, the centre retained powers of control and veto. The objective in some cases seems to have been to extend rather than relinquish control (Olowu 1995). Current decentralisation programmes suffer the same fate. In Uganda, for example, there has been transfer of power to the grassroots level, but there has been no transfer of the capabilities required to control resources and use them to transform rural communities. This has led merely to the transfer of resources to the local level and with them the decentralisation of mismanagement and corruption (Watt et al. 2000). In Ethiopia, despite the federal model, the centre has retained effective control of natural resources at the local level.[20]

In practice, therefore, decentralisation has remained difficult to achieve. This is more so where devolution of power to rural communities is required. First, the central leadership in most countries survives politically on the support of the rural communities. To relinquish power to the local elite would render the central elite more dependent on the latter. Second, since colonial times the centre has always been tempted to want to control natural resources, especially land and water. Colonial and post-colonial systems of power have tended to survive on the institutionalisation of relations of subordination between wealthy and powerful communities and powerless and poor communities. In this equation rural communities are kept weak as part of that traditional power structure needed for the survival of relations-based alienation, domination, exploitation and subordination (Hoppers 1998: 41–48).

Decentralisation programmes have also failed to transfer actual control of land resources to local people. Central governments in many African countries still act like the colonial governments which retained overall control over land and purported to manage local land resources in trust for communities. Local people are thus given communal tenure, which makes ownership amorphous and strips community lands of their commercial value. Concepts of trusteeship imply immaturity of the communities for which the land is held. Already the future of communal land tenure has stimulated fruitful and continuing debates. The system is still supported by many researchers, and some arguments that have been advanced are persuasive. One of them is that customary land tenure has its own value at the local level, which does not have to be commercial at the national level (Quan 2000). Another is that the registration of titles in rural areas would be prohibitively expensive.

Platteau (2000: 57) cautioned that informal arrangements may be more secure than is normally acknowledged and tenure under arrangements that seem precarious may be quite dependable. But customary tenure is constrained by restrictions on transactions and cannot be used as collateral to obtain credit from formal financial institutions. As a result it confines rural people to informal land and credit markets, limiting access to productive capital and leaving them only the options of social capital, loan sharking and usury. They become even more vulnerable if they end up selling land cheaply to the local elite, which has access to systems of formalising tenure and awarding titles.[21]

In practice, therefore, decentralisation has meant, for many countries in Africa, simply a transfer from the central to the local sphere and in that vein a shift of responsibilities from state institutions to households or informal social support systems. The current withdrawal of the state from development activism to passive management of poverty alleviation schemes is permeated by the pretence of the state as a facilitator of development. In urban areas states are indeed facilitating new avenues of wealth creation for a lucky and happy few. But states have withdrawn completely from the rural areas.

In the last decade theories of social capital have gained prominence in development discourse.[22] In looking for ways to fill the glaring gap left by the state, development agencies have discovered that rural areas have always survived on networks based on kinship, neighbourhood, mutual aid and collective labour. These have in fact sustained rural societies for centuries. "Social capital" has thus been elevated to a new approach that will enable the poor to be more integrated in their national economies. But in essence it is popular in development circles because it legitimises the withdrawal of the state from the rural arena. The concept of social capital is most commonly applied in sectors of the poor and not to corporate and formal sectors because the needs of the latter are taken care of by formal organisations including banks and insurance companies as well as cushioning policies such as subsidies and tax relief.

Decentralisation policies resemble broad programmes for the institutionalisation of the "social capital approach". In the post-adjustment era, decentralisation has become a policy mechanism for experimenting with theories and practices of social capital beyond the grassroots level. Responsibilities are transferred to the local level without transferring the necessary resources. In turn, local authorities leave residents to depend on their traditional systems of survival. Such policies have effectively become a mechanism for decentralising poverty.

5. REGULATION AND RURAL POVERTY

Since the colonial period the rural sectors in Africa have been managed by intrusion. In order to stimulate commodity exchange relations, colonial governments adopted migrant tax policies to create a need for money. This triggered waves of labour migration. Regulation was the main means of establishing the then-emerging markets. The poll tax continued to be the main instrument for ensuring those who could work went out and obtained employment at least to be able to pay taxes. The local marketplaces served as control centres, since anybody who went to those markets to sell or purchase goods had to have a valid tax certificate. Those netted for being without a certificate were subjected to forced labour, in addition to being obliged to pay their taxes. Another mechanism used was the centralisation of major services at a district level. To access facilities including secondary schools, hospitals and licensing authorities, people had to travel to the district centres. On the way they encountered roadblocks that were used to check tax credentials. Therefore, to travel one had to have a clean tax record or face compulsory public labour when and where required.

These mechanisms for controlling rural populations continued after independence. Regulation is still a main determinant of the relations of control, subordination and exploitation. Three major intrusions or interventions sustain these relations. First is production and marketing intrusion. Second is land de-privatisation, and third is deregulation and liberalisation. Tax policies during and after the colonial period created a double dilemma. There was, and always has been, a constant demand for labour in the urban centres and in the large-scale agricultural sectors. Though labour was needed in the mines, plantations and public offices, these sectors had to be assured of cheap food supplies. The rural areas, while losing their most dynamic young workers, were thus saddled with the task of producing food for themselves and the other sectors. The same sectors are still relied upon today to grow the export crops – such as coffee, cotton, cocoa, gum and pyrethrum – that are not in the domains of the large-scale public and commercial farm sectors in most countries.

The capacity of agricultural activities to benefit rural smallholder producers depends essentially on the fertility of the land used, the tools of production, the amount and variety of crops grown, technical support from experts and the price obtained for produce. During the colonial period local authorities and chiefs were used to coerce peasants to produce and sell their products at very low prices. The chiefs ensured that every family planted a minimum acreage of food and cash crops. This strategy became counterproductive when it served as a rallying point for nationalist agitation. In the post-colonial era, the same methods were revived. They were launched through forced collectivisation in Mali in the early 1960s and in Tanzania and Ethiopia in the 1970s (Hyden 1979). Militias and rural youth brigades were the instruments of the coercion. In the post-independence period peasants have been kept vulnerable by several interventions. For instance, marketing boards and cooperatives were given a monopoly in produce purchase and marketing. The marketing boards, wrongly called "commercialisation and stabilisation offices" in francophone countries,[23] were actually among the most effective agents of destabilisation and suppression of free commercialisation of peasant production. They ensured that rural farmers through their cooperatives received payments in such small instalments that they could not form a basis for any reinvestment activities.

The marketing boards focused on managing prices and did nothing to help peasants improve their productivity. Though many governments did institute extension programmes, the language in which the extension experts were taught was alien to the rural poor. Indeed, dialogue failure between the "experts" and the rural poor is a reoccurring theme in African development. This was further complicated by the information itself, which was predominantly focused on cash crops and livestock, which are the domains large-scale commercial farmers in the public or private domains. It said little about the food crops or small livestock on which smallholder producers depend (IDRC 1989).

For rural families, agricultural productivity certainly continues to be limited by the preponderance of illiteracy among farmers and their low levels of education, compounded by dependence on child and adult unskilled labour. Low productivity is also due to the lack of labour-saving techniques and technology, proper harvesting and storage facilities and broad knowledge about the environment. The role of the state in perpetuating these constraints is predicated on its preferential allocation of resources in favour of urban areas.

Price instability also perpetuates rural poverty. Commodity prices fall at harvest time and rise at planting time. Yet peasant are forced to sell most of their crops at harvest time in order to meet their reproductive and other needs. By the time prices rise they have little left to sell and end up pledging future produce to obtain loans from cooperatives or intermediate buyers. These cycles entrench poverty and indebtedness, leading in some cases to suicide. Within these complex price structures are market constraints. Most markets were traditionally controlled by cooperatives and marketing boards. These bodies decided when and where to buy crops and when to pay the peasants. In some countries marketing boards have been dissolved without having paid long-outstanding arrears to peasants. In Tanzania the practice was to give promissory notes to farmers, but for years these notes were not honoured (Naali 1986).

The marketing boards continued the colonial mechanisms for exploiting the rural poor. They used price fixing as a barrier to competition, leading to price stagnation and falling peasant production. They also became institutions for income control, as they determined not only how much but when and in what instalments to pay. They facilitated bureaucratic exploitation, since most of their proceeds went into government coffers and bureaucrats' salaries. In short, they became instruments for alienating peasants from their produce, for oppression and exploitation, thereby institutionalising rural poverty. As Dijkstra (1997: 35) observed, the purpose was to keep prices low and pacify urban communities either by sustaining low wages in urban areas or supporting industrial sectors through cheap supply of raw materials.

By controlling market outlets the state in practice constrained rural communities' capacity to produce more than they could consume. Though the recent wave of deregulation has opened markets, new forms of intrusion have emerged. In the place of marketing board bureaucrats and cooperative agents of the state, wholesalers, negotiators, brokers, facilitators, speculators, usurers, shortcut experts, patrons, disposers, protectors, overseers and retailers have entered the scene. Most of these have become well-linked with local leaders and the local literati. Meanwhile, the regulatory frameworks that once gave a monopoly to the cooperatives and marketing boards has not been replaced with new marketing regulations. The peasants are left to deal with multiple actors, most of whom enrich themselves on producers' lack of knowledge and information.

The two case studies of the wood workers and the charcoal traders in Tanzania summarised in the box (overleaf) illustrate not only the impact of deregulation without controls but also the new networks of exploitation that have evolved as governments have retreated from the rural marketplace.

Africa's rural areas have thus seen the worst of the two systems of local governance. The marketing boards and cooperatives were used to control and exploit the rural communities be they farmers or pastoralists. In the era of deregulation, peasants have been exposed to numerous actors and their intermediaries, as seen in the case of the charcoal makers. Both before and after market liberalisation the rural communities remain the losers. The cooperatives and marketing boards fixed prices and paid producers insufficient amounts and in unfavourable instalments. The new traders are many and they pay on the spot, but they pay very little since so many of them are involved in the marketing chain. On the other extreme, the deregulation has left no control at all and the local communities bear witness to the plundering of their resources as illustrated by the case of the wood looters of the Munene forest in Tanzania.

6. CONCLUSION

Rural communities in Africa have been bypassed by the major development thrusts of the last three decades. The human development approaches of the late 1960s and 1970s based on African socialism and humanism were used to mobilise rural communities to support the consolidation of state power over all forms of production and organisation. The human capital theories concentrated on improving schools in urban areas and on civil service reforms. The human resources approach mistook human resources for education and left health and rural productivity out of its visor. It focused more on the establishment of polytechnics, civil service training centres and institutes of development management. Although in some countries rural development training colleges, or "folk development colleges" as they were called, were established, both these and the polytechnics failed to integrate local systems of production into their curricula. They therefore remained removed from the daily ways of rural communities. The basic needs approach was based on perceived needs of the poor and, as indicated earlier, failed to bring prosperity to rural areas.

Box 1. *Deregulation Without Controls: Two Case Studies from Tanzania*

The Wood Workers of Munene Forest

The beauty of the rainforest adorns the Kagera river on the Tanzania-Uganda border. On the Tanzanian side, just beyond Kyaka, a large forest reserve stretches out to the borders of Tanzania, Uganda and Rwanda. The greenery of the forest lends an air of sanctity and gives credibility to the conservation programmes active here. Yet unknown to many, the forest has for decades been exploited by unlicensed operators. Officials discovered in June 2003 that these operators had entered the forest using obscure paths and had even installed small factories in which they made furniture and engaged in large-scale lumbering. Local residents have for years depended on these activities for employment and timber needs. In the absence of a road network one wonders how such a large operation could have continued for so long.

When the operations were discovered they were found to have installed generators and were cutting wood indiscriminately using advanced equipment and techniques. Because the local people survived from employment in the factories they had helped to conceal the operations from authorities. In July 2003 the regional commissioner for Kagera region who is a brigadier decided to enter the forest and flush out the operators. Equipment and timber were confiscated, but a few days later the operators hit back, killing the local administrator who led the authorities to their operations. These operators were unlicensed, paid no taxes and used the local resources to enrich themselves with nothing going to the local authorities. In the absence of clear regulations, this may not be the only case in the region whereby rural communities keep and maintain local biodiversity which is expropriated by others, within or outside of the law.

The Charcoal Makers and Traders of Kisarawe District

The charcoal business in the Kisarawe district of Tanzania's coastal region testifies to the impact of deregulation and lack of marketing control at the local level. Charcoal is the main source of energy for most people in Dar-es-Salaam. Kisarawe is rich in forest resources. The local people and those from other communities earn their livelihoods by making and selling charcoal. Local authorities do not regulate the exploitation of forest resources. Most charcoal makers operate without guidance on forest management or renewal. After the charcoal is produced the operators wait for buyers, most of whom come from Dar-es-Salaam. Competition among the charcoal makers is stiff, which already reduces their bargaining ability.

At the top of the charcoal marketing chain are the wholesalers who remain in Dar-es-Salaam and rarely interact with the producers. But there are many brokers who link the buyers with local producers. These brokers are local and have their own local suppliers whom they keep away from the buyers. Between the brokers and the producers are local negotiators who connect the producers with the brokers or, more rarely, directly with buyers. After the charcoal is bought there are facilitators whose task is to help the buyers ferry the charcoal through the system without paying duties or taxes. The facilitators organise young people with bicycles to transport the charcoal via footpaths in the bush to avoid roadblocks at which duties would have to be paid or proof given that they had been paid. The ferrying is done by paid shortcut experts and is carried out after dark. The same experts are used to load the charcoal quickly on waiting lorries or sometimes railway wagons. In Dar-es-Salaam the charcoal is quickly taken off of the wagons and delivered to the wholesalers. The off-loaders are the last operators in the chain. All these operations take place quickly and at night. Each actor in the chain makes some profit. The communities gain very little in the form of revenues; the local economy merely serves the interests of the various links in the market chain.

Neither did the structural adjustment that followed bring the announced improvement for the agricultural sector. It did bring a weakening of the state's capacity to govern the developmental process, which further weakened the position of the rural poor.

The current approach, centred on good governance, has introduced some novel ideas that could help rural communities become respected rather then manipulable participants in their national economies. It introduces participation, agrarian reform, employment generation, human security, population stabilisation, technological change and poverty alleviation as main goals of development. If this new approach is to succeed it must avoid the pitfalls of the past. The other paradigms failed because of the total absence of actual participation by rural people in decisions that were to shape and guide their lives and livelihoods. The rural people were and still are occasionally mobilised to vote or perform public services, such as road construction, on a voluntary basis. Women and youths are mobilised to sing and demonstrate in support of the groups in power or to provide a semblance of democracy and political tolerance during elections. This has been mistaken for popular participation. Actual popular participation requires communities to govern themselves through institutions they are familiar with and systems that reflect their own culture and aspirations. It requires their total involvement in decision-making at all levels and through all the relevant groups. To be meaningful to them, participation has to strengthen their own institutions, ensure a stable growing and peaceful society and provide for human security, which is necessary for community development.

None of the major development policy paradigms seems to have seriously addressed the issue of rural poverty. This lends credence to the idea that the plight of the poor can be better understood in the context of the delight that plight generates in the sectors of prosperity. In order to address the root causes of poverty in the rural areas a few challenges must be addressed.

First are challenges in the area of economic policy. Low agricultural productivity is a function of stagnant cultivation methods, small acreages, lack of inputs, declining soil fertility, low skills for agricultural production and the retention of obsolete methods of animal and crop management. Herein, lack of access to fertile land, credit and markets are major factors. Of the same order are problems related to rural markets. Either they are closed as in the past or they are flooded and chaotic as in many areas now. In either case producer proceeds are diminished. Price fixing has been a perennial problem for rural communities. During the era of state marketing rural producers were not allowed to organise outside state-controlled cooperatives. Now that states allow other actors, the time is ripe for rural communities to organise to be able to play a role in the determination of prices. The chaos that persists at the moment thrives on the lack of such organisation and even competition among producers at the grassroots level. If these producers organised and set uniform or relative prices they would increase their proceeds and survive the impact of chaotic deregulation.

The second set of challenges is in the area of social policy. The education system still reinforces rural poverty. School enrolment trends are negative for rural areas and for rural women in particular. It is also important to note that the number of secondary schools has not increased in many countries and the terminal level of education has remained the primary school level. At this level the skills acquired are basic and incapable of empowering individuals or communities to be dynamic actors in the world economy. This skills level would not enable school leavers to access and absorb information on production, health and survival. Hence, school systems must be improved, to absorb more people and to raise the terminal level of education for all to the secondary school level.

Together with education is the health policy challenge. If the rural areas are expected to continue producing over 70 per cent of the food and raw materials required, the rural population must be supported in remaining healthy. Rural communities still suffer many diseases of poverty that impinge on their capacity to produce optimally. Ill-health is so common in rural areas that when a person does not fall sick for a long time people fear that he or she may lose immunity and when illness does strike will not survive. Diseases of poverty arise from poor housing, lack of clean water and sanitation, inadequate nutrition, and sharing water and food resources with wild animals. Women, who are the main producers of food and cash crops, have their own extra burden of diseases that has to be separately addressed. These relate to maternal health, child health care, household health and other types of care and diseases to which women are more vulnerable than men, such as vector- and water-borne diseases to which those involved in agriculture are more exposed and HIV/AIDS to which women are biologically more vulnerable.

Finally, good governance is in danger, since in Africa local governance is often weak, in particular rural capacity and policy implementation. This is crucial in development processes in many less developed countries. But in Africa, where the rural population has been ignored, exploited and abused for decades, it is even more so.

NOTES

1. When the tin mining operations declined in the Kilembe region of Uganda, the road infrastructure deteriorated. When the Mpanda copper mines were closed in Tanzania, the railway that served that fertile area was closed. The same was done in Nigeria: when oil discovered in the south replaced the tin mining industry on the Jos Plateau, the railway system disappeared.

2. Figures on the gross agricultural product of many African countries reflects cash crop farming communities and peasant sales of primary commodities that reach the urban and international markets. The food crops of peasants and their small livestock are normally reported tangentially if at all. But this also impacts government allocations to support these activities, which in most cases is very low.

3. On the broader picture of minority alienation by the sate in Africa see Dzingirai (2003: 243–264). On the plight of the San people see Taylor (2002: 467–488).

4. For a detailed account on the history and current state of nomads in Africa see Azarya (1996).

5. Detailed studies on the takeover of land from the Maasai include Parkipuny (1975) and Anacleti (1975).

6. When government officials in Uganda go to Karamoja, people there normally ask "How is Uganda?" In the Turkana region of Kenya people talk about "events in Kenya" because they do not see themselves as part of that country. In both cases these populations have been so neglected that the feeling seems mutual.

7. The author was on the advisory team for this project.

8. In Tanzania some people in the rural areas still vote for Nyerere, even four years after his death. In Uganda they vote for Museveni by ticking alongside his picture, in some cases for fear that the ruling

party will know if they vote otherwise. Most intellectuals as well the general population in Ethiopia are still dominated by socialist ideology more than a decade after the end of socialism because the system of education has not changed and there is no alternative ideology.

9. Income is still considered the primary measure, although the poverty indicators, for example, in UNDP (1996: 40–41) include non-income measures such as health, education, food and nutrition, gender inequalities, population and environment and politics and conflict.

10. Examples include Burkina Faso, Ethiopia, Guinea Conakry, Mali, Mozambique and Tanzania.

11. See also World Bank (1996), Chapter 8.

12. UNDP (1995: 29–33 and 101–104)

13. Most of the young men engaged in petty trade in Tanzania come from matrilineal societies where they cannot own land and have no future. Most of the women engaged in petty trade and the sex industry in East Africa come from patrilineal societies where women have no ownership or inheritance rights and are treated as chattels that can be sold to other families for reproduction and as sources of labour, care and comfort. Indeed, these systems drive many youths to migrate to cities.

14. In June 2003 there was a riot in Mgeta district in Tanzania. It was a new district and at an inaugural gathering the new district commissioner told public that he would execute his mandate from the regional headquarters because the living conditions in the district were below his standards. The local people responded by rioting.

15. Government of Zambia (2002: 33–36); Government of Burkina Faso (2002); United Republic of Tanzania (2003: 62); Government of Mozambique (2001: 4).

16. In a remote village in Bukoba district in Tanzania, villagers wanted the local militia to search the house of a suspected burglar. They were told to pay a "feet fee" for the militia who would have to walk to the suspect's house and a "mouth fee" for village chair who would have to handle the case.

17. The Government of Tanzania set the deadline of 16 August for all district councils to stop levying the abolished taxes. But the information only reached the councils in the towns and not the villages. Up to September some village people had not heard about the new tax policies and were still paying duties. The money collected was being appropriated by corrupt leaders.

18. Safety standards for these attendants are low and this increases the risk of HIV infection.

19. Olowu (2001) emphasises devolution as a means for such recognition.

20. This expansionary character and tendency of the African elite is discussed by Harrison (2002).

21. See Maganga (2002), Odgaard (2002) and Mathieu et al. (2002).

22. For a detailed critique of the theories of social capital see Molyneux (2002).

23. For a general assessment of these boards see Dijkstra (1997: 23–37).

REFERENCES

Aina, Tade Akin (1995) 'International non-metropolitan migration and the development process in Africa' in Jonathan Barker and Tade Akin Aina (Eds) *The Migration Process in Africa*. Uppsala: Nordiska Afrikainstitutet.

Anacleti, A.O. (1975) 'Pastroralism and development: Economic changes in pastoral industry 1750–1961', MA dissertation, University of Dar-es-Salaam.

Azarya, Victor (1996) *Nomads and the State in Africa: The Political Roots of Marginality*. Leiden: African Studies Centre.

Callagy, Thomas M. (1984) *The State–Society Struggle: Zaire in Comparative Perspective*. New York: Columbia University Press.

Dijkstra, Tjalling (1997) *Trading the Fruits of the Land: Horticultural Marketing Channels in Kenya*, Leiden: African Studies Center.

Dzingirai, Vupenyu (2003) 'The new scramble for the African countryside', *Development and Change* 34 (2): 243–264.

Government of Burkina Faso (2000) *Poverty Reduction Strategy Paper*. Ouagadougou: Government of Burkina Faso.

Government of Malawi (2002) *Malawi Poverty Reduction Strategy Paper*. Lilongwe: Government of Malawi.

Government of Mozambique (2001) *Action Plan for the Reduction of Absolute Poverty 2001–2005*. Maputo: Government of Mozambique.

Government of Zambia (2002) *Zambia Poverty Reduction Strategy Paper*. Lusaka: Government of Zambia.

Harrison, Elizabeth (2002) 'The problem of locals: Partnership and participation in Ethiopia', *Development and Change* 33 (4): 587–610.

Hodgson, Dorothy L. and Richard A. Schroeder (2002) 'Dilemmas of counter-mapping community resources in Tanzania', *Development and Change* 33 (1): 79–100.

Hoppers, Catherine A. Odora (1998) *Structural Violence as a Constraint to African Policy Formation in the 1990s: Repositioning Education in International Relations*. Stockholm: Institute of International Education, Stockholm University.

Hyden, Goran (1979) 'Administration and public policy', in Joel Barkan and John J. Okumu (Eds) *Politics and Public Policy in Kenya and Tanzania*, pp. 93–116. New York: Praeger Publishers.

IDRC (1989) *Sharing Knowledge for Development: IDRC Information Strategy for Africa*. Ottawa: International Development Research Centre.

Maclure, Richard (1997) *Overlooked and Undervalued: A Synthesis of ERNWACA Reviews on the State of Educational Research in West and Central Africa*. Washington, D.C.: US Agency for International Development.

Maganga, Faustin (2002) 'The interplay between formal and informal systems of managing resource conflicts: Some evidence from south-western Tanzania', *The European Journal of Development Research* 14 (Special Issue): 51–70.

Mathieu, Paul, Mahamadou Zongo and Lacinan Paré (2002) 'Monetary land transactions in western Burkina Faso: Commoditisation, papers and ambiguities', *The European Journal of Development Research* 14 (Special Issue): 109–128.

Ministry of Foreign Affairs (1999) *Co-financing Between the Netherlands and the World Bank 1975–1996*. The Hague: Royal Netherlands Government.

Molyneux, Maxine (2002) 'Gender and the silences of social capital: Lessons from Latin America', *Development and Change* 33 (2): 167–188.

Naali, Shamshad (1986) 'State Control Over Cooperative Societies and Agricultural Marketing Boards', in Issa G. Shivji (Ed.) *The State and the Working People in Tanzania*. Dakar: Codesria.

Odgaard, Rie (2002) 'Scrambling for land in Tanzania: Processes of formalisation and legitimisation of land rights', *The European Journal of Development Research* 14 (Special Issue): 71–88.

Olowu, Dele (1995) 'The failure of current decentralization programmes in Africa', in James Wunsch and Dele Olowu *The Failure of the Centralized State: Institutions and Self-Governance in Africa,* pp. 74–95. San Francisco: Institute for Contemporary Studies.

Olowu, Dele (2001) 'African decentralization policies and practices: From 1980s and beyond'. ISS Working Paper Series No. 334. The Hague: Institute of Social Studies.

Parkipuny, L.M. (1975) 'Maasai predicament: Beyond pastoralism', MA dissertation, University of Dar-es-Salaam.

Platteau, Jean-Philippe (2000) 'Does Africa need land reform?' in Camilla Toulmin and Julian Quan (Eds) *Evolving Land Rights, Policy and Tenure in Africa*, pp. 51–74. London: UK Department for International Development, International Institute for Environment and Development and Natural Resources Institute.

Taylor, Michael (2002) 'The shaping of the San livelihood strategies: Government policy and popular values', *Development and Change* 33 (3): 467–488.

Quan, Julian (2000) 'Land tenure, economic growth and poverty in sub-Saharan Africa', in Camilla Toulmin and Julian Quan (Eds) *Evolving Land Rights, Policy and Tenure in Africa*, pp. 31–50. London: UK Department for International Development, International Institute for Environment and Development and Natural Resources Institute.

United Republic of Tanzania (2003) *Poverty Reduction Strategy: The Second Progress Report 2001–02*. Dar-es-Salaam: Government of Tanzania.

Scarrit, J.R and Susan McMillan (1995) 'Protest and rebellion in Africa: Explaining conflict between ethnic minorities and the state in the 1980s', *Comparative Political Studies* 28 (3): 325–349.

UNCED (1987) *Our Common Future*. The World Commission on Environment and Development (Brundtland Commission). Oxford: Oxford University Press.

UNDP (1993) *Human Development Report*. New York: Oxford University Press for the United Nations Development Programme.

UNDP (1994) *Human Development Report*. New York: Oxford University Press for the United Nations Development Programme.

UNDP (1995) *Human Development Report*. New York: Oxford University Press for the United Nations Development Programme.

UNDP (1996) *Human Development Report*. New York: Oxford University Press for the United Nations Development Programme.

UNDP (1998) *Human Development Report*. New York: Oxford University Press for the United Nations Development Programme.

UNDP (2000) *Human Development Report*. New York: Oxford University Press for the United Nations Development Programme.

UNDP (2002) *Poverty and Human Development Report on Tanzania*. Dar-es-Salaam: United Nations Development Programme.

Watt, David, Rachel Flanary and Robin Theobald (2000) 'Democratisation or the democratisation of corruption: The case of Uganda', in Alan Doing and Robin Theobald (Eds) *Corruption and Democratization*, pp. 37–64. London: Frank Cass.

World Bank (1996) *World Development Report 1996*. New York: Oxford University Press for the World Bank.

World Bank (2000) *World Development Report 1999/2000*. New York: Oxford University Press for the World Bank.

ZCTU (1993) 'Beyond Structural Adjustment'. Harare: Zimbabwe Trade Union Congress.

CHAPTER 10

LOCAL GOVERNANCE HYBRIDS: ENABLING POLICIES AND CITIZEN APPROACHES TO POVERTY REDUCTION

A. H. J. (BERT) HELMSING

1. INTRODUCTION

Development policy has increasingly focused on poverty reduction since the early 1990s.[1] More recently, the international community has formulated new millennium goals (eight in total), each of which centres on a different dimension of poverty.[2] While, as the millennium goals demonstrate, there is a growing consensus on objectives, there is less agreement on how to achieve these goals. Development policy is shifting away from conventional state-driven interventionism. The policy discourse is moving towards an emphasis on creating "enabling environments". But conceptions of the enabling environment differ.

Under structural adjustment, neoliberal viewpoints on enablement concentrated on creating what was called a "level playing field", where state intervention and regulation were reduced as much as possible. This chapter adopts a different notion of enablement and "enabling policies", namely policies that enable actors to make the most effective contribution towards solving their own problems. Our approach to enablement, in contrast with the neoliberal approach, does not call for *reduced* regulation, but for *better* regulation. The state has a crucial role to play as regulator of the economy and of society. Though the poor are a weak player in many respects, the state can improve their positioning in the market, for example, by protecting minimal rights and improving their access to resources. Through these and other supportive policies, the state can strengthen the position of the poor and their relations with other civil society actors and with the state itself. In this conception the state becomes less important for the poor as a direct provider of basic services and welfare, but it remains essential for a different reason, namely as regulator.

It is also necessary to reconsider the notion of the poor as a passive target group. If enablement is about creating conditions for the poor to make their greatest contribution

177

towards solving their own problems, then the notion of the poor as a passive target group ceases to be relevant. This chapter proposes the notion of the *poor as citizens* who have a *right* to develop their own initiatives (supported by public and private actors) and to be involved in defining the policies that shape their livelihoods. This can be argued on legal and moral grounds. Yet there are also pragmatic-instrumental reasons for the policy involvement of the poor; that is, the advantages that are usually ascribed to a participatory project approach.

In sum, this chapter discusses a governance hybrid, one which centres on the binomium "enabling policies–citizen approach" to poverty reduction. What would be the contours of this approach? Section 2 briefly develops its two components. Section 3 examines six constituent elements in more detail by drawing on and learning from a number of cases. The final section considers limitations and draws some conclusions.

2. ENABLEMENT AND CITIZENSHIP

The United Nations Centre for Human Settlements (UNCHS) first enunciated, in 1988, the "enabling" approach of government in its *Global Strategy for Shelter to the Year 2000* (Helmsing 1997, 2002a). That strategy acknowledged the fact that despite efforts to provide shelter, a growing proportion of the urban population was living in slum and squatter settlements, especially in developing countries. It urged governments to concentrate less on direct intervention and more on the creation of incentives and facilitating measures to enable housing and other urban services to be provided by households themselves, community organisations, NGOs and businesses. In this way the full potential of all actors involved in shelter production and improvement would be mobilised (UNCHS 1990: 8).

This new conception was based on the recognition, supported by ample evidence, that conventional state policies on housing and slum and settlement improvement were (i) quantitatively insignificant in relation to the scale and growth of the housing and settlement problems, especially in developing countries, and (ii) qualitatively inadequate in targeting, reaching and providing affordable solutions to the poor.

Subsequently others, such as the World Bank and United Nations Development Programme (UNDP), adopted and adapted the "enablement" notion and incorporated it into their own policy agendas. The World Bank subsumed it under the "good governance" banner and emphasised, above all, market-enabling policies. UNDP adopted a broader stance, incorporating participation, social justice and sustainability (UNDP 1991, 1993). Since the early 1990s a number of authors have made further contributions to the general concept as well as to substantive areas of policy. Notably Burgess, Carmona and Kolstee (1994 and 1997) distinguished market, community and political enablement. More recently Smith (2000) examined different strands of enabling practices by local governments.

With regard to the question of enablement of the poor as citizens, in terms of poor people's own initiatives and policy involvement, it is useful to differentiate roles, for instance, the poor as consumers, as workers and as producers. Thus poverty reduction strategies and interventions, instead of addressing "poverty" in general, need to relate to the specific problems and needs of the poor in these different roles. Discussing poverty in generic terms, as is so often done, is not very helpful when it comes to identifying and elaborating possible areas for poverty reduction strategies. Looking at the poor in their different roles helps to bring analyses closer to the operational level. What kinds of actors and grassroots organisations are involved? What sorts of initiatives are being undertaken and what support is needed? These variables will differ for project interventions and policies and regulatory frameworks governing consumption, wage work and production. Here the central focus is whether the "citizenship approach" addresses the special problems of poverty reduction and offers new possibilities.

This chapter proposes replacing the notion of the poor *as a target group* for state policies and programmes with the notion of the poor *as citizens* who have a right to their own initiatives. They thus deserve support in their involvement in defining the policies that shape their livelihoods. This can be argued on both legal and moral grounds. But there are also pragmatic-instrumental reasons for involvement of the poor in policymaking. These are the advantages usually ascribed to a participatory approach in development projects: information and knowledge gains; closer links between needs and demands and responses; and greater efficiency and lower costs, as those who actively participate tend to be more fully prepared to contribute their own resources. A higher degree of ownership and sustainability can also be among the gains, since those who are involved in the design and execution of projects are often more willing to share the costs of operations and maintenance. Such well-known arguments in favour of a participatory approach at the project level may apply to policymaking – indeed, probably with even greater validity. Policies are of a higher order; they encompass and orient projects and thus, by definition, they should have a far greater outreach and effect. In fact, it is the "macro" nature of this "citizenship approach" at the policy level which renders it, at least potentially, a powerful tool when it comes to the design and implementation of poverty reduction strategies.

Purposeful action is needed to promote the inclusion of the poor in society. This means to enable the poor to develop their own initiatives at the level of concrete interventions and to perform their role as citizens at the policy and regulatory level. Such an approach brings out the civic involvement of the poor rather than their role as passive recipients of benefits or, in the best of cases, as participants in the design of projects and programmes sponsored by third parties but which are "targeted" at them. These third parties usually perceive the poor as having specific needs for inputs and support in their roles as workers, consumers and producers. The citizen approach explicitly recognises the poor as capable and entitled to intervene, not just at the project level but also at a higher, more general level: in the identification of objectives and priorities for policy and in monitoring and evaluating the implementation of such policies. Both public and private agencies should embrace such an approach and actively support the initiatives and actions of the organised poor. The latter is the flipside of enablement.

"Citizenship" in this sense refers to (i) the systematic enablement of the poor, their communities and organisations at the level of their own initiatives and (ii) the role of the poor and their organisations in setting the policy and regulatory environment of the markets concerned, whether in consumption, wage labour or production. Hence, the citizenship approach represents an enablement strategy at the level of direct initiatives and at the same time a "political loop" through which the empowered poor endeavour to exercise their potential for indirect actions, basically via state or public authorities and rule-making. In more concrete terms, the citizenship approach enables the poor to take direct action, for example, via consumer cooperatives, unions and producer associations, but also by indirectly influencing specific policies and defining objectives, resources, programmes and conditions of access. In short, we are dealing with the enablement of the organised poor to intervene – with government and private support – both directly and indirectly in those sectors and markets which are of interest to them (see Wils and Helmsing 2001). This is not only relevant for employment and income generation, but especially for basic services of which the delivery has been liberalised, privatised or contracted out.

Such a citizenship approach is not yet common. It approximates yet still differs from the recent emphasis on social policies. Social policies, too, are connected with citizenship. Indeed, they are based on plebiscitarian principles which acknowledge the entitlement of all citizens including the poor to social security and a set of basic services. Yet from this perspective the poor appear as citizen-beneficiaries of collective programmes, rather than as active citizen-participants entitled to develop their own initiatives and help define policies. It is this active involvement of the poor which matters, particularly when it comes to strategies to reduce poverty.

Note also that the citizenship approach is distinct from "community development" approaches. The latter tend to ghetto-ise the poor in that they adopt perceptions and managerial practices which isolate "the poor" from mainstream economy and society, mostly under the protection of an umbrella poverty fund and special institutional arrangements. The citizenship approach seeks to enable the poor to advance their interests and to protect and claim their rights vis-à-vis other civil and economic actors. The state, as economic and social regulator, shoulders responsibility for economic and political stability and thus has a special role to play here. As citizens the poor can exercise claims and lobby for change through their relation with the local and central government. The citizen approach also distinguishes itself from the "empowerment approach", which focuses primarily on the poor and their organisations and much less on their interactions with other actors.

3. CONSTITUENT ELEMENTS OF NEW GOVERNANCE HYBRIDS FOR POVERTY REDUCTION

If we define governance as 'another style of governing' (Stoker 1999) then the binomium "enabling policies–citizen approach" to poverty reduction may be seen as another local governance type or "hybrid", which would be characterised by a number of constituent elements. Without being exhaustive, we consider the following elements:

- The poor become actors in their own right; policy formulation is brought down to the local context and is built up from there *without* becoming exclusively local. This re-affirms the *importance of decentralisation and local governance* as a framework for enabling citizen approaches.

- *Enabling policies are different from interventionism in a procedural sense;* actors that seek to enable the poor will have different relations, procedures and structures to plan, administer and finance their enabling actions.
- *Enabling policies are substantively different* from state interventionist poverty reduction policies.
- *Enabling policies use other instruments*, such as leveraging of resources, convergence in policies and actions, localising tendering and impacts, and institution and capacity building.
- Citizen approaches require *organisation.* Community and functional organisation of the poor and especially *second and higher order organisation* acquire crucial importance.
- In order to strengthen the position and positioning of the organised poor, different types of *allies* may play an important role.

The sections below elaborate on these constituent elements. The discussion is based on research carried out in the context of the UN-Habitat Community Development Programme which operated in 22 locations in seven countries[3] and on seven case studies commissioned specifically to explore the citizen approach.[4] The work done so far is no more than exploratory.

3.1 Decentralisation and Local Governance as a Framework for Enabling Citizen Approaches

Decentralisation has become a central issue in development cooperation and the debate surrounding it. Most developing countries are actively pursuing decentralisation policies that devolve functions and responsibilities to local governments. All multilateral and many bilateral agencies dedicate resources to this theme. The phenomenon has greatly increased in complexity. In earlier waves of decentralisation the problematique was restricted to the organisation of the public sector. The comparative static question, "which level of government is most appropriate", in the last decade of the previous century, became *secondary* to a more fundamental questioning of the relations between society and state. The underlying causal factors of the current wave of decentralisation and governance are examined elsewhere (Helmsing 2002a). The discussion has since broadened (Litvack et al. 1998, World Bank 1998). It has been argued that part of the state's lost legitimacy may be regained through decentralisation to and the democratisation of local-level government. This could be achieved, for example, via the institutionalisation of elected and representative local governments, direct election of mayors, the formation of local government sub-structures at village and neighbourhood levels, and via citizen participation in decision-making, planning and budgeting. In time these have become important dimensions of decentralisation.

From the above it may be clear that poverty reduction has not been among the major factors or forces shaping decentralisation. Nonetheless, decentralisation provides an opportunity to adopt new approaches to poverty reduction. The arguments in favour of decentralisation, especially potential improvements in the responsiveness of local governments, are important here. In addition, it should be noted that decentralisation offers *new public spaces* for the organised poor to influence policies and programmes that affect their livelihoods. A good example is in Bolivia, where decentralisation and popular participation laws of 1994 and 1995 created a framework for poor people's participation in public affairs and in public investment planning. Under this new governance hybrid more than 13,500 district organisations were created (called *Organizaciones Territoriales de Base* or OTBs). Many of these were based on pre-existing informal, ethnic or traditional organisations (VPEPP 2000). OTBs play an important role in demand generation and articulation. A large majority of the country's 314 municipalities now have "public vigilance committees" as well, which monitor public expenditures. Subsequently, the Law on National Dialogue of 2001 determined that funds which come available for poverty reduction under HIPC (Highly Indebted Poor Countries) II, would be channelled to the new municipal governments. Many countries have implemented extensive national dialogues when designing poverty reduction strategies. However, Bolivia is one of the first to decentralise poverty reduction funds, thereby bringing decision-making and implementation closer to the poor.

Decentralisation, if well-resourced, can make a difference, especially when other actors enable the poor to claim decentralised resources. The well-known experience of Porto Alegre (Brazil) in participatory budgeting was made possible in part by the considerable increase in intergovernmental transfers to the municipal level (Chavez 2002).

Decentralisation may be necessary, but it certainly is not a sufficient condition for poverty reduction. On the one hand, "elite capture", parasitic bureaucratic behaviour and clientelist politics may prevent the organised poor from benefiting from new opportunities. Even more so, our research on enabling policies in the UN-Habitat Community Development Programme showed that deconcentrated sector agencies can also be enabling if they succeed in substantially changing their administrative, financial and planning procedures and arrangements (Helmsing 2002a, Wils and Helmsing 1997).

3.2 Enabling Policies are Substantively Different

The central argument behind the notion of enabling policies is that government is incapable of addressing the needs of the poor through conventional direct intervention in markets and through public sector solutions. Instead, it plays an indispensable role in creating conditions for other actors, especially the poor themselves, to contribute towards poverty reduction. When government itself ceases to be the agent directly responsible for delivery, then incentives to and regulation of other actors becomes central.

The liberalisation of basic services has made the issue of regulation, in order to protect the poor, even more acute. Even if governments are not directly responsible for implementation they still can and should be held accountable for the end result. Seen from the other side, the state can help the organised poor to strengthen their position vis-à-vis other actors in civil society and the economy. This is the core of enabling policies.

Thus, for example, in housing the emphasis has shifted from public sites and services schemes to questions of regulation, organisation and capacity building. This includes (i) regulation of land and housing markets so as to improve access by the poor and reduce tenure insecurity for housing tenants (after all, most urban poor are renters); (ii) establishment of norms and standards for housing construction and materials; (iii) regulation of the construction industry. Community mortgage programmes, with all their limitations, are examples of how governments can create incentives and finance improved access for the poor (Berner 1999).

It should be recognised that government failure to deliver services to the poor can be described as a *governance* failure. This represents a whole new set of problems about which we know relatively little and for which there are, as yet, few solutions (Helmsing 2002a).

With regard to enabling policies for poverty reduction it is important to think of the poor as consumers, producers, workers and citizens, as indicated earlier. Thus, liberalisation of basic services brings about the need to protect the consumer rights of the poor. This is evident, for example, in contracts concerning access to privatised services and pricing that affects the poor. The poor, as producers, may be involved in the construction, maintenance and retailing of basic services. As workers, the focus of enabling policies could be on the organisation of the informal sector and on minimal workers rights. The poor, as citizens, may be represented on regulatory boards to ensure that their interests are taken into account.

Apart from regulation, it is important to consider the question of incentives and how and to whom incentives are directed. What actors are in the best position to assist the poor and what motivates them to act out that role? How can government structure incentives so as to maximise effectiveness?

Poverty reduction through enabling local *producers* has its own characteristics and presents its own class of issues. Policies and initiatives have moved from individual approaches to local economic development in which the position and positioning of local producers is framed in the context of the local economy and relevant industry. These efforts focus on building a *combination* of individual and social asset bases of local producers. This involves investing in community productive organisations, introducing knowledge and skills on (new) products and process technologies, reducing entry barriers to markets and increasing competitiveness by organising "collective efficiency" and joint enterprises (e.g. in processing or marketing) and linking local producers to external markets (Helmsing 2002b, 2003). In most cases external actors play a key role, interacting with local organisations.

In a rural water supply scheme in Gujarat (India) local women were trained as masons and, with the assistance of the Self-Employed Women's Association of India (SEWA), were able to form cooperative enterprises. In her study of this experience Van Wijk examined the gender impacts of both improved water supply and the income generated through the new enterprises (Van Wijk et al. 2002). Bebbington (2002) analysed how local producers in Salinas in highland Ecuador succeeded in improving their economic position thanks to external assistance in product innovation and production organisation. Rarely, however, do local organisations of poor producers succeed autonomously in transforming their local economy, though there are exceptions (e.g. the Otavalo indigenous economic networks, see Bebbington 2002).

An aim of the Uni-Trabalho (university-work) network, founded by the Brazilian Confederation of Trade Unions and universities, is to provide "incubators" for cooperative enterprises. This is an important step towards new types of enabling citizenship approaches. The incubators are primarily oriented to human capital development, providing knowledge resources on enterprise development and cooperativism. According to Pegler (2002), in 2001 there were some 33 incubators operating across the main regions of Brazil involving some 13,000 poor workers organised in 137 cooperatives and 109 project groups. Over time the incubators have become entry points for other supporting actors in the areas of finance and markets. This is well illustrated in Pegler's case study of Nova Amafrutas, a processing operation for perennial fruit in the Amazon (ibid.). In this case, several hundred factory workers and several thousand outgrowers were threatened with unemployment following the withdrawal of a multinational company. A variety of supporting actors came to assist, including the Confederation of Trade Unions, the state government (which facilitated land access) and the Dutch Inter-Church Organisation for Development Cooperation (ICCO). ICCO financed a crucial transition phase, co-financed an *escola densa* (a "learning by learning" and "learning by doing" project) and facilitated contacts with a new international buyer and an international financial institution. These subsequent and necessary actions to make Nova Amafrutas commercially viable also introduced new dilemmas. Greater risks were involved in "going global", and power balances were shifting as well, bringing the risk that local participants might become marginalised while other actors, higher up in the chain gained control. The role of countervailing allies (like ICCO) becomes all the more important in such a context, but how long-term can and should be their commitment? In this regard, Bebbington (2002) raised the question of the extent to which the enablement of poor local producers in (global) markets is at all feasible and sustainable.

3.3 Enabling Policies are Procedurally Different

Enablement requires governments to conduct their affairs in a fundamentally different way. Instead of self-contained, hierarchical bureaucratic processes, mediated by more or less democratically elected politicians, enabling governments seek to involve other actors in the formulation and implementation of policies and programmes.

Seen from the other side, the citizen approach implies that communities take part in decision-making affecting their own livelihoods and are involved in managing programmes to improve their livelihoods and human settlement conditions. Government facilitates community participation and management rather than intervening directly and unilaterally. In short, government enables communities to take their own responsibility, deriving its own actions from community initiatives. Government enablement may thus be redefined as (local) government creating appropriate legal, administrative, financial and public planning frameworks to facilitate neighbourhood communities to (i) organise into community-based organisations, (ii) manage community-level affairs and (iii) undertake collective action. Yet, if government enablement is to take root, it should be reflected in the management and administration of government's own affairs. The evaluation of the UN-Habitat Community Development Programme experience in seven countries showed how difficult it is to institutionalise enabling practices and also pointed out the problem of disabling attitudes (Wils and Helmsing 1997).

One of the challenges in operationalising enablement is finding indicators by which to identify and assess the degree of enablement in particular areas. We have addressed this problem elsewhere (Helmsing 2002a). Indeed, our research in the UN-Habitat Community Development Programme showed the adoption of enabling practices to be low and rather uneven. The greatest progress has been made in adopting enabling practices in planning, but that too is very partial in actual practice within government administration and management. Such enabling practices in planning have taken the form of changes in internal government procedures, political representation of community-based organisations and actually allowing community-based organisations to wholly or partially manage public funds. While it is relatively easy to incorporate enablement into public planning, it is much more difficult to change the ways and means by which government administration and finance operate (Wils and Helmsing 1997).

The Porto Alegre case mentioned earlier is by now one of the best-known examples of a "local governance hybrid" in which communities play a major role in defining local public priorities and programmes. The local government reorganised its own institutions and procedures to facilitate participatory budgeting, creating a community relations commission, a planning cabinet, thematic and regional coordinators and sub-local administrative centres that support the process in neighbourhoods. It also modified its annual planning and budgeting cycle to incorporate, in two rounds, the results of civil society deliberations. Furthermore, local government acts on the outcomes of the process and communicates these actions to those involved. Although the "deliberative space" of the participatory budgeting process itself has *not* been institutionalised, local government has clearly adopted enabling policies to facilitate community voice (Chavez 2002).

3.4 Enabling Policies Use Other Instruments

Our review of different cases showed that enabling policies use various instruments. Participatory planning, monitoring and evaluation systems, such as those implemented in Porto Alegre, Gujarat, Santa Cruz (Bolivia) and Limón (Honduras), form the core of enabling policies and citizen approaches. Leveraging resources is a second instrument. Enabling policies cannot be driven by one actor. In essence, they always involve some kind of "quid pro quo", and leveraging of resources is one way of realising this. Seeking convergence among actors and stakeholders is a third instrument. Enabling policies often require the creation and strengthening of new local institutions. Institution building and training or capacity building therefore belong to the repertoire of enabling instruments. Last but not least, local tendering can be an important catalyst for enabling citizen approaches. Each of these instruments is elaborated below.

3.4.1 Participatory Planning, Budgeting, Monitoring and Evaluation
A considerable literature has accumulated evaluating the experiences of community-based planning in rural and urban settings in developing countries (see e.g. Helmsing and Wekwete 1990, Korten and Klaus 1984, Uphoff 1986). Lessons learned relate to the sequencing of the planning process, the roles of community members in identifying and prioritising needs, the formulation and ownership of action plans and the role of external facilitators. Issues found to be important are the degree of participatory needs identification, how needs are articulated and priorities set and the provision of resources and commitment of various actors. Experiences show that synchronising planning is easier than synchronising budgets and implementation.

In structuring enabling citizen approaches, democratic processes are needed for selecting community members to be involved, ensuring balanced representation of gender and generations and separate elaboration of diagnosis and priorities prior to assemblies for decision-making (Wils and Helmsing 1997, Oosterhout 2002). Selection and decision-making cannot simply be left to community leaders.

One indicator of effective enablement is the degree to which community-based planning, monitoring and evaluation systems are synchronised with the procedures and practices of the principal enabling actors, notably municipal governments, sector agencies and NGOs. In the past, participatory planning was often undertaken in isolation from such actors. The annual participatory budgeting cycle of Porto Alegre demonstrates how advanced synchronisation and integration can become (Chavez 2002). The participatory planning, monitoring and evaluation system implemented by the NGO Christian Commission on Development (CCD) in Honduras ensures that targets and indicators are jointly defined so that all parties concerned (the municipality, NGOs and communities) can monitor their own progress and that of others (Wils 2002).

The use of participatory planning, monitoring and evaluation systems may be justified on the grounds that participation is not a means but an end in democratic decentralisation. But even as a "means", participatory planning, monitoring and evaluation may bring important effectiveness and efficiency gains in the form of improved targeting of public investment (reducing costly mistakes or "white elephants"), greater mobilisation of resources and increased efficiency in basic services. In an analysis of the participatory methods used by CCD, Wils (2002) showed that compared to "traditional" NGO approaches, more resources were generated and more projects completed within a shorter timeframe.

3.4.2 Leveraging Resources
The practices examined in the UN-Habitat programme case studies provide arguments for the leveraging of funds as an enabling instrument. Leveraging has a number of effects. First, it multiplies the resources available to finance programmes and activities. Second, it spreads risk amongst individual contributors. Although community contributions may burden poor people's budgets, such contributions are generally recognised as a way to create ownership and ensure that proposed initiatives are really wanted. Especially for more distant donors, the very fact that local people contribute spreads risk. This might be particularly important in situations where the

contributor does not directly control implementation. Third, leveraging contributes to horizontal accountability. It becomes more difficult for any of the contributors to default on their commitments (except of course when the number of contributors becomes very large). Thus, the organised poor, who tend to be in the more dependent position, can seek the support of other contributors to pressure a defaulting one to meet its obligations. A potential disadvantage of leveraging is that it may require longer lead times and contribute to efficiency losses in the narrow sense.

The Poverty Alleviation Programme (PAP) in Santa Cruz, Bolivia, is one example in which leveraging has been applied as a tool. The programme has not only been effective in leveraging community contributions, it also succeeded in leveraging local government funds, as well as funds from other donors and national funding agencies. This draws other actors into an enabling policy mode, thus helping to mainstream the approach. Across all project categories, the PAP financed 56 per cent of the activities, local government financed 21 per cent, communities provided 12 per cent and other actors financed 10 per cent. Within the PAP's small project category, community contributions varied between 45 and 50 per cent (Oosterhout 2002).

Leveraging of resources was also effective in Limón, Honduras, contributing to a rapid expansion of basic services for the poor (Wils 2002). Communities operating in a participatory mode with CCD and the municipal government achieved in 16 months results similar to those other communities accomplished in 10 to 12 years using traditional NGO approaches, even under better resource conditions (e.g. staffing). In the case of Kerala, India, the leveraging of resources that became available to local government under the new decentralisation policy greatly expanded the community's capacity to expand its sanitation programme (Van Wijk et al. 2002).

3.4.3 Convergence Among Actors

Convergence is another important instrument. Poverty is a multi-dimensional phenomenon, and insecurity and vulnerability have many aspects. Approaches must therefore be flexible and avoid pre-selecting one or another sector(s) through which programmes are to be implemented. Integrated analysis is needed. In the past this idea was often translated into "integrated" action in which one agency undertook a range of related programmes. Experiences with such programmes and (multi-sectoral) agencies have been disappointing however. Agencies often came to function as temporary bureaucratic bypasses of existing sectoral offices and, as such, had built-in sustainability problems. In addition, integrated programme agencies often lost focus and efficiency.

Enabling approaches are more concerned with seeking to "buy in" or enlist existing agencies and departments, expanding the range of effective responses to citizens' needs by re-orienting programmes through processes of convergence. For example, settlement upgrading often requires a combined response from physical planning departments, land registries, public works and field offices for basic services. Recognition that there is no single owner of a problem is central.

The critical question on convergence is at what level? In Porto Alegre, the participatory budgeting process exemplifies convergence at the city-wide level. In many other instances, convergence has been achieved at the more modest but nonetheless important neighbourhood level. The Bangalore Urban Poverty Project (BUPP) in India started by organising city-wide convergence of public and non-governmental actors, but bureaucratic struggles turned out to require inordinate amounts of time (De Wit 2002). In the PAP-Santa Cruz experience, convergence was achieved slowly over time. Though bureaucrats were often reluctant to change their attitudes and practices, they could be convinced by a combination of positive examples and local pressure. Several authors have noted the twin forces of "leading by example" and "political pressure from un-served communities" (arising from the implied demonstration effect). In Santa Cruz, such processes contributed to the incorporation of other districts into the poverty alleviation programme (Oosterhout 2002). In Kerala they helped extend a community-based sanitation programme to other *panchayats* (Van Wijk et al. 2002).

3.4.4 Institution Building and Capacity Building
Enabling citizenship approaches assume that the poor are organised so they can become an actor in shaping policies and programmes. In some countries, communities are well endowed with social capital and have thriving organisations. In others, past repression or neglect has led to a decline in associational life.

Enabling citizenship approaches requires up-front efforts and investments in institution building and capacity building. Though such investments may be difficult to finance and recuperate, primarily because of their public character, they can contribute to improved public decision-making and "democratic governance". This is an argument for government and donors to absorb such costs, rather than their being borne by the population alone.

Community skills are central to the use of participatory planning, monitoring and evaluation systems. Capacity-building workshops, both at the start of and during the enablement process, enhance social learning. In some cases quite a bit of time must be invested in building community organisation and management skills.

Institution and capacity building within local government is equally important. As several case studies demonstrate, enabling policies require capable local government. Participatory budgeting exists by virtue of a committed and professional local bureaucracy (Chavez 2002). Research by the UN-Habitat Community Development Programme found that senior local government officials are often far less "enabling" than their field officials. Hence, apart from creating enabling structures and procedures, there is a need for awareness-building and adjustment of attitudes (Wils and Helmsing 1997).

3.4.5 Localising Tendering Processes
Localising spending not only impacts poverty reduction but also stimulates local collective action. Localising public tendering, however, can be a sensitive issue. In many countries legislation emphasises making tendering competitive and complex, in order to control corruption. This often reduces the scope for governments to make tendering accessible to local producers. Furthermore, well-established industries (notably large construction companies which often have ties to local political elites)

may oppose any localisation that favours small local enterprises on the grounds that it would be "unfair" competition. Looking at it from another perspective, there are many forms of "regulated" competition in the construction industry. An additional political argument might be added to these: localising small tenders could increase the poverty reduction impacts of projects and through local collective action stimulate "democratic governance". In Santa Cruz, Bolivia, the Municipal Ordinance of Exception created a legal avenue to award competitive tenders for small projects to enterprises originating in the same or adjacent neighbourhoods where the projects were to be implemented (Oosterhout 2002)

Sri Lanka has also innovated in this area with its so-called "community construction contracts" whereby the construction of sanitation and infrastructure works and different types of buildings is contracted out via community development councils (UNCHS 1994). The engineering departments of local government produce the technical designs and specifications. In essence, the communities choose from a menu of options. Local NGOs assist the councils in preparing and undertaking the works. The use of community construction contracts has now spread to other Asian countries, attesting to their success as enabling municipal practices (CityNet 1997).

Evidence from both the Santa Cruz and Sri Lanka cases show that local micro-enterprise or community-contracted works are considerably cheaper (25–56 per cent) than similar projects carried out under standard local government tendering procedures. Furthermore, the works are completed more quickly and with better quality. Localising tendering processes not only makes economic sense, it has other benefits as well. It increases the propensity of communities to leverage their own resources, as these flow back into the community in the form of work for local businesses.

3.5 Second and Higher Order Organisations of the Poor

The general approach of the UN-Habitat Community Development Programme was to achieve sustainable human settlement improvement and poverty alleviation by empowering communities. This was to be done by strengthening the organisation of communities in locally based organisations and by endowing households with the knowledge and skills to participate in collective action, to manage community projects and to relate to external agents, especially government. Research done through this programme by Wils and Helmsing (1997) stressed, first, the importance of *community organisation*. Organisation and democratic representation of different interests within the community is a necessary first step. Particularly in the area of housing and settlement improvement, community organisations at the level of neighbourhoods are a primary unit of organisation, as this is the lowest level at which collective needs can be expressed. Second, community participation, as measured in terms of frequency of and attendance at community assemblies and meetings set by other actors, turned out *not* to be a decisive factor in community empowerment. The main purpose of these sorts of meetings tends to be to communicate other actors' plans and intentions, seeking to legitimise initiatives and extract information from communities. Instead, the creation and operation of *community management* in terms of planning, monitoring and evaluation systems and community-based construction, maintenance and operation of projects proved decisive. Planning, monitoring and evaluation helped to instrumentalise people's participation and bottom-up planning.

   In contrast with commonly held views, community organisation is not necessarily more difficult in urban areas (as compared to rural) nor does it require a particular maturity and duration of community life. Both the UN-Habitat programme and PAP-Santa Cruz studies found that much depends on the scope for collective action. If collective action brings quick results, it can trigger further collective action and the organisation that may be needed for it. In the case of PAP, even where communities were only recently established, initial collective action triggered social capital in the form of *juntas vecinales* (neighbourhood committees). The case of Gujarat, India, showed that enabling actions themselves can generate new forms of community organisation. This coincides with the conclusion by Evans that 'social capital makes collective action possible, but collective action is an important source of social capital' (Evans 2002: 225). In a similar vein, Douglass argued that 'the sense of efficacy gained through the successful execution of projects was central to enhancing community cohesion and capacity for future collective action' (Douglass et al. 2002: 225). To this one could add that the new ways of organising a relationship with the poor, notably the use of participatory and bottom-up methods, can have the (un)intended effect of helping to strengthen community organisation thanks to new generations of (well-trained and often young) community leaders, something which was observed in the PAP-Santa Cruz case. This does not mean that community organisation is easy and all-inclusive. Slum communities tend to be socially stratified and the very poorest are often difficult to incorporate.

   Second- and third-tier community organisations can play a major role in strengthening the position of community groups. Associations of grassroots groups and federations of associations seem to have several advantages. *First*, greater numbers raise voice. Apex organisations can yield more than proportional influence. This generally acknowledged fact is also borne out by our case studies. *Second*, associations facilitate sharing of information and experiences. *Third*, thanks to their larger size and scale of operation, associations can undertake functions that are not feasible at the community level. Second- and third-tier organisations can also strengthen the autonomy of community-based organisations vis-à-vis the state as well as the market. The PAP in Santa Cruz stimulated the formation of such a second-order federation by inviting community organisations to be represented on its municipal steering committee. The city-wide federation of *juntas vecinales* was able to enter public policy arenas that were previously beyond the reach of the poor. This applied to the municipal council as well as to the Santa Cruz Civic Forum. The *fourth* and final point is that community organisations become training grounds with wide application. In both Bolivia and Ecuador, indigenous organisations have acquired more political clout by increasing scale. In Ecuador second-tier organisations succeeded in wielding greater political influence such that indigenous leaders have started to occupy positions as municipal councillors and even as provincial prefects (Bebbington 2002). In Bolivia their influence is growing, although many factors are involved. In elections in 2002 "indigenous" parties gained important positions in the national parliament. In other words, social learning is taking place alongside a move from the local to higher levels of influence.

A different aspect concerns recognition and institutionalisation. Though recognition is crucial we may ask whether institutionalisation is needed. The argument in favour is that institutionalisation (especially legalisation) creates a new legal resource that active communities can claim. Wils and Helmsing (1997) reached this conclusion, yet found that the legalisation is often a challenge. Without it, however, community initiatives risk dependence on a temporarily favourable situation, like a committed mayor. Evans (2002) supported this view. Others see danger in institutionalisation, as this may detract from the authenticity of community initiatives as social movements, Chavez (2002) argued this position in relation to participatory budgeting in Porto Alegre. In many African countries, community organisations are created as part of so-called "mixed" local government structures. With legalisation comes centralised control (see Helmsing 1999). Thus, much depends on *how* legalisation takes place, whether it protects the authenticity of the community organisation and whether it is accompanied by political or bureaucratic controls.

In Ecuador, two federations of indigenous peoples succeeded in bringing issues such as bilingual education, land and the national constitution as a pluri-ethnic state onto public policy agendas at both the national and local levels. With the collaboration of other allies, the federations achieved changes in key provisions concerning land reform. However, as observed by Bebbington (2002), the higher the level of policy formulation the more difficult it is for organisations of the poor to influence policies. This brings us to the strategic importance of allies.

3.6 Strategic Importance of Allies

We have seen how community organisation, especially second-tier organisation, is crucial in enabling citizen approaches. However, communities rarely have sufficient organisational resources, knowledge and clout to go it alone. Community-based organisations need allies. There are several kinds of allies. Most well-known are NGOs, but there is also scope for others, such as trade unions and political parties.

3.6.1 NGOs as Allies

Both advocacy and development NGOs are powerful potential allies. They are often in a position to provide knowledge, ideas and perspectives that can connect livelihood issues to broader policy and to institutions. NGOs can help communities claim their rights and convey their claims and proposals in a manner understandable by other actors. They can also provide links to other support organisations.

At the same time we should recognise that NGOs have become part of the aid chain and of the "social service contractor" industry. It is sometimes difficult to distinguish self-serving intermediary NGOs from NGOs genuinely interested in and committed to poverty reduction. Few NGOs have managed to avoid paternalistic self-centred or NGO-led participatory approaches which in the end may become an obstacle to change. NGOs themselves must assume an enabling posture. The experiences of CCD in Honduras demonstrate how an NGO can enable both communities and local governments by offering participatory planning, monitoring and evaluation methodologies that serve to bridge efforts of both sides, mobilise resources and produce faster results (Wils 2002). NGOs that concentrate on supporting community collective action and enterprises (like the Indian Society for the Promotion of Area Resource Centres) stand a better chance to succeed than NGOs which themselves are service providers.

Also, the cases of PAP-Santa Cruz and Gujarat and Kerala show that NGOs can have a comparative advantage over government in relating to communities. They are often more successful in improving community organisation and management skills and can provide technical inputs on substantive issues such as community-based planning, construction, environment, sanitation and health.

3.6.2 Trade Unions as Allies

In several countries traditional trade unions are playing an enabling role and as such can be an important ally for community-based organisations. As Schiphorst (2002) argued, traditional trade unions restrict themselves to the formal sector, which has been shrinking in many countries in both absolute and relative terms. For a number of reasons it is difficult for such unions to attract members from the informal sector. In particular, a large majority of informal (self-employed) workers are women, while unions are usually male dominated. Moreover, informal enterprises tend to be unlicensed and difficult to locate, and their owners hesitate to become union members so as to avoid becoming too visible. Neither do informal workers have need of formal sector union services, such as collective bargaining. Thus, instead of seeking to enlist informal sector workers, traditional unions and the International Labour Organisation are now seeking to assist informal sector workers to organise themselves. In essence this is an enabling policy. The unions share their own organisational experiences, provide support and assist newly formed unions morally and financially during the difficult start-up phase. They may also lobby municipal governments to ease certain regulations.

Informal sector unions tend to be community based. Successful ones have advocated a mixture of issues related to the poor as workers and consumers. Activism for basic services serves both ends, as many informal workers engage in home-based activities. Priority issues may be settlement improvement, physical security in slum areas and public health (Schiphorst 2002). SEWA in Gujarat is such a union. It assists local women who have learned construction skills through NGOs to organise themselves in small enterprises and become independent contractors for the Gujarat Water Supply and Sanitation Board in a donor-funded rural water supply scheme (Van Wijk et al. 2002). The women began by forming a union, which then became a SEWA affiliate. The Brazilian case of Uni-Trabalho described by Pegler (2002) is an example whereby a confederation of trade unions sought out a new role for itself, namely to support informal and cooperative enterprises. By building alliances with other actors, especially universities, it positioned itself to make technical expertise available and disseminate it further through "incubators".

3.6.3 Allies in the Local Bureaucracy

Field officials of community development departments rally behind an enabling approach more easily than higher level executives. This is because any approach that externalises part of the public decision-making process may reduce the power of the bureaucracy – something senior executives can be rather sensitive about. It might therefore be difficult to convince the city engineer or treasurer of the virtues of new local governance hybrids. In the 23 local governments surveyed by the UN-Habitat Community Development Programme, senior executives were consistently more public-sector centred and less oriented towards community enablement (Helmsing 2002a, Wils and Helmsing 1997).

To mainstream poverty reduction strategies, broad "buy in" is needed by the local bureaucracy. Public works departments, planning and finance offices and land registration bureaus need to incorporate the concerns of poor citizens into their policies, plans and daily practices. A welfare office responsible for community and neighbourhood development cannot be the sole actor, but it can be a strong ally within government. The Porto Alegre experience of participatory budgeting shows the crucial role of a committed local government administration that supports citizen approaches with innovative procedures and processes (Chavez 2002).

3.6.4 Allies in Political Parties

Political parties and politicians are often seen as antagonistic towards community approaches. When community organisations do engage with political parties, they often fall victim to partisan rivalries and conflicts. For their part, politicians and political parties may seek to co-opt community leaders in order to organise vote banks and clientelist networks. This often has divisive impacts that weaken citizen approaches. NGOs and experts therefore often advocate preserving autonomy. For enabling citizen approaches, such an option is not an attractive one.

Politicians may view municipal or city-wide participatory approaches as a threat, perceiving them as pre-empting their role in defining collective and public preferences and reducing their role in public decision-making. This is beside the fact that participatory processes may reduce opportunities for patronage, clientelism and fraud. Nonetheless, as rightly argued by Evans (2002), politicians are, for communities, the primary interlocutor with agencies that deliver infrastructure and services. As a way out of the dilemma, Evans advocated a greater role for opposition parties. They may be less instrumental in providing infrastructure and services, but they can open political spaces for communities and social movements to participate in municipal or other local fora; they can also advocate political proposals that are more enabling. In other words, opposition parties rooted in social movements are nowadays likely to adopt enabling legislation when they come to power. The New Democratic Front in Kerala, when it came into power, played a key role in implementing the new national-level decentralisation policy along with complementary state policies which increased the powers and the resource base of the *panchayats* (Van Wijk et al. 2002). This, in turn, made possible the leveraging of resources to expand the local sanitation programme.

The central challenge is not to substitute representative democracy with participatory democracy but to find the right balance between the two. The participatory budgeting process in Porto Alegre is an example in which such a balance has been found. Participatory budgeting is an extensive process of identifying public preferences and setting public priorities for sectors, programmes and projects. Yet in the end the municipal council decides. The Labour Party has been the primary interlocutor in this redistributive process. Its prime contribution is not only enabling policy but acceptance of the outcomes of participatory democracy as inputs into formal democratic local government. Elsewhere in Brazil, many parties, including conservative ones, have adopted participatory budgeting. This reinforces the view that such a process is no more than a public management technique. However, the proof of the pudding is in the eating: it depends on whether resources are allocated in a more redistributive manner and whether outcomes in terms of community preferences and priorities are adopted and implemented by local government.

4. LIMITATIONS AND CONCLUDING OBSERVATIONS

Can enabling citizen approaches be adopted anywhere? Or are there particular conditions that need to be met? The case studies done as part of the UN-Habitat Community Development Programme generated a number of considerations. First, and especially in relation to basic services, national decentralisation policies are important insofar as these create overall enabling conditions, especially when accompanied by substantial resource transfers to municipal governments. Bolivia is a case in point, and the Porto Alegre and Kerala cases are illustrative in this respect. Second, political commitment on the part of key actors is crucial. This not only applies to domestic allies but also to donors. Sometimes even initially reluctant actors, such as the municipal government in the Santa Cruz case, can be pulled aboard. A strong and well-organised local government is a key interlocutor, particularly if it is willing *and* able to organise its own share of enabling citizenship governance and invest up-front in creating new institutions and in social learning.

As indicated earlier, a tradition of social movements and community organisation is instrumental in developing advanced forms of hybrid governance, as in the cases of Porto Alegre and Kerala. This does not mean that without such a tradition no advances can be made. Quite the contrary, the Bolivian and Honduran experiences show that participatory poverty reduction experiences can trigger collective action. Collective action, in turn, helps to build social capital. Obviously, there is some degree of path dependence. That is to say, the sophistication of institutions of hybrid governance depends on past capacities and experiences. Kerala and Porto Alegre both have a long tradition of social movements and comparatively high levels of literacy and education.

As that last statement signals, there can be limiting factors. A governance hybrid can develop in a number of directions. A distinction can be made between the demand side and the supply side of governance. The former is the political dimension of determining public policies and preferences, while the latter refers to the role of non-state actors in the delivery of basic services (Helmsing 2002a). The Porto Alegre experience is rather advanced on the demand side of governance while minimal on the supply side. In contrast, the Santa Cruz and Limón cases are, on the whole, more rudimentary (especially in terms of participatory methods of planning) on the demand side and strongly developed on the supply side of governance.

These governance hybrids may be more easily applied to basic services, such as water, sanitation and infrastructure than to income generation projects. In the case of income generation beyond the local market, governance hybrids may require enabling actions from a range of actors at higher levels. As the Nova Amafrutas case revealed, this can create tension between bottom-up empowerment and top-down enablement.

Of greater importance may be the absence or presence of powerful opposing forces. For example, in Ecuador land reform had already undermined the power of the local *hacendados* (landlords). This opened the way for indigenous second-tier organisations to acquire a position in local development and beyond (Bebbington 2002).

Last but not least, acceptance of the voice and interests of the poor by other domestic actors, and their positive response to it, is an important precondition for enablement to succeed. International cooperation and aid can play an important role in softening resource conflicts but, in the end, poverty reduction depends on the presence of a basic, even if tacit, solidarity with the poor.

NOTES

1. Many of the ideas in this paper were developed in work undertaken with Frits Wils. Our cross-disciplinary dialogues were a constant source of inspiration. Useful comments were provided by Nicholas Awortwi, Erhard Berner, Bamidele Olowu, Lee Pegler and Joop de Wit. Any errors and omission are mine.

2. The millennium goals include reduction of extreme poverty and hunger; achievement of universal primary education; promotion of gender equality and empowerment of women; reduction of child mortality; improvements in maternal health; combating HIV/AIDS; ensuring environmental sustainability including life improvements for slum dwellers in terms of water, sanitation and tenure security; and global partnership for development, including ODA targets, market access and debt sustainability.

3. These countries were Bolivia, Costa Rica, Ecuador, Ghana, Uganda, Zambia and Sri Lanka.

4. The Poverty Alleviation Programme (PAP) in Santa Cruz, Bolivia was a purposeful initiative to experiment with a new model of urban poverty reduction. It lasted less than four years, but had noteworthy impacts (Oosterhout 2002). It was a city-wide project that brought together for purposes of planning, administration and funding, the municipality, slum dwellers (including women) via their federation and elected vigilance committee, urban NGOs and bilateral donors. Porto Alegre's experience with participatory budgeting is another well-known city-wide experience. Chavez, who examined it in detail, argued that participatory budgeting is not merely a development management technique, but entails a set of conditions for it to work. According to Chavez (2002), local context is crucial for replication. The Porto Alegre experience contrasts with that of Santa Cruz (Bolivia), as Porto Alegre primarily focused on the poor as consumers, while in Santa Cruz the poor were also involved as producers, thereby increasing the poverty reduction impact and generating different dynamics. The experiences with community-based water supply and sanitation in Gujarat and Kerala, both in India, provide insight into potential gains of this approach, especially for women. These cases also show the role of a new type of trade union and give us an opportunity to examine how national enabling sectoral policies work out differently depending on the responses by state and local governments (Van Wijk et al. 2002).

   A paper by Frits Wils (2002) on participatory methods of planning, monitoring and evaluation by NGOs and the grassroots groups they work with complements this. It uses a case study in Honduras to show that participation by the poor, when extended along the entire project cycle, can generate important additional advantages. Compared with "traditional" NGO processes, enabling approaches are more efficient and sustainable.

A study by Bebbington (2002) on the significance of second and higher order organisations of indigenous peoples in Ecuador is of interest in general terms. The study goes beyond a project focus to examine and compare the relative gains from organisation of the poor in social and economic development. The paper also shows that economic development often requires a different approach than social development and social services. So, how should an organisation for the poor look? Who could be allies of the poor? Can traditional trade unions play such a role?

Schiphorst (2002) examined the roles and limitations of traditional trade unions. Poverty reduction through employment and income generation that seeks to move beyond mere subsistence generates its own class of problems and issues. The innovation network in Brazil based on universities and trade unions and the specific case of Amafrutas in the Amazon are interesting illustrations of increasing complexity when poverty reduction through economic promotion moves beyond the immediately local context. These cases show the tension between bottom-up empowerment and top-down enablement (Pegler 2002).

REFERENCES

Bebbington, A. (2002) 'Organisation, inclusion and citizenship: Policy and (some) economic gains of indigenous people's organisations in Ecuador." The Hague: Institute of Social Studies (In Mimeo).

Berner, E. (1999) 'Poverty alleviation and the eviction of the poorest: Towards urban land reform in the Philippines', *International Journal of Urban and Regional Research* 24 (3): 554–566.

Burgess, R., M. Carmona and Th. Kolstee (1994) 'Position paper on urban strategies and urban design prepared for the international seminar the hidden assignment', Publikatieburo Bouwkunde. Delft, The Netherlands: Technological University of Delft.

Burgess, R., M. Carmona and Th. Kolstee (Eds) (1997) *The Challenge of Sustainable Cities: Neoliberalism and Urban Strategies in Developing Countries.* London: Zed Books.

Chavez Miños, D. (2002) 'Porto Alegre, Brazil: A new, sustainable and replicable model of participatory and democratic governance?' The Hague: Institute of Social Studies (In Mimeo).

CityNet (1997) *Partnership for Local Action: A Sourcebook on Participatory Approaches to Shelter and Human Settlements Improvement for Local Government Officials.* Bangkok: UN-Habitat.

Douglass, M., Orathai Ard-am and Ik Ki Kim (2002) 'Urban poverty and the environment: Social capital and state-community synergy in Seoul and Bangkok', in P. Evans (ed.), *Liveable Cities? Urban Struggles for Livelihood and Sustainability*, pp. 31–66. Berkeley: University of California Press.

Evans, P. (2002) 'Political strategies for more liveable cities: Lessons from six cases of development and political transition', in P. Evans (ed.) *Liveable Cities? Urban Struggles for Livelihood and Sustainability*, pp. 222–247. Berkeley: University of California Press.

Helmsing, A.H.J. (2003) 'Local economic development: New generations of actors, policies and instruments', *Public Administration and Development*, 23 (1): 67–76.

Helmsing, A.H.J. (2002a) 'Decentralisation, enablement and local governance in low income countries', *Environment and Planning, C: Government and Policy* 20: 317–340.

Helmsing, A.H.J. (2002b) 'Partnerships, meso-institutions and learning: New local and regional economic development initiatives in Latin America', in I. Baud and J. Post (eds) *Re-aligning Actors in an Urbanizing World: Governance and Institutions from a Development Perspective*, pp. 79–101. Aldershot and Burlington: Ashgate.

Helmsing, A.H.J. (1999) 'Decentralisation and emerging patterns of local governance: A comparative analysis of Uganda, Zimbabwe and Zambia', Regional Development Studies Working Paper No. 4. Addis Ababa: Addis Ababa University.

Helmsing, A.H.J. (1997) 'Government enablement and planning', in: A.H.J. Helmsing and J. Guimaraes (eds) *Locality, State and Development: Essays in Honour of Jos G.M. Hilhorst*, pp. 107–130. The Hague: Institute of Social Studies.

Helmsing, A.H.J. and Wekwete, K.H. (Eds) (1990) *Subnational Planning in Southern & Eastern Africa: Approaches, Finance and Education.* Aldershot: Avebury, Gower Publishing Company.

Korten, D.C. and R. Klauss (1984) *People-Centered Development: Contributions Toward Theory and Planning Frameworks.* Connecticut: Kumarian Press.

Litvack, J, J. Ahmad and R. Bird (1998) 'Rethinking decentralisation in developing countries', Washington, D.C.: The World Bank (sector studies series PREM).

Ludeking, G. and Chr. Williams (1999) 'Poverty, participation and government enablement: A summary of the findings, lessons learned and recommendations of the UNCHS (Habitat) and the Institute of Social Studies Evaluation research (1996–98)'. Nairobi: UNCHS.

Oosterhout, F. (2002) 'An enabling citizenship approach to urban poverty alleviation: The case of PAP in Santa Cruz'. The Hague: Institute of Social Studies (In Mimeo).

Patel, S. (1998) 'Establishing a legal basis for community action: Lessons from the Society for the Promotion of Area Resource Centres (SPARC) and the Asian Coalition for Housing Rights (ACHR)'. Paper presented at the Community Development Programme meeting held at Cape Town, South Africa. March.

Pegler, L. (2002) 'Enabling citizenship in income generation: Two case studies from Brazil', The Hague: Institute of Social Studies (In Mimeo).

Schiphorst, F. (2002) 'Producers in the informal sector and their relationship with traditional trade unions'. The Hague: Institute of Social Studies (In Mimeo).

Smith, B. (2000) 'The concept of an "enabling" local authority', *Environment and Planning, C: Government and Policy*, 18: 79–94.

Stoker, G. (1999) 'The unintended cost and benefits of new management reform for British local government', in G. Stoker (ed.), *The New Management of British Local Governance*, pp. 1–21. London: McMillan.

UNCHS (1994) 'The community construction contrast system in Sri Lanka'. Nairobi: UNCHS-Habitat.

UNCHS (1990) *The Global Strategy for Shelter to the Year 2000.* Nairobi: UNCHS-Habitat.

UNCHS (1991) *Evaluating Experiences with Initiating Enabling Shelter Strategies.* Nairobi: UNCHS.

UNDP (1990) *Human Development Report 1990.* Oxford: Oxford University Press.

UNDP (1991) *Cities, People and Poverty: Urban Development Cooperation for the 1990.* Oxford: Oxford University Press.

UNDP (1993) *Human Development Report 1993.* Oxford: Oxford University Press.

UNDP (1997) *Human Development Report 1997.* Oxford: Oxford University Press.

Uphoff, N. (1986) *Local Institutional Development: An Analytical Sourcebook with Cases.* Connecticut: Kumarian Press.

VPEPP (Viceministerio de Planificación Estrategica y Participación Popular) (2000) *Participación Popular en Cifras, Volumen III.* La Paz: VPEPP.

Van Wijk, C., T. Mathew and K. Prathap Reddy (2002) 'Water supply, sanitation and rural poverty reduction in Gujarat and Kerala states'. The Hague: Institute of Social Studies (In Mimeo).

Wils, F. (2002) 'NGDOs and participatory planning monitoring and evaluation methods for enabling citizenship approaches: The case of Limón, Honduras'. The Hague: Institute of Social Studies (In Mimeo).

Wils, F and A.H.J. Helmsing (2001) 'Enabling communities and markets: Meanings, relationships and options in settlement improvement', ISS Working Paper No. 335. The Hague: Institute of Social Studies.

Wils, F. and A.H.J. Helmsing (1997) 'Shadow on the ground: The practical effectiveness of the community development programme'. The Hague: Institute of Social Studies Advisory Service.

Wit, J. de (2002) 'Urban poverty alleviation in Bangalore: Institutional and community level dilemmas', *Economic and Political Weekly*, September, 3,935–3,942.

Wit, J. de (2001) 'Partnerships, participation and patronage: Realities of urban poverty alleviation in Bangalore slums', ISS Working Paper No. 350. The Hague: Institute of Social Studies.

World Bank (1994) *World Development Report.* New York: Oxford University Press.

World Bank (1998) *World Development Report.* New York: Oxford University Press.

CHAPTER 11

CIVIC ENGAGEMENT, SOCIAL ACCOUNTABILITY AND THE GOVERNANCE CRISIS

WILLIAM REUBEN[1]

1. DEFINING CIVIC ENGAGEMENT

Civic engagement is the participation of private actors in the public sphere with the aim of influencing decision-making or pursuing common goals. Participation is conducted through direct and indirect interactions of civil society organisations and citizens at large with government, multilateral institutions and business establishments. Engagement of citizens and citizens' organisations in public policy debate or in delivering public services and citizens contributing to the management of public goods is a critical factor in making development policy and actions responsive to the needs and aspirations of the people affected.

Throughout the developing world, as the twenty-first century begins, civic engagement has become a realistic and even necessary option in the development process and a key to achieving sustainable development. This first decade of the century will witness unprecedented direct citizen participation in the development process. Citizens' active involvement in designing, implementing and monitoring economic and political reforms has never been as possible as it is today. The main factors behind this phenomenon are recognition of poverty as the central development challenge, democratisation and political pluralism, internal and external pressures for good governance, transparency in policymaking, decentralisation and technological advances (e.g. satellite communications, the Internet, and the knowledge economy).

2. ASPECTS OF CIVIC ENGAGEMENT

2.1 Civic Engagement and Development

Development theory increasingly accepts that civil society, like the market and the state, plays a crucial role in development. Recent development literature coincides in attributing the engagement of civil society in development to the promotion of state

Max Spoor (ed.), Globalisation, Poverty and Conflict, 199–216.
© 2004 *Kluwer Academic Publishers. Printed in the Netherlands.*

and market transparency and social equity.[2] Civic engagement contributes valuable insight to the policy arena. 'In many countries civil society has been responsible for securing from the political system major social changes (ranging from the abolition of slavery to the creation of social safety nets)' (Whaites 2002: 13). Market failures and segmentation and state ineffectiveness can be partially attributed to a lack of transparency and information flows. When civil society engages in development processes, particularly in mobilising the interests of poor and marginalised people, it plays the role of "democratising development dialogue" by giving access to information and voice to those who are not in decision-making positions (Kanbur and Squire 2001). Empirical evidence shows a strong correlation between famines and lack of political freedom, transparency and people's inability to defend their basic entitlements (Dreze and Hussain 1995). The argument that democratic information flows, strong systems of public accountability and civil society scrutiny play crucial roles in sustainable development and in building resilient economic institutions gained greater support after the Russian and East Asian financial crises.

The costs of civic engagement can be significant, however, for both governments and civil society. Engaging primary and secondary stakeholders can be time- and resource-intensive and may limit government decisiveness. Consultations and other mechanisms of participation can introduce new tensions, such as competition among stakeholders with different interests, or a rise in expectations which cannot be met by the government or by the specific project or task at hand. Nonetheless, most development agencies note that the benefits of participation outweigh the costs.[3]

2.2 Civil Society

The existence, strength and shape of civil society are, in the long run, determinant factors of civic engagement. Civic engagement cannot take place without the presence of active citizenship, and active citizenship cannot exist in a sustained and effective way in the absence of civil society, that is to say, in the absence of private organisations mobilised on behalf of public interests and values. Civil society is more than just the aggregate of civic organisations. Like the state and the market it is a specific sphere of social interaction, which has its own rationale and dynamics. Civil society materialised with the emergence of citizenship and citizens' organisations with relative autonomy from the state and with a different rationale from business establishments. Like state institutions and unlike business establishments, they mobilise on behalf of collective values and public interests. Like business entities and unlike the state, they have a private nature. It is this specificity that confers to civil society a particular social dynamism and a distinct role in modern social change.

Generally speaking, civil society is the result of the emergence of organisations pursuing collective goals outside of the state. In the history of the West, the rise of civil society has corresponded with the emergence of independent cities ("burgs") and therefore of autonomous citizens' organisations. Artisans' guilds, professional associations and labour unions were probably the first expressions of citizens' organisations in the West. These, from an independent standpoint, engaged in public debate and organised their interests to protect themselves from the absolute power of the state, or to manage common goods and undertake public action.

Several studies examine the origin of civil society in various cultural and historical contexts. For instance, Ibrahim found reminiscences of civil society in traditional Arab civil formations which share public space with political authorities. Leaders, elders and notables of craftsmen's guilds, *sufi* orders, merchant organisations and sects or *millets* performed several functions in the overall governance of pre-modern Arab society, like advising or influencing rulers, mediating inter-communal conflicts and managing intra-communal affairs (Ibrahim 1998).

In a polemic paper, Fatton (1999) suggested that the dominant paradigm of voluntary association that singles out western civil society organisations does not necessarily work in contemporary Africa. 'If civil society was to be a useful heuristic tool in deciphering contemporary African history, it has to be conceptualised as the realm of collective solidarities generated by processes of class formation, ethnic "inventions" and religious "revelations".' Strong traditional links characterised by assigned status and belonging underpin collective action in African societies. According to Fowler, this phenomenon contributed to shape the unique character of African civil society, typified by the presence of "modern" organisations, like NGOs, alongside traditional groups, like community organisations in which assigned kinship and religious bounds alternate with personal interests and voluntary decisions to take collective action.

On similar lines, Holanda (1989) suggested that in Brazil, as in many other Latin American countries, the state that has emerged as a result of the hacienda system 'represents a continuity of the patriarchal family, and not a departure, a transgression from it'. Devoid of the separation between family and state, public and private, the men who held public office did not establish different parameters for their actions in the ambits of the public and private spheres. On the same path, Chaves, Silva and Dagnino (1994) observed that the origins of Brazilian civil society were rooted in a rural culture in which citizens' rights were perceived not as entitlements but as concessions of the *hacendados*. This culture has favoured patronage in power relationships between the state and civil society, and within civil society itself, where in many cases a clientelistic culture of donations, command and subservience still prevails.

Civil society is a diverse and changing phenomenon and as such its shape and composition, its importance, strength and engagement with public institutions and its relationship with the market change from region to region, from country to country, and from one historical period to the next. Thus, when considering civil society several facts are important to note:

- Civil society is neither monolithic nor static, but rather diverse and dynamic. Existing levels of social capital, political regimes, legal frameworks, cultural environments and external influences generate different forms and expressions of civil society. These further shape civil society's ability to influence public decisions affecting development.
- Not all forces present in civil society play a positive role in development. There are organisations that oppose social change and technological innovation; others

favour social or cultural segregation, and still others are linked to drug trafficking or economic mafias.

- As a whole, civil society is a determining factor in the development process, comparable to the market or the state. The strong engagement of civil society organisations in development policy formulation and implementation contributes to mobilising social forces for poverty reduction and to creating the required consensus for the accomplishment of sustainable development targets. It brings transparency to public action and creates effective environments for decentralised design and implementation of development policies and programmes (OED 2000).

2.3 Civil Society Diversity and Asymmetry

Although it is true that the state, the market and civil society are social spheres with their own rationale and relative autonomy, they can only be fully understood through their interrelations. That is to say, the shape of civil society and its role in social transformation can be appreciated only in conjunction with the roles played by the state and the market in the evolution of a particular society, and vice versa. Civic engagement can thus be understood only as the result of the specific political and economic background of a given country.

Given the dominant role of NGOs as development actors, international cooperation has given them special consideration in programmes, policy dialogue and operational work. Yet, NGOs comprise just one sub-set of civic engagement in the form of civil society organisations. As professional or voluntary organisations that advocate or provide services in favour of their members or others, NGOs are driven by not-for-profit motives. Their field of work spans economic and social development, welfare, emergency relief, environmental conservation, human rights protection and public advocacy. Their incomes usually come from international cooperation, governmental or private sources and/or from revenues generated by their services.

Beyond NGOs, farmers' organisations, trade unions, community groups, professional guilds, political parties, informal networks, faith-based associations, student and youth organisations, academic bodies and business chambers are other, different expressions of modern civil society. These organisations sometimes represent the interests and values of their own members; on other occasions they may express the interests of others, based on ethical, cultural, political, scientific, religious or philanthropic considerations.[4]

Relationships within civil society are heterogeneous and asymmetrical among its members. This is in part reflected in civil society organisations' different levels of access to power and economic resources (Fowler 2000). Lower levels of institutional development and influence are likely to characterise organisations of the poor. Local and poor people's organisations often suffer from "civic exclusion". Many hurdles prevent these organisations from engaging in public debate and public policy implementation. Lack of access to information, resources and political leverage may prevent organisations of the poor from achieving direct representation and engagement in development initiatives. There is a compelling need to pay special attention to these obstacles in order to empower the poor and create an enabling environment for their participation.

In recent years, civic engagement at the global level has been effected through transnational networks and alliances of civil society organisations.[5] In consequence, the boundaries between local, national and international civic engagement are becoming blurred and imprecise. This trend has prompted some analysts to speak of emerging transnational forms of civic engagement.

2.4 Representation, Voice and Accountability

Representation of civil society organisations and particularly of NGOs has gained ground as an issue in civic engagement. Because of the increased presence of democratically elected governments in transition and developing countries in the last decades, elected officials now systematically question the legitimacy of local and, above all, international NGOs which claim to represent the interests of the people and the poor of their countries (Edwards and Hulme 1992, Edwards 2000). They argue that the lack of clear representation leads to unaccountable organisations. The question is therefore, to whom are NGOs accountable if it is unclear who they represent?

In recent decades, civil society has undergone profound transformations everywhere in response to dramatic changes in social relationships and complex economic processes. One of the most profound transformations affecting the very structure of civil society has to do with the representation of social interests. The transition from societies organised around clearly defined productive sectors to societies organised around increasingly diffuse economic processes has changed the way social interests are represented. The representation of social interests has moved away from class-based organisational models to less clearly defined, micro-social models. Societies have moved from a civil society divided into large social interest blocs, based on productive status, to multilayered, highly diversified civil societies.

Yet this new civil society structure, rich in diversity and means of expressing the most varied interests, has not necessarily resulted in improved accountability. Pressure groups have become more diffused, and forging lasting alliances is more complicated. Contemporary civil society's new and dispersed structure presents complex challenges for "operationalising" the representation and accountability of civil society organisations when they engage in development advocacy.

Traditional sector or class-based civil society organisations are membership organisations such as labour unions, craftsmen's guilds, peasant unions and employers' associations. These have clear representational structures based on internal election processes. Leadership is formal or is formalised by these mechanisms of representation. Formal internal elections and clear lines of accountability legitimise those who sit at the negotiating table, serve as spokespersons for trade association interests or make decisions on behalf of their constituencies. These processes are comparable to mechanisms for political representation and accountability in democratic states. In membership organisations, accountability is derived from representation, as authorities ought to be accountable to those who elect them. Hence, in this type of organisation, elections are the most important mechanism for holding authorities accountable.

Representation and accountability in the non-membership organisations that are involved in development advocacy are much more complicated. The worst mistake one could make would be to associate these structures with traditional models of social interest representation. Like most NGOs, these organisations are not membership based, they cannot profess to directly represent a specific constituency. The leaders of these organisations cannot claim to exercise a representation that was not conferred to them by those for whom they speak. Non-membership civil society organisations do not represent groups or specific segments of society. Rather they represent values, beliefs and lifestyles. When they engage in development advocacy they should be accountable to the constituencies from which they derive power and legitimacy: usually the poor and excluded groups in developing and transition countries. Like international or national development agencies, development NGOs face the challenge of building mechanisms of horizontal and downward accountability – accountability to the poor and excluded – and not just to their donors and owners. In this case, representation is derived from accountability. Accountability confers credibility and representation in these organisations (Reuben 2002).

2.5 Civic Engagement and the Market

Civic engagement vis-à-vis the market plays an important role in development. Civic engagement brings to private behaviour a socially ethical dimension that is not necessarily present in market relations, where profit is the dominant motivation. There are two fundamental ways by which civic engagement can assist markets in working for the poor. The first is by fighting market segmentation and exclusion. Micro-finance is a good example. Civil society organisations have been important in developing credit schemes to bring financial services to poor people who otherwise would be excluded from financial markets. Furthermore, in many countries civil society organisations have helped to bridge the gap between commercial banks and micro-businesses. However, as in micro-finance, there are other good examples where intervention by civil society organisations has been instrumental in overcoming market failure and segmentation. Peasant organisations have been at the core of making land markets accessible to the poor; farmers' unions, cooperatives and consumers' associations have played a critical role in breaking monopolistic control over food markets.

The second way in which civic engagement has helped markets to work for the poor is by exercising pressure on private corporations to incorporate a *social responsibility* dimension in their business. National and international NGOs, and consumers' associations in developed, transition and developing countries have contributed to the spread of new "business ethics" that take into consideration the social and environmental impacts of their industry. Campaigns against child and bonded labour and fair trade initiatives are examples of the positive impact of civic engagement over market dynamics.

2.6 Civic Engagement and the State

There is a fundamental misunderstanding regarding the role of civic engagement in development. Civil society and the state do not necessarily play antagonistic roles; neither can one replace the other. The existence of a strong civil society does not preclude the existence of a robust state. There is clear evidence that an effective and sound public sector depends very much on the existence of a dynamic civil society and strong citizen involvement in the public realm. Civic engagement in public policy formulation and implementation brings transparency and effectiveness to public institutions. By the same token, a robust public sector with the capacity to enforce the rule of law, set clear rules of engagement for civil society and promote sound public policy provides an appropriate environment for civic engagement. In the case of failed or weak states experiencing conditions of widespread conflict and social ungovernability, civil society organisations may offer an institutional basis for public service delivery and, in many cases, contribute to conflict resolution and reconciliation efforts. Links between national NGOs and their external partners or international cooperation agencies constitute vital gateways for emergency aid, external peace-building and reconstructive endeavours.

3. THE GOVERNANCE CRISIS

As a point of departure, this section addresses the assumption that any governance crisis expresses a fundamental contradiction between the citizenship and the state. The fact that the power exercised by the state, through law, coercion and the administration of public resources, is the result of the delegation, or take-over, of citizens' sovereignty permeates the entire range of tensions that characterises this relationship.

Although this contradiction has existed since the emergence of the state, it has taken on different shapes and varied in degree according to the characteristics of political regimes and the level of disjuncture between ruling institutions and citizens. The smaller the extent to which citizens feel represented and serviced by public institutions, the bigger the governance crisis. The greater the degree of separation between the actions of rulers and citizens' expectations and control over rulers' actions, the greater the governance crisis. However, these statements are not as simple as they appear, because they embed two distinct elements on the same side of the equation: expectations and representation. These do not necessarily march together. The majority of citizens may feel that their values are represented and respected by a given government, but that does not necessarily mean that their expectations are fulfilled. Representation responds to the existence and functioning of democratic institutions and mechanisms of control over those who exercise power. Fulfilment of expectations is linked to the capacity of public institutions to manage and deliver public goods.

The struggle between these two basic elements that underpin governance are well expressed by O'Donnell (2000) as the tension between the rational desire of citizens 'to live under political arrangements that furnish some basic public goods and collective solutions' and 'the dangers that lurk behind the great power, including coercive power, that must be constituted if such goods and solutions are to be furnished'. In short, it is the tension between decisiveness in fulfilling expectations and accountability in holding public power under the control of those represented by the state. The point of encroachment of these two contradictory elements defines the type of regime and therefore the likely political solutions to address the governance dilemma. Tyrannies will tilt the balance in one direction, deadlock regimes in the other. Still other regimes will use a more balanced mixture of the two elements to work out the governance equation (Figure 1.)

However, there is no permanent equilibrium to this equation. The changing conditions of the relationship between the government and civil society permanently incline the balance in one direction or the other. The relationship is determined by many contextual factors, a fundamental example of which is the political regime and the type of approach a specific government has adopted towards organised interests expressed in civil society. Manor (1999) identified seven possible types of strategies governments may adopt.

- *Laissez-faire:* A rather passive approach which does not promote strong engagement with civil society but which may enable the organisation of citizens in independent civil society organisations.
- *Combination of conflictive and harmonic relationships:* An approach that seeks the creation of divisions within civil society by establishing alliances with some groups while confronting others which government regards with suspicion or obvious antipathy.
- *Repression of all manifestations of citizens' organised interests:* An approach usually adopted by autocratic governments which creates a restrictive environment for the emergence of open expressions of civil interests.
- *Co-optation approach:* Governments seek to co-opt some or all interests, drawing them into relationships of dependency in order to control them, to stimulate a dependent and corporatist foundation for civil society.
- *Patronage approach:* The aim here is to control citizens' diverse interests through distribution of goods, services and funds. This is a variation of the previous approach. Patronage systems vary in the degree to which they serve the

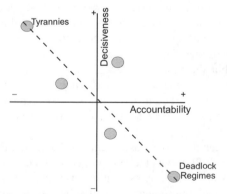

Figure 1. *Political Regimes and Governance Solutions*

interests of a particular political body or personality, their level of decentralisation and extension. This approach usually divides citizens' interests along clientelistic alliances.

- *Proactive engagement:* Governments seek to mobilise all or the majority of organised interests in order to build political consensus. This approach may create a climate of strong citizen engagement in public debate and action, although it can surpass the boundaries of independent and critical mobilisation.
- *Civic mobilisation through nationalistic appeals:* A variation of the previous approach. Governments seek to mobilise citizens' support by appealing to national causes like nationalistic values and identities.

Usually governments employ a mixture of these approaches, creating or limiting in different ways and degrees spaces of engagement with organised citizens and therefore shaping the relationship between public institutions and civil society. However, civil society organisations also have their own agency in defining their relationship with the state and other development actors like political parties, business organisations and international institutions, utilising different forms and approaches to civic engagement. Four main types of civic engagement strategies can be identified on the side of civil society.

- *Confrontation strategy:* Civil society organisations perceive the government as the main obstacle for achieving their objectives. National and international campaigns led by civil society coalitions are used to denounce what they perceive as an illegitimate government that needs to be replaced to create political and social conditions to build a new national consensus. The confrontation strategy usually gives way to social and political unrest and therefore to the narrowing of civic freedoms.
- *Parallel track strategy:* Civil society organisations decide not to engage with the government and establish a set of parallel services that they themselves deliver directly to their clients and constituencies. There is limited or no engagement between civil society organisations and public institutions, and on many occasions attempts at either compromise or confrontation are deliberately fudged. Often this strategy is made possible by large flows of external funding and the presence of permissive and weak governments. Competition for external resources and local influence usually characterises the relationship between governments (or their bureaucratic bodies) and civil society organisations.
- *Selective collaboration:* This strategy combines collaboration on some fronts with critical distance, or even confrontation, on others. Collaboration is usually in the areas where there is a convergence of perspectives and aims with a set of government policies or specific government institutions and officials. Yet areas of confrontation may relate to specific policy issues or public figures. This strategy usually leads to complex tensions within civil society and between civil society organisations and the government, though it also opens dynamic spaces for negotiation and constructive engagement.
- *Full endorsement:* Civil society organisations fully engage and endorse government objectives and policies. This situation characterises the political and

social climate in the aftermath of deep political and social crises and the
emergence of national unity governments. In this case, many of the leading
figures in government held positions in civil society organisations before they
became elected or appointed to a public position. These are usually transition
situations that, in time, shift to one of the previously mentioned strategies after a
"grace period" or to conditions of enduring clientelist deals and patronage.

Given the heterogeneous character of civil society, these strategies coexist and create
all sorts of tensions between different civil society organisations and between civil
society organisations and the government. The complex mix of government
approaches to civil society and civic engagement strategies frames the specific
formula of the governance equation. Nevertheless, existing legal frameworks and
institutional and policy settings, as well as engagement practices and mechanisms,
determine the environment for more or less constructive and effective civic
engagement and governance solutions.

4. SOCIAL ACCOUNTABILITY

4.1 Governance and Social Accountability

To what extent can decisiveness and accountability, the two elements in permanent
tension in conformance with the governance equation, be reconciled to enhance policy
effectiveness and sustainability? In other words, is it possible to increase decisiveness
and therefore achieve efficiency in delivering and managing public goods and services
in order to meet the expectations of populations and at the same time expand
accountability and therefore achieve higher levels of legitimacy and credibility of
public institutions? Any answer to these questions would have to consider the
decision-making processes and mechanisms in place and the systems by which
governments are held accountable.

Delegation of representation prevails in accountability systems relying on purely
horizontal controls (O'Donnell 2000). National comptroller bureaus, ombudsman
offices and procurement units are appointed either by the legislative or executive
branches of the state. Their legitimacy depends on the credibility of public institutions
at large.

In countries characterised by a climate of distrust in public institutions, these
horizontal control mechanisms generally enjoy limited or no public confidence. In
democratic regimes where effective and credible processes of delegation of authority
prevail, the same mechanisms of control may enjoy higher levels of public confidence
and therefore contribute to improved citizen trust in public institutions. In fact, in
effective democracies, these mechanisms incorporate a combination of downstream
and upstream accountability through downwards transparent reporting and upwards
legitimate delegation of authority. However, in both situations, strict horizontal
controls result in increased transaction costs, working against efficient and timely
government policy implementation and delivery of public goods .

Another set of accountability mechanisms can be identified that are vertical in nature and characterised by the exercise of direct participation of civil society organisations and citizens at large. International development institutions have identified a number of such mechanisms from a wide range of innovative experiences in developed and developing countries. In particular, the Social Development Department of the World Bank has identified various types of social accountability mechanisms that can be applied at different stages of the policy sequence (Wagle and Shah 2002). These mechanisms operate along the budget and public expenditure cycle, understanding that budgets and their execution reflect actual policy decisions and their implementation. The assumption is that even though budgets and expenditures are cloaked in "the mystical veils of public accounting", they are dressed down in the rhetoric attached to policy statements and declarations. In that sense, they are more trackable and subject to public scrutiny.

Vertical mechanisms may lead to a virtuous cycle of accountability and development effectiveness. There is evidence that civic engagement in vertical accountability contributes to improving targeting of poverty reduction strategies and enhancing the quality of services delivered. Concurrently, anti-corruption controls set up by social accountability mechanisms reduce resource "leakages" at the local level, which are usually outside the control of traditional public auditing systems.

4.2 Social Accountability in the Budget Cycle

Social accountability in budget processes spans four key interventions that may or may not follow one another in a cyclic path: budget formulation, budget analysis and demystification, expenditure tracking and delivery monitoring.

The first intervention relates to *budget formulation*. When revenue and expenditure estimates are being prepared, processes of civic engagement can influence how expenditure proposals are made – to which sectors and in which amounts. Budget formulation processes usually involve the finance ministry, the cabinet and sector ministries. Opening the processes to the public helps governments present policy and resource options so that participation does not result merely in a wish list of conflicting demands or unrealistic expectations. Examples include the participatory budget formulation process now used in some 70 Brazilian municipalities (Box 1). This case demonstrates how participation of organised communities and sectors in budget formulation can lead to increasing government effectiveness. Civic engagement can also involve preparation of alternative budgets to exert pressure for changes in official budgets. The alternative federal budget drawn up by the Canadian Centre on Policy Alternatives is a case in point.

The second intervention is *budget analysis and demystification*. This begins when the budget is in draft stage and spans its discussion and approval by the legislature, including evaluation of the impact and implications of different policy proposals and allocations. Many times such analyses evaluate previous budgets and track changes in allocations for specific expenditures. The civil society organisations that perform the analytical work generally adopt a specific focus (gender, poverty, children, etc.), which guides their critical evaluation of allocations and the political economy of

---

Box 1. *Porto Alegre: Participatory Budgeting in Brazil*

Brazil is Latin America's largest country, both in terms of land mass and population. When it enacted its democratic constitution in 1988, after 20 years of dictatorship, it also had one of the largest proportions of poor people among comparable middle-income countries.

With the new constitution came reform. As part of that reform Porto Alegre, the largest industrial city in Rio Grande do Sul with 1.3 million inhabitants, began to develop a participatory budgeting model based on collective debate and choice-making.

The Porto Alegre participatory budgeting process spans an entire calendar year and includes assessments of past years' budget expenditures, an election of regional delegates, several presentations of community demands, a participatory reconciliation of needs and demands, the preparation and submission of a budget proposal to the local congress, the preparation of regional and thematic investment plans and finally monitoring and follow-up meetings. The results accredited to the participatory budgeting process are indicative of the impact that civil society and citizen involvement can have on local development:

- the number of households with access to water services rose from 80 per cent to 98 per cent from 1989 to 1996;
- the number of children enrolled in public schools doubled;
- earnings and expenditure were brought into balance;
- in poorer neighbourhoods, 30 kilometres of roads have been paved annually since 1989;
- increased transparency and reduced corruption;
- responsible and balanced municipal fiscal management;
- due to transparency affecting citizens' motivation to pay taxes, revenue increased by almost half.

Over 70 Brazilian cities now employ the Porto Alegre model of participatory budgeting.

---

budget formulation. Here, civic engagement provides a vehicle for demystifying the seemingly technical content of the budget, as well as for raising general awareness and even informing and training parliamentarians to pressure the executive for pro-poor changes. Among the numerous examples of civic engagement to improve budget literacy are the following: initiatives of the Institute for Economic Affairs in Kenya, DISHA (Developing Initiatives for Social and Human Action) in India and IDASA (the Institute for Democracy in South Africa), and now the International Budget Project, which is a network of budget analysis practitioners.

Tracking relates to the period after budget appropriations are debated and approved by parliament, when allocations are disbursed to ministries and the tiers below. Because of institutional weaknesses, such as opportunities for rent-seeking, patronage and leakages, disbursed amounts may not reach the intended beneficiaries in countries where the link between planning and implementation is tenuous. By identifying the elusive bureaucratic channels through which funds flow, civic groups can highlight bottlenecks in the flow of resources and other systemic deficiencies (Wagle and Shah 2002). This stage may involve triangulation of information received from independent enquiries, funds released by finance ministries, accounts submitted by line agencies to the exchequer and the establishment of call centres to register and channel complaints to an ombudsman office. Examples of tracking systems are Uganda's World Bank-funded public expenditure tracking system, the G-Watch project in the Philippines, the *Majdoor Kisaan Shakti Sangathan* of Rajasthan, India, and the more recent social monitoring initiative in Argentina (Box 2). Each of these initiatives

---

Box 2. *Argentina: Social Monitoring*

In Argentina, the World Bank, the Inter-American Development Bank (IDB) and the Government of Argentina, with technical support from the United Nations Development Programme (UNDP), are coming together to promote a culture of greater public transparency and accountability as one of several responses to the serious economic crisis.

The initiative began in March 2002, when teams from the World Bank, IDB, the Argentinean government and UNDP collaborated with civil society organisations and NGOs to design a participatory monitoring mechanism that would capitalise on the breadth of Argentinean civil society by creating a space for them to hold their government accountable for social expenditures. The design process continued through April 2002. A workshop promoted by the Catholic University in Buenos Aires brought together 200 representatives of NGOs and civil society organisations to discuss experiences and approaches to participatory governance. The result was a World Bank and IDB-funded initiative which focuses on the involvement of civil society in the monitoring of expenditures, access, quality of service and transparency in World Bank and IDB-funded social emergency programs. A coalition of civil society organisations and NGOs inform beneficiaries of their rights and entitlements regarding the social emergency services, as well as surveying beneficiary populations, gathering and disseminating information and providing regular feedback to government agencies and the public via mass media awareness campaigns. The system will be complemented by call centres run by a national consumers' association that will screen and register complaints from the public and transfer information to the ombudsman office and to the respective implementing agency for follow up.

A number of outcomes of the social monitoring programme are expected:
- improved quality and outreach of social emergency services;
- increased trust in and credibility of public institutions;
- enhanced capacity of civil society and beneficiaries to promote and ensure oversight of financial resources;
- reduction of clientelism and political patronage of social emergency programs.

---

demonstrates how tracking can be used to upset the status quo of harmful bureaucratic and political arrangements by facilitating a more transparent, fair and effective use of public funds for the poor.

The fourth type of intervention is *performance evaluation*, where the performance of selected publicly funded agencies (sometimes an entire government) is assessed. One of the more analytically robust and powerful instruments for evaluating public performance draws on the private-sector practice of soliciting feedback from citizens and compiling "report cards". The methodology involves administration of a survey to determine quality of, access to and satisfaction with such services. There are two practices: one in which an independent group conducts the survey for a government or donor agency; another where citizen groups conduct the surveys. The Public Affairs Centre in Bangalore, India (Box 3), and the World Bank-funded Filipino Report Card 2000, among other examples, offer useful precedents.

The report cards usually identify key constraints the poor face in accessing public services, views about the quality and adequacy of services and treatment received in interactions with public service providers. Civic input at this stage can influence both the internal reform measures of the evaluated government agencies and external evaluators, such as the report of an auditor general to the parliament. An interesting feature of report cards as a mechanism to monitor and evaluate performance of public services is that they incite healthy competition between delivery agencies within

---

Box 3. *Bangalore, Karnataka, India: Citizen Report Cards*

Initiated in 1993 by a small advisory group of local leaders, the innovative citizen report card monitoring scheme involves collecting direct feedback from users on the quality, efficiency and adequacy of government-provided services. Since its first implementation in Bangalore in 1993, the report card scheme has been institutionalised as one of the core functions of the Public Affairs Centre, a non-profit society established in 1994 with the goal of improving governance in India by strengthening civil society institutions in their interactions with the state. The report card exercise was repeated in 1999 and has been replicated in at least five other Indian cities, as well as at the state level in Karnataka.

Lessons learned from the report card process speak to the impact that civil society can have on the monitoring of service delivery. In terms of civil society growth, the number of groups interested in social monitoring has grown from 30 in 1993 to over 200 to date. Due to this growth and to the wide dissemination of report card findings (most of which were negative) through mass media, public agencies have been forced to, at least, listen and react to citizen concerns. In addition, the report card scheme provides a tangible and quantified assessment of the extent to which the citizenry is (dis)satisfied with the government's performance as a provider of services, allowing for inter-agency comparisons and triggering internal reforms.

---

public monopolies, especially when a system of incentives and disincentives (budget allocations and bonuses subject to results and public acknowledgments) is linked to the mechanism.

Evidence so far indicates that with the use of appropriate mechanisms, social accountability may become a promising alternative for enhancing controls over public institutions and the use of public resources. Such accountability can render institutions more effective and responsive to the needs and aspirations of the poor. Nonetheless, enforcement of vertical control mechanisms is generally weak. Usually such mechanisms lack the formal power required to make their moral authority effective. It is thus worth exploring ways by which vertical and horizontal accountability might be combined in circumstances where horizontal control mechanisms lack credibility and public trust. The social monitoring programme in Argentina is an example of such a combination, in this case of a civil society coalition that runs the monitoring system and an ombudsman. The capacity of report cards to influence an auditor general's report to the parliament is another example of this wholesome connection, though results from these experiences remain to be seen. Another challenging aspect for systems of social accountability is creation of the enabling environments necessary for sustained and effective civic engagement. Social accountability does not take place in a vacuum. It requires a set of regulatory, institutional and political conditions.

5. STRENGTHENING CIVIC ENGAGEMENT AND SOCIAL ACCOUNTABILITY

There is an ongoing debate about the conditions that enable effective and sustained civic engagement and social accountability and about which organisations can contribute to creating those conditions. However, almost all parties agree that context and civil society capacity to exercise social agency are the two most important factors. In other words, effective civic engagement requires an enabling environment and institutional capacities in civil society organisations to influence public debate, hold public institutions accountable and provide services.

An enabling environment for civic engagement and social accountability has three basic dimensions: (i) the regulatory framework, (ii) the political and institutional setting and (iii) the civic culture of a given society. The civic culture affects the way that civil society organisations and public institutions engage in policy dialogue, advance systems of social and public accountability and cooperate in the provision of public services. It comprises existing participatory practices as well as experiences and values among citizens and their organisations, including the footprint of political history, customary religious values about the role of individuals or groups of individuals (i.e. women) in public life, prevailing role models and other culturally defined factors. The regulatory framework includes the legal provisions and practices regulating citizens' participation. The political and institutional settings involve ideological statements, political will and institutional arrangements vis-à-vis civic engagement. All these dimensions determine the freedom of citizens to associate; their ability to mobilise financial resources to fulfil the objectives of their organisations; their access to voice, media and the Internet; their access to information; and the existence of spaces and rules of engagement for negotiation and public debate. Together, these factors synthesise the complexity of conditions that affect the ability of civil society organisations and governments to engage in public debate and in systems of social accountability (Table 1).

Multilateral and bilateral agencies can play a positive role by providing their client governments with the financial resources and technical assistance to transform their

Table 1. Conditions Affecting Civic Engagement and Social Accountability

Factors	Regulatory Environment	Political/Institutional Setting	Civic Culture
Association	Freedom of association	Recognition	Social capital
Resources	Tax systems, fund-raising and procurement regulations	Government grants, private funds, contracting, other transfers	Social philanthropy (the culture of giving)
Voice	Freedom of expression, media and ICT-related laws	Political use of public media	Communication practices (use of media by different social groups)
Information	Freedom of information, rights to access public information, public disclosure regulations	Information disclosure policies and practices, ability to disseminate public information	Information networks, word of mouth
Negotiation	Legally established dialogue spaces (e.g. referendums, lobby regulations, public forums)	Political will, customary policy dialogues and social accountability mechanisms, capacities of parliaments and local and national governments to engage	Tolerance, listening practices

regulatory frameworks and institutional set-ups into more friendly and helpful environments for social accountability. Though mass media and cultural activities can contribute to creating a "civic culture", the cultural dimension requires long-term social transformations which cannot be easily modelled by public interventions.

Building the capacity of civil society organisations can strengthen civic engagement and support social accountability. Indeed, the presence of resilient, responsive and credible organisations is essential for civic engagement. Such organisations should be financially sustainable (which does not necessarily mean financially self-sufficient), efficiently managed and adequately skilled. To build such capable organisations requires the development of flexibility and the ability to learn from and communicate with international constituencies and partners through the establishment of regional and global networks. But, as stated in previous sections, widespread credibility is perhaps the key for civil society organisations to play a role in social accountability. The old conundrum of *qui custodet custodes* deserves particular attention when civil society engages in social accountability initiatives. Voluntary reporting of accounts, deals and activities should become the norm among the civil society organisations participating in public debate and social accountability processes. Transparent management of representation and acceptance that the poor and excluded groups do not necessarily enjoy the same access to resources, information, voice and negotiation as other established organisations, are the bottom line of any substantive step towards creating sustainable and effective systems of social accountability.

Major efforts are needed in supporting enabling environments for civic engagement. At the same time, capacity building within civil society should focus on helping the poor and other excluded groups to develop their presence in processes of policymaking and service delivery. As indicated in previous sections, civic inclusion should be at the core of any strategy geared towards increasing social accountability and the effectiveness of civic engagement.

NOTES

1. The author of this paper is the Civil Society Coordinator of the Social Development Department at the World Bank. The findings, interpretations, judgments and conclusions expressed in this paper are those of the author. They should not be attributed to the World Bank. Michelle Levy-Benitez contributed with the case studies.

2. See for example North (1990) Sen (1999: Ch. 6), Hoff and Stiglitz (2001), Appadurai (2001), Grindle (2001), and Stiglitz (1997).

3. As far as the World Bank is concerned, a significant number of studies endorse these conclusions. For example, in World Bank (1999) it was concluded that a majority of projects studied showed potential for success because their preparation and early implementation were highly participatory. In World Bank (1998a) the authors found in one study that government agencies which actively sought to encourage involvement of beneficiaries achieved a 62 per cent success rate in their projects, while those that did not achieved just a 10 per cent success rate. More recently see also OED (2001).

4. Given the wide diversity of civil society organisations, it is not easy and perhaps not useful to develop a general typology of these institutions. As all classification efforts, they make sense subject to the objective they pursue. Many typologies follow different criteria to classify civil society organisations. Some emphasise the nature of their membership or their constituency (i.e. membership/

non-membership; intermediary/community based; social sectors/class). Others focus on their objectives (i.e. advocacy/service delivery). Still others emphasise their level of intervention (local/national/international/global). See for example World Bank (1998b) and World Bank (2000).

5. For example, the International Campaign to Ban Landmines, transnational networks such as the Earth Summit and Beijing, the World Commission on Dams, the Jubilee 2000 Coalition, the World Social Forum and protests against the World Trade Organization, the World Bank and the International Monetary Fund. See Edwards (1999: 177–185), Keck and Sikkink (2000), Florini (2000) and O'Brien et al. (2000).

REFERENCES

Appadurai, Adjun (2001) *Deep Democracy*. Chicago: University of Chicago (In Mimeo).
Dreze, Sen and Athar Hussain (1995) *The Political Economy of Hunger: Selected Essays*. London: Clarendon Press.
Edwards, M. and Hulme, D. (Eds) (1992) *Making a Difference: NGOs and Development in a Changing World*. London: Earthscan.
Edwards, Michael (1999) *Future Positive: International Co-operation in the 21st Century*. London: Earthscan.
Edwards, Michael (2000) *Democratizing Global Governance: Rights and Responsibilities of NGOs*. London: The Foreign Policy Centre.
Fatton, Robert (1999) 'Civil society revisited: Africa in the new millennium', *West Africa Review* 1 (1) July <http://www.icaap.org/iuicode?101.1.1.5.
Florini Ann (Ed.) (2000) *The Third Force: The Rise of Transnational Civil Society*. Washington: Carnegie Endowment for International Peace.
Fowler, Alan (2000) 'Civil society, NGOs and social development: Changing the rules of the game'. UNRISD Occasional Paper No. 1. Geneva: United Nations Research Institute for Social Development.
Grindle, Merilee (2001) 'The quest of the political: The political economy of development policymaking', in Gerard Meier and Joseph Stiglitz (Eds) *Frontiers of Development Economics: The Future in Perspective*, pp. 345–380. New York: Oxford University Press for the World Bank.
Habermas J. (2001) *The Postnational Costellation: Political Essays*. Translated, edited and with an introduction by Max Pensky. Cambridge: MIT Press.
Hoff, Karla and Joseph Stiglitz (2001) 'Modern economic theory and development', in Gerard Meier and Joseph Stiglitz (Eds) *Frontiers of Development Economics: The Future in Perspective*, pp. 389–459. New York: Oxford University Press for the World Bank.
Holanda, Sérgio Buarque (1989) *Raízes do Brasil* (21st Edition). Rio de Janeiro: José Olympio Publishing House.
Ibrahim, Saad Eddin (1998) 'Populism, Islam, and civil society in the Arab world', in John Burbidge (Ed.) *Beyond Prince and Merchant: Citizen Participation and the Rise of Civil Society*. New York: PACT Publications.
Kanbur, Ravi and Squire, Lyn (2001) 'The evolution of thinking about poverty: Exploring the interactions', in Gerard Meier and Joseph Stiglitz (Eds) *Frontiers of Development Economics: The Future in Perspective*, pp. 183–226. New York: Oxford University Press for the World Bank.
Keck, Margareth and Katheryn Sikkink (2000) *Activist Beyond Borders: Advocacy Networks in International Politics*. New York: Cornell University Press.
Manor, James (1999) 'Civil society and governance: A concept paper', Sussex: IDS. 26 August 1999. <http://www.ids.ac.uk/ids/civsoc/public.doc>
Meier, Gerard and Joseph Stiglitz (Eds) (2001) *Frontiers of Development Economics: The Future in Perspective*. New York: Oxford University Press for the World Bank.
North, Douglass (1990) *Institutions, Institutional Change and Economic Performance*. Cambridge: Cambridge University Press.
O'Brien, Robert, Anne Marie Goetz, Jan Aart Scholte and Marc Williams (Eds) (2000) *Contesting Global Governance: Multilateral Economic Institutions and Global Social Movements*. Cambridge Studies in International Studies. Cambridge: Cambridge University Press.

O'Donnell, Guillermo (2000) 'Further thoughts on horizontal accountability'. Paper presented at the
    workshop Political Institutions, Accountability, and Democratic Governance in Latin America.
    University of Notre Dame, Kellogg Institute, 8–9 May.
OED (2000) *Annual Review of Development Effectiveness: Towards a Comprehensive Development
    Strategy*. World Bank Operations Evaluation Department. Washington, D.C.: World Bank.
OED (2001) *Participation Process Review*. World Bank Operations Evaluation Department. Washington,
    D.C.: World Bank.
Reuben, William (2002) 'Tell me, who are those guys; to whom are they accountable? Dilemmas for civil
    society accountability and representation'. Paper presented at Cornell University, New York.
    February.
Sales, Teresa (1994) 'Raízes da desigualdade social na cultura política Brasileira', *Revista Brasileira de
    Ciências Sociais*. São Paulo (25): 26–37.
Sen, Amartya (1999) *Development as Freedom*. New York: Anchor Books.
Stiglitz, Joseph (1997) *Whither Socialism?* Cambridge: MIT Press.
Teixeira Chaves, Claudia Almeida Silva and Evelina Dagnino (1994) *Civil Society in Brazil*.
    <http://www.ids.ac.uk/ids/civsoc/docs/brazilwhole.doc>
Wagle, Swarnim and Parmesh Shah (2002) 'Participation in public expenditure systems: An issues paper'.
    Participation and Civic Engagement Group. Social Development Department. The World Bank.
    <http://www.worldbank.org/participation/webfiles/pem.pdf>
Whaites, Alan (2002) *Development Dilemma: NGO Challenges and Ambiguities*. London: The World
    Vision.
World Bank (1998a) *Assessing Aid: What Works, What Doesn't and Why*. Policy Research Report. New
    York: Oxford University Press for the World Bank.
World Bank (1998b) 'The Bank's relations with NGOs: Issues and directions'. Social Development Paper.
    Washington, D.C.: World Bank.
World Bank (1999) 'Non-governmental organizations in World Bank-supported projects', Précis No. 177.
    Washington, D.C.: World Bank Operations Evaluation Department.
World Bank (2000) *Consultations with Civil Society Organizations: General Guidelines for World Bank
    Staff*. NGO and Civil Society Unit. Washington, D.C.: World Bank.

CHAPTER 12

BLURRING THE STATE-PRIVATE DIVIDE: FLEX ORGANISATIONS AND THE DECLINE OF ACCOUNTABILITY

JANINE R. WEDEL[1]

'In a world of multiple, diffused authority, each of us shares Pinocchio's problems; our individual consciences are our only guide.' Susan Strange (1996), *The Retreat of the State*

1. INTRODUCTION

Spurred by two decades of deregulation, public-private partnerships and a worldwide movement towards privatisation, non-state actors now fulfil functions once reserved for government. Moreover, the inclination to blur the "state" and "private"[2] spheres now enjoys global acceptance. An international vernacular of "privatisation", "civil society", "non-governmental organisations" and other catch terms de-emphasising the state is parroted from Washington to Warsaw to Wellington.

The blurring of the traditional boundaries of state and private encompasses long-practices of governance "reorganisation" such as the outsourcing and subcontracting of government work. It includes state-private hybrids such as the well-entrenched quasi-non-governmental organisations (often called "QUANGOs")[3] of the United Kingdom and the "parastatals" of Third World countries. Such reorganisation is being intensified in many places (witness the increased outsourcing in the United States). But the acceleration of this familiar practice fails to convey the processes under way in governments, which are increasingly labyrinths of interconnected state and private structures. Indeed, in some settings, governance is reordering itself in ways that standard vocabularies are ill-equipped to characterise.

For example, the "flex organisations" I observed in Eastern Europe can switch their status – from state to private – according to the situation, strategically manoeuvring to best access state, private and international resources. These organisations play multiple, sometimes conflicting, and ambiguous roles that overlap with both government and business, enabling them to bypass the constraints governing either type of activity.

217

Max Spoor (ed.), Globalisation, Poverty and Conflict, 217–235.
© *2004 Kluwer Academic Publishers. Printed in the Netherlands.*

These organisations violate the rules and regulations of both the state and the private
sphere – the state regulations governing accountability and the private rules governing
competition. Yet they drive and implement policy in crucial areas and wield more
influence in governance than the otherwise relevant state organisations, rendering state
agencies ineffectual.

Although concepts such as "conflict of interest" and "revolving door" describe
boundary-crossing between the state and private organisations and the ramifications of
such fluidity at the individual level, little vocabulary in English effectively captures
such activities at the group or organisational level. This gives rise to questions such as
how can we best explore "flex organising" and its implications. Although
boundary-crossing may be found in a variety of contexts, it is logical that it is most
intense in countries which are divesting state-owned resources, as they "transition"
away from central planning. What can we learn about governance in contexts where
such patterns are most intense? What issues of accountability arise in such cases and
what do they portend for governance generally?

2. RETREAT OF THE STATE?

The blurring of "state" and "private" has been structured by conscious policies
emanating mainly from the United States and the United Kingdom and spurred around
the globe by the politics and programmes of the international financial institutions.
Proponents of the "new governance" popularised in the United States, the United
Kingdom and New Zealand in the early 1990s have advocated 'glory in the blurring of
public and private', a call for the privatisation of public purposes (Osborne and
Gaebler 1992).

It was not always that way. In the mid-1900s, it was 'unthinkable', as Scholte
(2000) noted, 'that private agencies would implement public policy'. But that was
before neoliberal thinking encouraged governments to outsource work to the private
sector, especially the supply of social services. The result is that deregulation,
devolution, privatisation and public-private partnerships have laid claim to
governance.

The neoliberal agenda, shaped by American political decisions of the 1970s and
early 1980s, encouraged, a decade later, what Strange (1998) called the 'retreat of the
state'. With this retreat, non-state actors with economic and political power became
increasingly important forces in international relations.[4] In a similar vein, Matthews
(1997) sketched a fragmentation of power that mitigates against state centrism. These
developments opened up space for actors to play new roles. They also set into motion a
world of opportunities for brokers, even as they helped shape states and relations
among states.

Although Strange's 'retreat of the state' accurately captures a diffusion of
authority, it misses a critical continuity. Globalisation has encouraged some
fundamental shifts in the state, yet, as contended by Scholte, 'states have remained
crucial to governance... Neither the diffusion of public-sector authority nor the growth
of private-sector regulation has displaced bureaucracy as the underlying principle of
modern administration' (Scholte 2000: 132–133). Although bureaucracy has become

'multilayered and more diffuse', he wrote, it still provides the chief continuity in governance (ibid.: 5). But if Scholte is right, where does "flex organising" fit in? How does bureaucracy respond to flex organising and vice-versa?

3. TWO DISCONNECTS

The diffusion of authority highlights two crucial disconnects. The first disconnect points to a dearth of compelling inquiry: the gap between rhetoric and models of governance on the one hand and their actual implementation on the other hand. For example, in the United States the practice of "government by third party" is masked, as Guttman expressed, by 'the fiction that the official workforce is in control, whereas the bipartisan policy has been to grow government through a *private* workforce'.[5] Indeed, the federal government today writes pay checks for millions more contract and grant employees than for civil servants (Light 1999).

A global discourse on privatisation, public-private partnerships and "good governance" glosses over the diversity of context-framed practices and patterns of "actually existing governance".[6] It is tempting to cast a positive light on "modern" public-private partnerships as opposed to "traditional" ones. Indeed, the greater role of NGOs and businesses in the workings of states is often hailed as a millennial model of governance. Yet "modern" organisational forms can have traits in common with "traditional" means of rule. The circumstances that enable NGOs such as Human Rights Watch and the Sierra Club to work effectively across borders also may facilitate the operations of transnational organised criminal and terrorist groups. This resemblance may be jarring because we tend to think more in terms of the groups' stated positions than of the ways in which they organise and are organised by their environments. Yet many groups are arranged flexibly to enable a wide range of activity and to serve the purposes of their members, but not necessarily anyone else. Thus, "mafias" and NGOs may be equally unaccountable to voters. Tellingly, although NGOs in Anglo-Saxon countries tend to conjure up images of public virtue and outreach, in some parts of the world they are seen as selfish usurpers of resources.

Another disconnect lies between mechanisms of accountability and governance in a process of government reorganising itself. Legal frameworks have not always kept abreast of trends in governance. Guttman contended this especially to be the case in the United States, which is seen as a model of "reinvented government". He argued that accountability mechanisms are out of sync with long-standing practices of outsourcing and subcontracting. While "private" employees may deliver services ranging from the management of nuclear weapons and the space programme to the development of government budgets and policies, the laws in place to protect taxpayer-citizens from official abuse often do not apply to non-governmental employees who perform governmental services.[7]

If this is the case in relatively stable societies, it is easy to imagine much greater problems of accountability in rapidly changing ones, in a world freed from Cold War restraints. When self-starting actors – operating under all manner of organisational rubrics – take matters into their own hands and organise themselves and others both at home and abroad, the results can be beneficial or harmful to citizens. The actors themselves may become the restraints and watchdogs, but unelected and owing little accountability to others; perhaps looking to make a profit, they may also remind us of the need for publicly accountable watchdogs.

4. SOCIAL ORGANISATION AND NETWORKS OF GOVERNANCE

How, then, might we study arenas of governance permeated by complex entanglements of formal and informal state and private structures and processes? A focus on the social organisation of governance can serve as a counterweight to the ideologised discourses of privatisation, public-private partnerships and the like.[8] A social organisational approach that explores how actors and organisations are interconnected can illuminate the structures and processes of governance that ground, order, appropriate and give it direction. This approach can provide a snapshot of the workings of governance in specific arenas.

Social network analysis, which focuses on social relations rather than the characteristics of actors,[9] is integral to the social organisational approach. Network analysis is powerful both 'as an orienting idea' and as a 'specific body of methods', as Scott (1991) put it. By examining relationships among formal and informal institutions, organisations and individuals, it is ideally suited to map mixes of organisational forms, the changing, overlapping and multiple roles that actors within them may play and the ambiguities surrounding them.

Further, network analysis provides a path around the methodological and theoretical conundrum of macro versus micro. It can connect the local or regional with the national and the local, regional or national with the international. Examination of networks enables us to look at these levels in a single study frame.[10]

Network analysis thus enables us to explore a number of relevant questions: How do various network and role structures shape decision-making and policy processes? To what extent do state authorities in a given context have the capacity to remain separate from the organisations and agendas of informal groups such as mafias, clans or lobbyists? In other words, to what extent do particular network configurations merely penetrate the state, and to what extent do they reorganise it? Addressing such questions can uncover constellations of actors, activities and influences that shape policy decisions, their implementation and governance itself.[11]

We cannot answer these questions in abstract or in general. To do so requires ethnographic grounding in the contexts in which they operate and empirical investigation of the social networks and activities of individuals and organisations. This is the case whether we are sorting out the networks that gird corruption or terrorism, the networks linking Enron to Arthur Andersen and to government organisations, or the networks connecting Human Rights Watch to the Ford Foundation and the United Nations.[12]

Based on such research, we then can establish patterns and theories. For example, the incentive structures in a given system encourage multiple and conflicting roles in governance; or an ever-increasing diffusion of governance creates more opportunities for conflicts of interest.

5. BOUNDARY CROSSING IN POST-SOCIALIST STATES

It is logical that the state-private nexus was the heartbeat of change in societies divesting themselves of state-owned resources. The very collapse of a centrally planned economy implies movement away from the state. The countries of Central and Eastern Europe and the former Soviet Union provide particularly rich contexts in which to explore the organisation of governance. Moreover, the international financial institutions, western governments and aid agencies have accelerated this movement with their aggressive promotion of privatisation, deregulation and other free-market projects in the region. As a result, political-economic influence overwhelmingly has resided in the 'control of the interface between public and private', as World Bank economist put it.[13] From the distribution and management of resources and privatisation, to the development of "civil society" and NGOs, much of the action has been at the state-private nexus.

Individuals and groups vying for influence have positioned themselves at this nexus. At precarious points in the change process, old systems of social relations, such as the informal groups and networks that functioned under communism and helped to ensure stability, have become crucial instruments of change. In the political, legal, administrative, economic and social flux that accompanied the collapse of communism, many individuals were empowered by the erosion of the centralised state and enticed by new opportunities for acquiring resources and wielding influence.

"Institutional nomads", a term coined by Kaminski and Kurczewska (1994), are members of Polish informal social circles who have come together to achieve concrete goals. They do so by putting their fingers into a kitchenfull of pies – government, politics, business, foundations and non-governmental and international organisations – and pooling their resources to best serve the interests of the group. Institutional nomads owe their primary loyalty to their fellow nomads, rather than to the formal positions they occupy or the institutions with which they are associated. Skapska stressed that vested interests are at stake in the circulation of nomads among institutions. Whether they come from a former opposition or communist party milieu, and whether they were workers or directors, 'members willy-nilly must stay loyal and collaborate'.[14]

Similar to the idea of institutional nomads, the Russian *clan*, as Kryshtanovskaya (1997) analysed it, is an informal group of elites whose members promote their mutual political, financial and strategic interests. In the Russian context, clans are grounded in long-standing association and incentives to act together, not kinship or genealogical units, as in the classic anthropological definition. Kryshtanovskaya explained it as follows:

> A clan is based on informal relations between its members, and has no registered structure. Its members can be dispersed, but have their men everywhere. They are united by a community of views and loyalty to an idea or a leader... But the head of a clan cannot be pensioned off. He has his men everywhere, his influence is dispersed and not always noticeable. Today he can be in the spotlight, and tomorrow he can retreat into the shadow. He can become the country's top leader, but prefers to remain his grey cardinal. Unlike the leaders of other elite groups, he does not give his undivided attention to any one organisation (ibid.).

Clans, institutional nomads and other such informal groups strategically place their members in as many different spheres (state and private, bureaucratic and market, legal and illegal) as possible, bridging old and new, formal and informal, in order to best access resources and advantages for the group. The most adept, and sometimes the most ethically challenged, actors play the boundaries, skilfully blending, equivocating, mediating and otherwise working the spheres — all the while using ambiguity to their advantage.

Thus, far from disappearing, informal systems played a pivotal role in many reform processes of the 1990s.[15] Such systems became integrated with state-sponsored reforms and helped shape their outcomes. By providing unrestrained opportunities for insiders to acquire resources, some reforms fostered the proliferation and entrenchment of informal groups and networks, including those linked to organised crime. For example, in Russia there was mass *grabitisation* of state-owned enterprises, as many Russians came to call the privatisation that was linked *en masse* to organised crime (Wedel 2001: 138–142). The "reforms" were more about wealth confiscation than wealth creation; and the incentive system encouraged looting, asset stripping and capital flight.[16] Billionaire oligarchs were created virtually overnight.

It is important to emphasise that the activities of informal networks and groups help to organise the state itself, not only to divert state resources and functions. Yurchak (2002) studied how actors shape the state sphere. He documented two separate spheres *within* the Russian state – the "officialised-public" and the "personalised-public".[17] He argued that when the Soviet Union fell apart it was principally the officialised-public sphere of the state, with its institutions, laws and ideologies, that succumbed to crisis. Yurchak observes,

> the personalized-public sphere expanded into new areas of everyday life, and many of its relations and understandings became even more important…the state's personalized-public sphere did not collapse but rather re-adapted to the new situation much better than was obvious at the time (Yurchak 2002: 311).

5.1 Flex Organising

I first discovered what I have come to call "flex organisations" while studying foreign economic aid to Russia in the 1990s. Flex organisations in Russia emerged exactly at the state-private nexus. Their defining feature is that they switch their status situationally – from state to private spheres – back and forth, thereby enabling their members to selectively bypass the constraints governing both spheres. They are defined as flex organisations (Wedel 2001: 145–153, 156, 172), in recognition of their impressively adaptable, chameleon-like, multifunctional character.[18] They are Janus-faced in that they keep changing their facade. They are also referred to as governmental non-governmental organisations (GNGOs).

Flex organising depends on the state-private distinction. It plays on the existence of separate spheres, each of which is subject to different rules and opportunities and neither of which is prepared to govern boundary-crossing. The inter-changeability inherent in flex organising enables actors to avoid state directives of accountability while skirting private codes of competition.

Flex organisations are empowered by informal groups or networks such as clans and institutional nomads. Their influence derives in significant part from their ability to access the resources and advantages from one sphere for use in another. In flex organising, spheres within and around the state are malleable and fluid. They are situationally and even fleetingly activated, deactivated and otherwise moulded by the actors operating under various configurations of state and private rubrics who employ state-ness and private-ness strategically to achieve the goals of their group. The capacity of individuals and organisations to manoeuvre, traverse and blur activities in different spheres (state, private and sometimes international) in fact underwrites whatever success they achieve. It is precisely the ability to equivocate that affords flex organisations their influence and resilience.

Flex organisations may share aspects of the same standing as state organisations, and/or they may even be NGOs. Whatever the specifics of their legal standing in a particular country, they have been set up by high-ranking officials and depend on the coercive powers of the state and continued access to and personal relationships with these officials. There may even be overlap between officials, for example, from a particular ministry and the leadership of a flex organisation that is legally an NGO. Such officials then play dual roles, representing and empowering both the "state" and "private" organisations.

After the Soviet Union was dismantled, Russia embarked on a course of economic reform with the help of the international financial institutions and western donor organisations. Flex organisations, the vehicles through which economic reforms were to take place, became prime recipients of US and other western foreign aid funds (Wedel 2001: Chp. 4). They were created around, and run by, a small, interlocking group of transnational actors made up of representatives from Russia (Anatoly Chubais and the so-called Chubais clan) and the United States (a group of aid-funded advisers associated with Harvard University and the Harvard Institute for International Development). Flex organising was central to the influence of this Chubais-Harvard group. The Russian Privatisation Centre, the donors' flagship organisation (and a flex organisation in my terms), is a case in point. It received hundreds of millions of dollars in aid from bilateral donors and loans from the international financial institutions.

To start my study of economic aid, I cast a wide net, as I had done in my earlier investigation of aid to Central Europe. A large part of my fieldwork consisted of exploring connections – among the Chubais clan, representatives of the Harvard group that monopolised US economic aid to Russia, and American and Russian officials. Although the individuals and groups involved in this case were sometimes located at different sites, they were always connected by the aid process or by actual social networks.[19]

I began by conducting interviews with individuals connected with the economic aid effort in a variety of organisations. I visited the organisations that received economic aid, such as the Russian Privatisation Centre, the Federal Commission on Securities and Capital Markets (also known as the "Russian SEC"), the Institute for Law-Based Economy (ILBE) and the Resource Secretariat. The individuals interviewed typically gave me their organisations' statements of purpose, often nicely presented for westerners. I also asked "Who runs this organisation?" "Who founded it?" "Who is on the board of directors?" "Who visits here?" and "Whose word counts?"

After as few as a dozen interviews in a half dozen different organisations, I was able to piece together the beginnings of a rough social network chart on who was connected with whom and in what capacities. I found the same set of names coming up – no matter which organisation was in question. The more interviewing I did, the more I began to understand that, despite the organisations' different functions, the same people ran them. Thus, despite the fact that there were several organisations, ostensibly engaged in different parts of the economic reform agenda, the same tight-knit group of interconnected individuals had set them up (together with the foreign aid establishment) and appeared to be running them, along with the Russian reform agenda and significant parts of the Russian government. These individuals were additionally connected with each other in a variety of capacities, including business and romance.

Such interlocking networks highlight a fundamental feature of flex organisations, the *empowerment of the organisation by a group* which uses the organisation to further the group's goals. The apparent interconnectedness of the Chubais-Harvard actors led me to examine their network structures more systematically, according to several characteristics. One such characteristic is the *single-stranded* versus *multiplex* nature of networks. Single-stranded means that the relationship between two people is based on only one role – that they know each other in only one capacity. Multiplexity, on the other hand, means that they have more than one role vis-à-vis one another.

In the United States-Russia case, actors knew one another in a variety of capacities; their networks were highly multiplex. With regard to interconnectedness through formal organisations (only), multiplexity is manifested in the following description of some ways in which two individuals were wedded to each other:

- Person A is connected to Person B through the Russian SEC, because A controls the ILBE, which funds the Russian SEC, and B is both vice-director of the board and managing director of the Russian SEC;
- A is connected to B through the Russian Privatisation Centre, because A is on its board of directors, while B is its deputy chairperson;
- In addition, A is connected to B through the State Property Committee, because A is senior legal advisor, while B is deputy chairperson.

In addition to (and using) their formal organisations, the same two individuals were connected to each other through business transactions, including the following:

- Person A is connected to B, who in two of his roles is deputy chairperson of the board and managing director of the Russian SEC. B arranges for A's wife's company – a little known mutual fund – to be the first licensed fund in Russia, over and above the applications of Credit Suisse First Boston and other big players;
- B, in another of his capacities – his role on the Gore-Chernomyrdin Commission – arranges for A's wife to become a member of a commission working group.

Another fundamental feature of flex organisations is their *ability to shift agency* – the flexibility after which they are named. An archetypal flex organisation, the Russian

Privatisation Centre switched its status situationally. Legally it was non-profit and non-governmental, but it was established by a Russian presidential decree. Though the Centre was an NGO, it helped carry out government policy related to inflation and other macroeconomic issues and also negotiated with and received loans from international financial institutions *on behalf of the Russian government*, when typically the international financial institutions lend only to governments.

Further, the Russian Privatisation Centre received aid from the international financial institutions because it was run by members of the Chubais clan, who also played key roles in the Russian government. (This introduced still more ambiguity between state and private roles and responsibilities.) As an NGO, the Centre received tens of millions of dollars from western foundations that generally support NGOs. As a state organisation, it received hundreds of millions of dollars from international financial institutions, which, as stated above, principally lend to governments.

Yet another key feature of flex organisations, *deniability*, flows from their shifting agency. Deniability means that, because actors and organisations can change their agency, they always have an "out". They can evade culpability for actions that might be questioned by voices of one of the parties by claiming that their actions were in the service of the other. Deniability is "institutionalised" in that it is built into the very structure of shifting and multiple agency.

For example, because Harvard's director in Moscow, Jonathan Hay, was given signatory authority over certain privatisation activities, he could, if questioned by US investigators, legitimately claim that he conducted those activities "as a Russian" and thus, with respect to those transactions should not be constrained by US norms or regulations. Likewise, if the Russian Privatisation Centre came under fire for its activities as a state organisation, it could legitimately claim to be a private one.

Another feature of flex organisations is the *propensity to bypass otherwise relevant institutions*, such as those of government (executive, judiciary or legislative). The Russian Privatisation Centre appears to have been set up precisely to circumvent such institutions. It bypassed the democratically elected parliament and the Russian government agency formally responsible for privatisation. Indeed, according to documents from Russia's Chamber of Accounts, the Russian Privatisation Centre wielded more control over certain privatisation directives than did the government privatisation agency.[20] Two Centre officials[21] were in fact authorised by the Russians to sign privatisation decisions on Russia's behalf. Thus did a Russian and an American, both of them officially working for a *private* entity, come to act as representatives of the Russian state.

Flex organisations, understandably, call to mind the notion of conflict of interest. But they serve to *obfuscate* conflict of interest. Unlike a lawyer who represents a client who has embezzled funds from a bank on the one hand, and represents the bank on the other – a clear conflict of interest – in flex organisations, the roles are ambiguous. In conflict of interest, an actor can deny the facts, but not the conflict if the facts are true. But with flex organisations, it is not always clear what the conflicts are because structures are themselves ambiguous. An actor can plausibly deny responsibility and get away with it. The difference lies in the ability of a flex organisation to exploit the ambiguity.

Thus, flex organisations are neither quasi-non-governmental organisations (QUANGOs) nor GONGOs (government-organised NGOs). Unlike flex organisations, QUANGOs and GONGOs depend on continued personal relationships with and access to high government officials. Unlike flex organisations, which switch their status back and forth between state and private, QUANGOs and GONGOs imply static relationships and thus cannot enable deniability, as do flex organisations by definition. Whereas QUANGOs and GONGOs can come into conflicts of interest with state organisations, flex organisations cannot because they camouflage any such conflicts.

5.2 Polish Agencies and Targeted Funds

During the mid-1990s in Poland information began surfacing – albeit reluctantly – on the existence of state-private hybrid organisations called agencies (*agencje*) and targeted funds (*fundusze celowej*). Although such organisations lack the inherently situational quality of flex organisations, the defining feature of agencies and targeted funds is their indistinct responsibilities and functions (Kaminsky 1997: 100). These organisations do not have the same legal status as state bodies, but they use state resources and rely on the coercive powers of the state administration. They have broad prerogatives that are supported by administrative sanctions and are subject to limited public accountability. In the words of Piotr Kownacki, deputy director of NIK (Supreme Chamber of Control), Poland's chief auditing body, they are part and parcel of the 'privatisation of the functions of the state', and they represent 'areas of the state in which the state is responsible but has no control'.[22] From the point of view of the state administration, these entities are "public", not state. On the other hand, clients associate them with the state.

Agencies and targeted funds have come to play a major role in the organisation of Polish governance and in the collection and disbursement of public funds. Some one-fourth of the state budget was allocated to them in 2001.[23] In addition, some agencies and targeted funds are or have been authorised by the state to variously conduct and receive monies from commercial activities, invest in the stock market, start new companies, spawn new agencies and manage foreign aid funds.

With regard to targeted funds, a NIK report calls them 'corruption-causing'. The NIK noted the 'excessive discretion' that the funds enjoy in their use of public resources, which is more than that of state organisations. It also lamented the 'lack of current controls' over the activities of targeted funds (NIK 2000: 45). One example is the Fund for the Rehabilitation of Disabled People (PFRON), which is supposed to subsidise the employment of handicapped individuals. Considerable discretion is built into every level of decision-making, from whether PFRON subsidises a particular workplace, to the amount of the subsidy, to whether the workplace further distributes the funds to its disabled employees (ibid.: 46).

Agencies also possess a lot of leeway. They have been created in all ministries with control over property. These include the ministries of transportation, economy, agriculture, treasury and defence, according to NIK Deputy Director Kownacki.[24] Agencies are set up by state officials, often attached to their ministries or state organisations and funded by the state budget. The minister typically appoints an agency's supervisory board, with these selections often based on political connections.[25] Some 10 to 15 per cent of an agency's profits can be allocated to "social" purposes: If the agency accrues profits, those profits go to the board and are sometimes funnelled into political campaigns. On the other hand, any losses are covered by the state budget.[26]

Agricultural agencies offer a case in point. With so much property under their control, including state farms inherited from the communist past, agencies have begun 'to represent [their] own interests, not those of the state', according to Piotr Kownacki. He observed that 'most of the money is taken by intermediaries' and the state has very little control over this process.[27]

Coal mining and arms also are dominated by agencies and present myriad opportunities for corruption.[28] The coal industry, for example, appears to be dominated by a group of *institutional nomads* who simultaneously hold and circulate in positions of government, various agencies and business. Collectively, the nomads organise themselves to cover all the bases by being involved in as many influential administrative, business and political positions as possible relevant to their success in the industry, regardless of which political parties are in power (Gadowska 2002).

Some agencies and targeted funds have become vehicles through which foreign aid is distributed (though aid organisations generally did not initiate the entities in Poland as they did with the Russian flex organisations described earlier). Notable examples are the European Union's Special Accession Programme for Agriculture and Rural Development (SAPARD) to restructure Polish agriculture and some EU programmes to improve environment and transportation.[29]

The number of agencies and targeted funds grew throughout the 1990s[30] – in a country that enjoyed the reputation of a transition "success story" – with its entry into NATO, pending accession to the EU and, during the mid- to late 1990s, the fastest economic growth rate in Europe. With a structure of governance in which accountability- challenged entities played a significant part, the result was that 'much tax-payer money flows to private hands on a large scale', as former NIK Director Lech Kaczynski expressed.[31] Kaminski (1997: 100) assessed that 'the real aim of the [agencies and targeted funds] is to transfer public means to private individuals or organisations or to create funds within the public sector which can then be intercepted by the initiating parties' (Kaminski 1997: 100). A number of analysts have linked the entities to the financing of political campaigns, although this has not been established by the relevant government regulatory bodies. As Stefanowicz has maintained, 'There is a silent truth between political parties. No financial report has ever disclosed how much political support is allocated to political campaigns [through agencies and similar entities].'[32]

Until the early 2000s, these entities managed to sidestep any substantial media spotlight. Only a few analysts, journalists and, notably, the NIK, tracked limited parts of what constituted a huge portion of the state or public budget. Since then, however, media coverage of NIK findings and journalists' reports of cases of state funds lining private pockets have become extensive. At the same time, public attention has been drawn to an (unrelated) parliamentary investigation of a high-level political scandal that has been broadcast live on television.

Public attention to corruption is among the factors that may have contributed to increased scrutiny of governance and public officials and the introduction of new anti-corruption regulations. In 2001, the parliament enacted a law limiting the number of agencies and targeted funds. In 2003, however, the NIK reported that these entities continued to result in losses to the state budget.[33] At this juncture, it is difficult to assess their future in the Polish state-public sphere.

Like flex organisations in Russia, Polish agencies and targeted funds are not holdovers from the communist period. They were, however, enabled by the breakdown of the command structure of the centrally planned state, which privileged a network-based organisation of governance and business, as described earlier. It is important to bear in mind that the bodies were made legally possible by legislation enacted after the fall of communism. As discussed earlier, the western models of deregulation, decentralisation and privatisation that were promoted in the region also provided inspiration.

6. THE DEEP AND WIDESPREAD UNACCOUNTABLE STATE?

What are the implications for states of a proliferation of flex organisations or similarly ambiguous and unaccountable organisational forms? The net effect may actually be the *enlargement* of the state sphere. Kaminski (1996: 4) argued that Poland's targeted funds and agencies have resulted in 'an indirect enlargement of the dominion of the "state" for which the state is responsible but has no control'. 'This has been accomplished', he wrote, 'through the founding of institutions that in appearance are private, but in fact are part of the [appropriated] public domain'.

This empirical data is consistent with the claim that globalisation has 'yielded a key continuity in governance, namely that of bureaucratism' (Scholte 2000: 133). However, Scholte appears to have missed a crucial development when he suggested that 'governance is mainly conducted through large-scale, relatively permanent, formally organised, impersonally managed and hierarchically ordered decision-taking procedures'. On the contrary, such forms of organising (arenas of) governance as the flex organisations, agencies and targeted funds detailed here are in large part *informally* organised and *personally* managed. Yet these organisations are symbiotic, and at times synonymous with the structures and processes of "actually existing governance".

The crumbling of the command structure of the centrally planned state encouraged the development – or perhaps the continuation – of flex organising in the Russian context, even though the flex organisations of the 1990s were created by foreign aid organisations and Harvard University, and propelled by millions of dollars from the West. Flex organisations mimicked the dual system under communism, in which many state organisations had counterpart Communist Party organisations that wielded the prevailing influence.[34] The creation of 1990s flex organisations and the massive western underwriting they received may have encouraged the development of what I have called the "clan state", a state that is powered by competing, tight-knit and closed clans (Wedel 2001, 2003). E. W. Merry, a former US senior political State Department officer, regretted the US-sponsored creation of 'extra-constitutional institutions to end-run the legislature'. He added that 'many people in Moscow were comfortable with this, because it looked like the old communistic structure. It was just like home.'[35]

Yet, although flex organising may be most visible and prevalent in societies undergoing fundamental transformation, such forms of organising may be found in a variety of settings. Flex organisations appear to exist in developing country contexts[36] – not only in centrally planned economies in the process of "unplanning" themselves, as well as in developed countries, including the United States.

For example, an influential group made of up "neoconservatives" who have been active in formulating and implementing US policy towards Iraq and the Middle East, appear to operate at least in part through flex organising. With Richard Perle (who advised President George W. Bush on foreign policy and defence matters during the Bush campaign) as a central actor, members of this long-standing tight-knit group are connected with each other through government, business, lobbying, think-tanks and media organisations and activities, as well as through family and marriage ties. Members occupy key positions in the Bush administration's Iraq and Middle East policy, both within and outside the formal government.[37] The group works in part by bypassing otherwise relevant structures and processes of governance and supplanting them with their own. Cross-agency cliques reportedly enable the group to limit information and activities to its associates across agencies.[38]

Further, the Defence Policy Board, formerly headed by Perle, appears to be a flex organisation that can circumvent the Pentagon. With Perle as its chairman since mid-2001, the Board evolved from a little-known organisation to one with wide influence and power. Its structure affords Perle more influence than he might have had as a government official (he turned down a government position in the administration).[39] It allows Perle and others in his circle to retain their private business interests while holding a not-quite-public office that provides access to defence planning and top-secret intelligence.[40] Perle resigned from his position as chairman of the Defence Policy Board (but not his membership) in March 2003 amid accusations of conflicting interests. He is reported to have advised companies and their clients on business dealings using sensitive government information he was privy to through his position on the Board.[41] The Board and the Perle circle remain decisive in shaping key parts of US foreign policy.

The conditions that facilitate flex organising have become more prevalent in the international arena of retreated states and diffuse authority. These circumstances provide greater incentives for people to play multiple, conflicting roles that overlap government, business and non-governmental organisations and enable them to selectively bypass the constraints on these institutions. Because these circumstances call for a higher degree of flexibility and deniability, flex organising likely will become more common.

7. IMPLICATIONS FOR ACCOUNTABILITY

Can citizens take advantage of the benefits of governance by flex organising while mitigating potentially harmful effects? What are the implications for accountability when states rely on actors and organisations that can fulfil multiple, conflicting and ambiguous roles? What does the overlapping and situational nature of flex organising mean for accountability?

"Transparency", a term that occupies a prominent place in the anti-corruption discourse, may be achievable with regard to flex organisations, depending on the context. It is possible, at least in some settings, to shed light on the existence and at least some of the operations of flex organisations. The extent to which this illumination can occur depends on the ability of information that is independent of the flex organisations and those who control them. Yet, while transparency is potentially attainable, accountability is not.

When ambiguity is woven into the fabric of social organisation and the structure of relationships, accountability is difficult, if not impossible. Because organisations and the actors who empower them engage in representational juggling and can shift their allegiances to achieve their own objectives, flex organisations provide deniability. Flex organisations are *inherently unaccountable* because shifting agency builds deniability into them.

This raises the issue of corruption and current approaches to countering it (Wedel 2002).[42] The classic definition of corruption, such as that employed by the World Bank, is "the abuse of public office for private gain". Jowitt (1983: 293) argued that such approaches to corruption are weak because they rely on 'the difference between public and private aspects of social organisation' which 'makes it impossible to specify the existence and meaning of corruption in settings where no public-private distinction exists institutionally'. Indeed, as we see from the flex organisations, agencies and targeted funds here analysed, the state-private divide may be fluid, subdivided, overlapping or otherwise obscure. Anti-corruption approaches that rest on the state-private divide cannot counter flex organisations because, by definition, flex organisations shift their agency.

The flex organisations I charted in Russia were frequently used by actors to pursue their own private and group agendas. Yet it is conceivable that actors and groups can use flex organising to serve official or public purposes. But whether they do so is entirely up to them; any accountability depends on the actors. Thus, the only way to increase the accountability of flex organisations to citizens is to reduce their ambiguity.

Some analysts from developing countries have identified what appear to be flex organisations in their own countries. These organisations effectively use some official structures while circumventing others. Government officials rely on their positions to accomplish specific tasks that they (and others) see as in the public interest but that cannot be accomplished in the official sphere.

However, flex organisations, I would caution, are not equipped to encourage broad-based citizen participation. There is a tendency among some development practitioners to set up organisations to bypass bureaucracies that are seen as cumbersome and inefficient. This practice is problematic. As a report prepared for the World Bank on government-NGO relations argues, although it may be easier to set up separate entities outside civil service rules, there is then a 'danger of conflicts of interest, self-dealing, or improper personal enrichment' (World Bank 1997).

Additional caution is warranted because some flex organisations have the standing of NGOs, which the development community looks to solve a plethora of problems and tends to regard as effective vehicles for public outreach. Flex organisations are often invigorated by people who come together out of long-standing association, friendship and family connections. Such organisations often lack incentives that would encourage expansion beyond their originating circles. They are better equipped to continue as isolated groups, contributing to fragmented governance, rather than attracting new members on the basis of common interests.

In addition, the donor community often sees NGOs as exemplars of and vehicles for creating democracy and civil society and mediating between citizens and the state, especially in transitional societies. Yet development agencies clearly should not charge entities such as flex organisations with public outreach. The very structure of flex organisations and similar ways of organising governance mitigates against the sharing of information and resources with a wider public.

NOTES

1. I wish to thank Sylvette Cormeraie, Bryant Garth, John Harper, Alina Hussein, Antoni Kaminski, Bruce Kapferer, Max Spoor and Susan Tolchin for their generous help and invaluable comments.

2. Although I juxtapose "state" and "private" in this chapter, how these terms are used and the relationships among them are key questions for research. For an analysis of alternative historic views of relationships among public, private, state and market, see Weintraub and Kumar (1997).

3. The term QUANGO was coined more than two decades ago by journalist Caroline Moorehead, who called them 'quasi-autonomous nongovernmental organisations' (Bird 1998).

4. Strange elaborates on the development of the neoliberal agenda by delineating five crucial political choices, made mostly by the United States, from 1971 to 1985, which propelled the neoliberal project in finance. Strange's five choices are the following: (i) the 'extreme withdrawal' on the part of the United States 'from any intervention in foreign exchange markets'; (ii) the false but convincing claim that monetary reform remained a serious issue on the international policy agenda; (iii) the United States' 'confrontational strategy of an oil-consumers' coalition armed… with strategic stockpiles against any repetition of the 1973 oil price rise'; (iv) the "stonewalling strategy… against the Conference on International Economic Cooperation', which followed from the failure to negotiate with the Organisation of Petroleum-Exporting Countries; and (v) the bolstering of 'cooperation between central banks in their dual role as bank regulators and lenders of last resort' in response to two notable bank failures (Strange 1998: 6–7).

5. Personal communication with Dan Guttman, a legal analyst specialising in American governance, 3 June 2003.

6. This expression takes its reference from the term "actually existing socialism", which 'came into use, to distinguish its messy reality from its hopes and claims', as Verdery (1996: 4) put it. The term was originally coined by Bahro (1978).

7. For a discussion of the tension between accountability and autonomy of "private" government contractors, including legal decisions, see Guttman (2000). His article also outlines the kinds of conflicts of interests that arise between private employees and their public overseers (ibid.: 896–901).

8. Concepts such as privatisation have often played a part in the discourse on globalisation. Commenting on that discourse, Kalb, Van der Land, Staring and Van Steenbergen (2000) observed that the very 'neglect, denial, or even conscious repression, of institutional complexity, social relationships, contingency, and possible contradictions' is what made the concept of globalisation into the "ideological magnet" it became'. Shore and Wright (1997) placed this argument in a larger context. They wrote that the 'masking of the political under the cloak of neutrality is a key feature of modern power'. Although policies are typically clothed in the language of neutrality — ostensibly merely promoting effectiveness and efficiency — they are fundamentally political.

9. Pioneers in the field of social network analysis were John Barnes, Clyde Mitchell and Elizabeth Bott, all associated with the Department of Social Anthropology at Manchester University in the 1950s. They saw social structure as networks of relations and focused on 'the actual configuration of relations which arose from the exercise of conflict and power' (Scott 1991: 27). For an analysis of the contribution of the Manchester school to the development of social network theory see Scott (ibid.: 27–33).

10. Anthropology provides a theoretical framework for connecting levels and processes. Inda and Rosaldo (2002: 4–5) stressed that "[a]nthropology... is most concerned with the articulation of the global and the local, that is, with how globalising processes exist in the context of, and must come to terms with, the realities of particular societies, with their accumulated – that is to say historical – cultures and ways of life.... What anthropology offers that is often lacking in other disciplines is a concrete attentiveness to human agency, to the practices of everyday life, in short, to how subjects mediate the processes of globalization.'

11. Several network analysts have linked network structures to collective processes. For example, Laumann, Marsden and Prensky (1989: 62) pointed out that '[f]eatures of a network can be used... to show the consequences of individual level network processes at the level of the collectivity'. See also Marsden (1981).

12. As Dezalay and Garth (2002: 10) point out, 'Tracing the careers of particular individuals makes it obvious... that the world of foundations and that of human rights NGOs have always been very closely related; how through concrete networks and careers the World Bank interacts with local situations; and how corporate law firms or advocacy organisations modelled on those in the United States are brought to new terrains.'

13. Personal communication with World Bank economist Helen Sutch, 1 November 2001.

14. Personal communication with Grazyna Skapska, 14 October 2002.

15. In many contexts across Central and Eastern Europe and the former Soviet Union, groups that originally coalesced under communism (notably *nomenklatura*, the system under which responsible positions in all spheres of government had to be approved by the communist party) decisively shaped property relations. The *nomenklatura* had the power to accept or veto candidates for any state job and asserted a final voice over responsible positions in all spheres, from police and army posts to factory management and school principalships on the basis of party loyalty, not ability or qualifications. This created a tangle of loyalties and favouritisms that precluded broader political, economic and social participation. As communism was crumbling, many members of the *nomenklatura* traded in their political advantages for economic ones.

16. See, for example, Bivens and Bernstein (1998), Nelson and Kuzes (1994, 1995), Hedlund (1999), Klebnikov (2000) and Anne Williamson's congressional testimony before the House Committee on Banking and Financial Services (21 September 1999). Commenting on the US role in sponsoring the reforms, E. Wayne Merry, former chief political analyst at the US Embassy in Moscow, assessed that '[w]e created a virtual open shop for thievery at a national level and for capital flight in terms of hundreds of billions of dollars, and the raping of natural resources'
    Frontline "Return of the Czar" interview with E. Wayne Merry, PBS website: http://www.pbs.org\wgbh\pages\frontline\shows\yeltsin\interviews\merry.html.

17. Yurchak (1998). See also Bonnell and Gold (2002).

18. The concept bears some similarity to anthropologist Aihwa Ong's notion of "flexible citizenship" in the sense that social structures enable alterative and multiple presentations as actors operate in and respond to a diversity of situations (Ong 1999).

19. The process of following the source of policies, in this case, the donors, their policy prescriptions, rhetoric, and organisation of aid, through to those affected by the policies, in this case, the recipients, has been called "studying through", as in Shore and Wright (1997). Studying through entails tracing connections between different organisational and everyday worlds, even where actors in different sites do not know each other or share a moral universe.

20. Interview with and documents provided by Chamber of Accounts auditor Veniamin Sokolov, 31 May 1998. See State Property Committee Order No. 188 (which gave Jonathan Hay veto power over the Committee's projects), 5 October 1992.

21. These were the Russian Privatisation Centre's CEO from the Chubais clan (Maxim Boycko) and the Moscow representative (Jonathan Hay) of the Harvard Institute for International Development, which managed virtually the entire $350 million US economic aid portfolio to Russia. (Wedel 2001: 145–153).

22. Interview with Piotr Kownacki, Deputy Director of NIK, 26 July 1999.

23. Interview with NIK official Andrzej Lodyga, 24 July 2002.

25. Interview with Piotr Kownacki, Deputy Director of NIK, 26 July 1999.

26. Interviews with Jan Stefanowicz, 14 and 15 July 1999.

26. Interviews with Jan Stefanowicz, 14 and 15 July 1999.

27. Interview with Piotr Kownacki, Deputy Director of NIK, 26 July 1999.

28. Interview with Piotr Kownacki, Deputy Director of NIK, 26 July 1999.

29. Interview with NIK official Andrzej Lodyga, 24 July 2002.

30. Interviews with Jan Stefanowicz, 14 and 15 July 1999.

31. Interview with Lech Kaczynski, 14 July 1999.

32. Interviews with Jan Stefanowicz, 14 and 15 July 1999.

33. Report of the activities of the supreme chamber of control in 2002 (*Sprawozdanie z Dzialalnosci Najwyzszej Izby Kontroli w 2002 Roku*), Warsaw, Poland: Najwyzszej Izby Kontroli, June 2003, p. 127.

34. Such organisations were highly compatible with Russian practices regarding influence and ownership. A number of analysts have pointed out that *de facto* control and influence over property are more important than *de jure* ownership. For further analysis, see, for example, Anne Williamson's congressional testimony before the House Committee on Banking and Financial Services, 21 September 1999, and commentaries on *Johnson's Russia List* by Jerry F. Hough (No. 3051, 11 February 1999), S. Lawrence (No. 3072, 28 February 1999), and Edwin G. Dolan (No. 3073, 1 March 1999).

35. Interview with E. Wayne Merry, 23 May 2000.

36. This claim is based in part on discussions with observers of governance in developing countries. For example, in the past several years I have given talks at the World Bank and the Ford Foundation in which my discussion of flex organisations has prompted audience members to describe similar forms in their home countries. Officials and programme officers from a variety of countries, including Egypt and Nigeria, have identified what appear to be flex organisations.

37. See, for example, Seymour M. Hersh, 'Lunch with the chairman', *The New Yorker* 17 March 2003, http://www.newyorker.com; Elizabeth Drew, 'The neocons in power', *The New York Review of Books* 50 (10), 12 June 2003; and Sam Tenenhaus, 'Bush's brain trust', *Vanity Fair*, July 2003, pp. 114–169.

38. See, for example, Seymour M. Hersh, 'Selective intelligence', *The New Yorker* 18 May 2003, http://www.newyorker.com and Jim Lobe, 'Insider fires a broadside at Rumsfeld's office', *Asia Times* 7 August 2003, regarding the Pentagon's Office of Special Plans and cross-agency cliques.

39. See, for example, Seymour M. Hersh, 'Lunch with the chairman', *The New Yorker* 17 March 2003. <http://www.newyorker.com>

40. See Andre Verloy and Daniel Politi, 'Advisors of influence: Nine members of Defence Policy Board have ties to defence contractors', *The Public-I* 28 March 2003, http://www.publicintegrity.org and 'The Bush 100: Snapshot of professional and economic interests reveals close ties between government,

business', *Center for Public Integrity:*
<http://www.publicintegrity.org/cgi-bin/WhosWhoSearch.asp?Display=List&List=All>

41. See Ken Silverstein and Chuck Neubauer, 'Consulting and policy overlap', *Los Angeles Times* 7 May 2003.

42. Issues of corruption are explored in the author's paper on blurring the boundaries of the state-private divide. Presented at the European Association for Social Anthropology (EASA) Conference, Copenhagen, Denmark, 14–17 Aug. 2002.

REFERENCES

Bahro, Rudolph (1978) *The Alternative in Eastern Europe.* London: Verso.

Bird, Dennis L. (1998) 'Book review of *QUANGOs and Local Government: A Changing World*, edited by Howard Davis' (1998), *Contemporary Review* (July) 273 (1590): 45.

Bivens, Matt and Jonas Bernstein 'The Russia you never met', *Demokratizatsiya: The Journal of Post-Soviet Democratization* 6 (4): 613–647.

Bonnel, Victoria and Thomas Gold (Eds) (2002) *The New Entrepreneurs of Europe and Asia.* Armonk, N.Y.: M.E. Sharpe.

Daniel Guttman's (2000) 'Public and private service: The twentieth century culture of contracting out and the evolving law of diffused sovereignty', *Administrative Law Review* 52 (3): 901–908.

Dezalay, Yves and Bryant Garth (2002) *The Internationalization of Palace Wars: Lawyers, Economists and Contest to Transform Latin American States.* Chicago: University of Chicago Press.

Gadowska, Kaja (2002) *Zjawisko Klientelizmu Polityczno-Ekonomicznego: Systemowa Analiza Powiazan Sieciowych Na Przykladzie Przeksztalcen Sektora Gorniczego w Polsce.* Krakow, Poland: Uniwersytet Jagielonski, Wydzial Filozoficzny, Instytut Socjologii (Rozprawa Doktorska).

Hedlund, Stefan (1999) *Russia's "Market" Economy: A Bad Case of Predatory Capitalism.* London: UCL Press.

Inda, Jonathan Xavier and Renato Rosaldo (Eds) (2002) *The Anthropology of Globalization: A Reader.* Malden, Massachusetts: Blackwell Publishers.

Jowitt, Ken (1983) 'Soviet neotraditionalism: The political corruption of a Leninist regime', *Soviet Studies: A Quarterly Journal on the USSR and Eastern Europe* 35 (3): 275–297.

Kalb, Don, Marco van der Land, Richard Staring, Bart van Steenbergen and Nico Wilterdink (Eds) (2000) *The Ends of Globalization: Bringing Society Back In.* New York, NY: Rowman & Littlefield Publishers.

Kaminski, Antoni Z. and Joanna Kurczewska (1994) 'Main actors of transformation: The nomadic elites', in Eric Allardt and W. Wesolowski (Eds) *The General Outlines of Transformation*, pp. 132–153. Warsaw: IFIS PAN Publishing.

Kaminski, Antoni Z. (1996) 'The new Polish regime and the spectre of economic corruption'. Summary of a paper presented at the Woodrow Wilson International Center for Scholars, 3 April.

Kaminski, Antoni Z. (1997) 'Corruption under the post-communist transformation: The case of Poland', *Polish Sociological Review* 2 (8): 91–117.

Klebnikov, Paul (2000) *Godfather of the Kremlin: Boris Berezovsky and the Looting of Russia.* New York: Harcourt.

Kryshtanovskaya, Olga (1997) 'Illegal structures in Russia', *Trends in Organized Crime* 3 (1): 14–17.

Laumann, Edward O., Peter V. Marsden and David Prensky (1992) 'The boundary specification problem in network analysis', in Linton S. Freeman, Douglas R. White and A. Kimball Romney *Research Methods in Social Network Analysis*, pp. 61–87. Brunswick, New Jersey: Transaction Publishers.

Light, Paul (1999) *The True Size of Government.* Washington, D.C.: The Brookings Institution.

Marsden, Peter V. (1981) 'Introducing influence processes into a system of collective decisions', *American Journal of Sociology* 86: 1,203–1,235.

Mathews, Jessica (1997) 'Power shift', *Foreign Affairs* 76 (1): 50–66.

Nelson, Lynn D. and Irina Y. Kuzes (1994) *Property to the People: The Struggle for Radical Economic Reform in Russia.* Armonk, New York: M.E. Sharpe.

Nelson, Lynn D. and Irina Y. Kunes (1995) *Radical Reform in Yeltsin's Russia: Political, Economic, and Social Dimensions*. Armonk, New York: M.E. Sharpe.

NIK (2000) *Zagrozenie Korupcja w Swietle Badan Kontrolnych*. Warsaw, Poland: Najwyzsza Izba Kontroli, Departament Strategii Kontrolnej, March.

Ong, Aihwa (1999) *Flexible Citizenship: The Cultural Logics of Transnationality*. Durham, North Carolina: Duke University Press.

Osborne, David and Ted Gaebler (1992) *Reinventing Government: How the Entrepreneurial Spirit is Transforming the Public Sector*. Reading, Massachusetts: Addison Wesley.

Scholte, Jan Aart (2000) *Globalization: A Critical Introduction*. New York, New York: St. Martin's Press.

Scott, John (1991) *Social Network Analysis: A Handbook*. London: Sage Publications.

Shore, Cris and Susan Wright (Eds) (1997) *Anthropology of Policy: Critical Perspectives on Governance and Power*. New York: Routledge.

Strange, Susan (1996) *The Retreat of the State*. Cambridge: Cambridge University Press.

Strange, Susan (1998) *Mad Money: When Markets Outgrow Governments*. Ann Arbor, Michigan: University of Michigan Press.

Verdery, Katherine (1996) *What Was Socialism and What Comes Next*. Princeton, New Jersey: Princeton University Press.

Wedel, Janine (2001) 'Corruption and organized crime in post-communist states: New ways of manifesting old patterns', *Trends in Organized Crime* 7 (1) Fall: 3–61.

Wedel, Janine (2001) *Collision and Collusion: The Strange Case of Western Aid to Eastern Europe*. New York: Palgrave.

Wedel, Janine (2002) 'Blurring the boundaries of the state-private divide: Implications for corruption', Paper presented at the European Association for Social Anthropology (EASA) Conference, Copenhagen, Denmark, 14–17 August.

Wedel, Janine (2003), *The New Interface of State-Private Control and Its Implications for Corruption in Central and Eastern Europe and the Former Soviet Union*. National Institute of Justice (forthcoming).

Weintraub, Jeff (1997) 'The theory and politics of the public/private distinction', in Jeff Weintraub and Krishan Kumar (Eds), *Public and Private in Thought and Practice: Perspectives on a Grand Dichotomy*. Chicago: University of Chicago Press.

World Bank (1997) 'The World Bank and civil society: Handbook on good practices for laws relating to non-governmental organizations (discussion draft)'. Prepared for the World Bank by the International Center for Not-for-Profit Law. Available via the Internet: <www.worldbank.org>

Yurchak, Alexei (1998) 'Mafia, the state, and the new Russian business'. Paper presented at the American Anthropological Association Annual Meeting in Philadelphia, 4 December.

Yurchak, Alexei (2002) 'Entrepreneurial governmentality in postsocialist Russia', in Victoria Bonnell and Thomas Gold (Eds) *The New Entrepreneurs of Europe and Asia*. Armonk, New York: M.E. Sharpe.

PART III

RESOURCE DEGRADATION, INSTITUTIONS AND CONFLICT

CHAPTER 13

MULTI-LEVEL GOVERNANCE AND RESILIENCE OF SOCIAL-ECOLOGICAL SYSTEMS

ELINOR OSTROM AND MARCO A. JANSSEN[1]

Belief structures get transformed into society and economic structure by institutions – both formal rules and informal norms of behaviour. The relationship between mental models and institutions is an intimate one. Mental models are the internal representation that individual cognitive systems create to interpret the environment; institutions are the external (to the mind) mechanisms individuals create to structure and order the environment (North 1996: 348).

1. INTRODUCTION

The last half of the twentieth century witnessed major efforts on the part of developed countries to assist developing countries in speeding up the process of economic and political development and, within their own borders, to improve natural resources management. While enhancing economic development and protecting the environment do not at first appear closely related, the underlying beliefs of many policy analysts in both fields have notable similarities. Initiatives in each policy area have been based on mental models which hold that solutions to difficult and complex problems can only be generated by scientifically trained analysts and implemented by impartial, national-level officials. However, both policy areas have been subject to considerable failure. Although initiatives based on these shared mental models have been undertaken and funds allocated to correct perceived problems, in many cases little or no improvement has been achieved, even after vast sums were spent. Or worse, the problems increased in magnitude.

This chapter first reviews the belief structures, or mental models, of the policy analysts who have been influential in recommending government strategies to achieve sustainable economic development and sustainable resource development. The dominant mental models used in both policy areas rely primarily on command and control (see Ostrom 1989 for a critique). The chapter then explores the concept of social-ecological systems as complex adaptive systems that differ with regard to their

239

predictability and their level of resilience to internal and external shocks or weaknesses. The adaptive cycle is described as well as how disturbances at one scale may trigger problems at the same or other scales, particularly if no experimentation has occurred or repertoire of adaptive responses been developed. Section four looks at some long-established institutions and examines why they may have been more resilient than other institutions. Neither policy field has paid much attention to the course of institutional development during the past millennium which facilitated the immensely productive economic development of Western Europe and the United States, or to successful long-term resource regimes throughout the world that have avoided ecological surprises for centuries. The chapter concludes by urging scholars to move beyond the dominant approach and draw on research conducted with various forms of a complex, adaptive systems perspective.

2. THE BELIEF IN THE EFFICACY OF TOP-DOWN SOLUTIONS

2.1 Development Assistance Policies

Major efforts to reconstruct economies after World War II were initiated at the Bretton Woods Conference in July 1944 when the World Bank and the International Monetary Fund were established. After Harry S. Truman declared his "point four" programme of technical assistance to developing nations in early 1949, the United States and many European countries began to focus on the "Third World". In reflecting on the perspective of those asked to implement the point four programme, Stone (1992: 36) indicated that development was viewed as a 'process in which "modernization," industrialization, and GNP growth, achieved largely by means of public investments and comprehensive national planning, would lead to increasingly prosperous and contented free society'.

During the early 1960s, many scholars subscribed to the gap theory of the development process. 'The "gap theory" stressed the lack of certain vital resources as the main stumbling blocks on the road to development. There was a lack of capital, caused by inadequate savings, and a lack of foreign exchange, but also of knowledge, of entrepreneurial spirit and of leadership qualities' (Elgström 1992: 46). Bauer summarised the beliefs of those who invested their professional lives in development assistance:

> The advance of LDCs depends on ample supplies of capital to provide for infrastructure, for the rapid growth of manufacturing industry, and for the modernization of their economies and societies. The capital required cannot be generated in the LDCs themselves because of the inflexible and inexorable constraint of low incomes (the vicious circle of poverty and stagnation), reinforced by the international demonstration effect, and by the lack of privately profitable investment opportunities in poor countries with their inherently limited local markets.
>
> General backwardness, economic unresponsiveness, and lack of enterprise are well-neigh universal within the less developed world. Therefore, if significant economic advance is to be achieved, governments have an indispensable as well as a comprehensive role in carrying through the critical and large-scale changes necessary to break down the formidable obstacles to growth and to initiate and sustain the growth process (Bauer 1984: 27).

The obvious solution to many western scholars, given the theories that were in vogue in the 1960s, was to recommend generous amounts of foreign assistance that could provide the missing capital, create new infrastructure and provide technical assistance. For some, aid was also seen as a way of "stabilising" and rewarding countries that were allies of the West. The relative success of the Marshall Plan in Western Europe played a role in the belief that pumping capital into an economy leads to a recovery. Western Europe had been through a devastating but short war. It had not suffered the loss of long-established and slowly evolving institutions that colonialism had imposed on much of the developing world. Europe had lost massive amounts of physical capital, but retained much of its social capital.

During the 1960s, billions of dollars of grants and loans to developing countries were accompanied by copious advice on organising and reforming governments. The dominant academic thinking of the day was that centralised regimes were most effective for achieving rapid national unity, effective public administration and extensive economic growth (see Ostrom 1999, Peluso 1992). Scholars and policymakers paid little attention to the need to establish transparent procedures, accountable regimes, effective local government, fair and open court systems and an active civil society. Nor was much attention given to stimulating private enterprise and providing the public policies needed to support a dynamic market economy.

In the 1960s, many developing countries enjoyed relatively high growth rates and optimism abounded. By the early 1970s, however, not only were economic growth patterns unstable and uneven across countries, even more disturbing was that the process of growth often adversely affected the poorest. Corrupt regimes diverted foreign aid into private goods rather than making public investments. Mauro (1995) found, for example, a strong inverse relation between the level of corruption in a country during the 1960s and 1970s and its level of spending on education and health.

Negative assessments have repeatedly been made of the results of international assistance to many developing countries based on extensive reviews of the empirical evidence.[2] In *Improving Aid to Africa* (1996: 2), Van de Walle and Johnston noted that 'aid has not succeeded in fostering economic growth and poverty alleviation in most African countries. From 1980 to 1993, the continent's rate of economic growth was actually negative.' The problem of poverty has generally grown worse around the world. In 1996, for example, approximately 1.3 billion people in developing countries subsisted on less than one US dollar a day (World Bank 1996a). Even in Latin America, 'the number of poor, now 33 per cent of the total population, has failed to fall despite economic recovery' (Birdsall and Londoño 1997: 32). Despite some notable achievements many practitioners and observers now view development aid as ineffective at best, and counterproductive at worst (Edgren 1995, Elgström 1992, White 1992, 1998, 1999). Research has confirmed this conventional wisdom (Boone 1996, Dollar and Easterly 1999, Dollar and Svensson 1998).

While there are important exceptions (such as the overall national record of economic and social welfare in Botswana and Taiwan and smaller projects throughout the world), the performance of development assistance is not strongly positive. In an effort to assess the impact of aid on economic growth, for example, Burnside and Dollar (1997) examined the growth rate of per capita GNP (averaged over four-year periods) for 56 developing countries beginning in 1970–73 and ending in 1990–93. They found that overall levels of aid had *no* independent effect on GNP growth (see also Boone 1994). Further, many projects and programmes have turned out to be unsustainable in the sense that the recipient government is unable or unwilling to continue the effort after development loans or grants are discontinued. When the funds stop, so does the project.

In recent years, particular focus has been on the perverse incentives generated by some forms of development assistance. Rather than emphasising the lack of material or human resources, a number of analysts have pointed to ways in which the incentives of development aid undermine its effectiveness (Bates 1998, Catterson and Lindahl 1999, Killick et al. 1998). Their work indicates that no matter how well-intentioned the assistance, or how many resources are transferred, development will occur only if political and economic institutions generate incentives that facilitate individuals' achievement of development goals.

Many reasons can be cited for this lack of effectiveness and sustainability. One is the excessive faith in the neutrality of centralised governments. Another is the delivery of aid to authoritarian regimes through naïveté or Cold War strategic thinking (Ertman 1997). Still another is that the aid process itself creates a series of perverse incentives within recipient governments and within agencies responsible for disbursing large volumes of funds (Bräutigam 2000).[3] Levels of corruption grew dramatically, with high costs both in terms of achieving an open, fair and effective government administration and in terms of opportunities lost.[4] The most general factor affecting the success or failure of development aid, however, appears to be the lack of effective institutions at all scales to generate incentives enabling development actors – donors; national, regional, and local governments; NGOs; contractors; investors and the citizens of recipient governments – to cope effectively with diverse collective-action problems.

Collective-action problems exist whenever multiple actors are needed to obtain a jointly beneficial outcome but each actor has a short-term incentive to hold back a full contribution to the joint endeavour. Collective-action problems pervade all aspects of development and resource management. The mental models used in advising policymakers (and in teaching generations of graduate students), however, have yet to focus on the wide array of these problems, which nonetheless must be solved at diverse scales in order for economic agents to gain the trust and assurance they need to sustain economic growth. Indeed, a wide variety of institutional arrangements are needed to encourage economic development.

The centralised state has failed throughout Africa due to its overlooking the self-organising and self-government capabilities of African peoples (Wunsch and Olowu 1995). Instead of focusing on the diversely structured collective-action problems that exist in any complex, dynamic political economy, the predominant mental model used until recently in development assistance was that problems of development should be articulated and tackled by national-level governments of donor countries dealing directly with the sovereign national governments of recipient countries (see Ostrom et al. 2002 and Martens et al. 2002 for recent analyses of the role of institutions and incentives in development). This is particularly paradoxical since many of the scholars involved are economists who recognise the importance of multiple independent actors in a market setting. However, their theory has not addressed how institutions are developed to enhance the emergence of a free, open, competitive market. North stated it well:

> There is no mystery why the field of development has failed to develop during the five decades since the end of the Second World War. Neoclassical theory is simply an inappropriate tool to analyze and prescribe policies that will induce development. It is concerned with the operation of markets, not with how markets develop. How can one prescribe policies when one doesn't understand how economies develop? (North, 1996: 342)

2.2 Natural Resource Policies

During the past half century, a similar set of mental models has dominated the thinking of policy analysts examining a diverse set of problems related to natural resources. The early work of Gordon (1954: 124) focused attention on the problem of over-harvesting from an open access natural resource:

> Wealth that is free for all is valued by no one because he who is foolhardy enough to wait for its proper time of use will only find that it has been taken by another... The fish in the sea are valueless to the fisherman, because there is no assurance that they will be there for him tomorrow if they are left behind today.

When Hardin (1968) dramatised this logic in his famous article *The Tragedy of the Commons*, scholar after scholar proclaimed the necessity of "the" government stepping in (see e.g. Clark 1976, Dales 1968, Ehrenfield 1972, Ophuls 1973). The actual policies varied, but the uniform recommendation was that the initiative for change would come from a national government. Stillman (1975: 13) was among the early commentators to point out the puzzling inconsistency of those who recommended a strong central ruler and thus presumed 'the ruler will be a wise and ecologically aware altruist', while at the same time presuming that the users of natural resources were myopic, self-interested and ecologically unaware hedonists. But somehow, these 'wise and ecologically aware altruists' have repeatedly subsidised the over-extraction of forests and other resources under a diversity of regime structures (Repetto and Gillies 1988).

National governments are seen as needed to devise new rules, impose them on users and enforce new definitions of rights and duties. One of the major shared belief systems in natural resource conservation and management has been the acceptance of the single-species growth curves and the capacity to use scientific investigation to determine the maximum sustainable yield (MSY) of a fishery or other renewable resource. A governmental scientific agency is delegated authority to determine the MSY for a fishery or forest resource and then to assign permits to a level of harvesting that is as large as possible while assuring sustainable yields over the long term (Dolšak 2000, Tietenberg 2002).

In addition to presuming that resource users are themselves helpless to overcome the temptation to over-harvest resources, several other mental models have been widely adopted by those concerned with natural resources. The ecosystem management approach, which is widely taught and applied, urges policymakers to do their analysis at the highest feasible level in order to capture all of the inter-connections. Somehow a national government elected by a general public is implicitly presumed to represent society and thereby is seen as able to do analyses that consider all the effects of its decisions!

Another policy prescription frequently recommended to cope with natural resource problems is total government ownership or control. Thus, governments throughout the world have created various kinds of reserves in areas thought to be suffering from resource depletion or other types of environmental degradation. State agencies have been assigned responsibility for managing natural resources themselves or for contracting out such management under concession arrangements. One of the aims of these agencies has been to reduce the level of disturbance stemming from both human and non-human threats to the resource.

In a cogent critique of policies adopted in the name of increasing the sustainability of natural resources, Ascher (2001) identified a number of perverse processes that occur in the day-to-day implementation of these policies:

- *short-term considerations* stemming from multiple sources, such as short election cycles;
- *perverse learning patterns* arising 'because of oversimplification in the face of complexity, or because the lessons run counter to institutional interests';
- *increased depletion* stemming from a lack of monitoring and control after a national government has declared its ownership of a natural resource, such as a forest, but lacks the resources needed to protect what it has declared that it owns (see Bromley et al. 1992);
- *truncated approaches* serving agency interests by 'enshrining simple strategies at the cost of sustainable protection' (Bromley et al. 1992: 745).

Unfortunately, clearly wrong policies have been adopted in regard to many natural resource systems. In spite of repeated warnings by Newfoundland fishers that the size of the cod they were catching was steadily decreasing, the Canadian agency responsible for management of the extensive eastern coastal zone insisted that its "scientific" data showed evidence of a fishery that could withstand the high levels of withdrawals (Finlayson and McCay 1998). Thus, "scientific information" at a highly aggregated level trumped detailed information about the contents of catches by local fishers (National Research Council 1998). When the collapse in the cod fishery came, it came suddenly (as so many other environmental disasters have occurred) and led to the closure of the entire fishery for years with little evidence of a rebound.

Wilson et al. (2001) questioned the presumption made by many officials that large-scale fisheries should be managed only at a large scale. Recent studies have demonstrated the existence of sub-species of fish at smaller scales than that usually managed by national or international authorities. Ignoring these metapopulations can lead to a different form of overfishing than is usually discussed in contemporary textbooks. 'In particular, rather than overfishing simply by harvesting too many fish, it may be possible to overfish by inadvertently destroying the spatial structure of a population' (Wilson et al. 2001: 60). In a similar vein, many models of natural resources do not address space explicitly.

Multiple studies document the perverse effect of over-reliance on one level of government – usually a large, central regime. As Berkes (2002) documented, over-reliance on central regimes is not confined to centrally planned economies. It occurs in almost all countries where 'resource management functions have been taken over by a managerial elite' (Berkes 2002: 296). Berkes provided typical examples of the types of impacts that central institutions have had on local-level institutions around the world. In Canada, for example, he described the management strategies of the aboriginal hunters from the Arctic and Subarctic who have monitored caribou distributions and migration patterns for many centuries. He pointed out a fundamental difference between the mental models used by the Canadian government and those used by the aboriginals. Users of the aboriginal system do not search for a way of controlling the caribou by developing a self-conscious estimate of herd size and hunting limits. Rather, the hunters pay close attention to the fat content of the caribou they harvest. This provides them with a reliable, qualitative model of the trends – increasing or decreasing – of caribou health over time. When the caribou are seen as less healthy, the normative system of the hunters is to reduce hunting until the fat content of the caribou appears to rise. By using this qualitative model, the hunters learn the direction of change in which a population of wild animals is headed and can respond accordingly (see also Berkes 1999). Furthermore, the cost of this method is dramatically less than the cost of conducting a head count of a widely dispersed population.

Unfortunately, many indigenous knowledge systems and related institutions have been destroyed as a result of new rules imposed by external authorities (see e.g. Mwangi 2003). It is certainly not the case that all indigenous systems are as effective as modern systems based on extensive data collection and analysis. However, many of these systems are being destroyed in the name of protecting natural resources, without due consideration of whether the indigenous system contains mental models and low-cost heuristics worthy of further investigation.

2.3 Underlying Similarities in Policy Analysis Belief Systems

The belief structures underlying a substantial amount of the scholarly literature and policy advice related both to economic development and to natural resources management share several core assumptions. Both fields share a fundamental commitment to the use of scientific approaches to help elucidate the core variables involved in a process and how they are related. Analysts in both fields have developed relatively simple models of the underlying problems of interest.

Benefit-cost analyses are repeatedly used to demonstrate that the benefits to a developing country of building a road, electric power project or major waterworks will be greater than the costs involved. Models of the economic costs and benefits of various harvest levels are also often used in analysing the relative efficiency of taxes, permits or transferable quota systems.

Scholars in both fields are committed to a determination of an *optimal* policy intervention – frequently a technological fix – and its implementation by a unified regime (even when that regime allocates private, transferable rights). Finding an optimal policy intervention is consistent with a presumption that most of the relevant processes involve a single beneficial equilibrium that is calculable through analysis and achievable through clear policy interventions. The problem identified for policy analysis is to develop a model of the system, determine the variables that affect its performance, determine which variables can be positively affected by a policy change and then develop and implement policies that induce the system to perform at optimal levels. To do this requires the advice of an expert.

Scott (1998) characterised as "seeing like a state" the belief system underlying many efforts to solve problems centrally through expert advice while ignoring the interests, information and capabilities of others involved. Scott attributed the gross failure of much such problem-solving to a belief system that he calls "high modernism".[5] When governments with strong powers adopt this belief system where citizens have a weak voice, the results have been massive tragedies in the twentieth century:

> High modernism must not be confused with scientific practice. It was, as the term ideology implies, fundamentally a faith which borrowed, as it were, the legitimacy of science and technology. It was, accordingly, uncritical, unskeptical, and thus unscientifically optimistic about the possibilities for the comprehensive planning of human settlement and production. The carriers of high modernism tended to see rational order in highly visual aesthetic terms. An efficient and rational city, village, or farm was, to them, a city that looked regimented and orderly in a geometric sense (Scott 1997: 4).

Many policy analysts argue that Scott is not criticising them but rather the authoritarian rulers of Communist Russia and many African countries. Extraordinary faith in the capacity of simple models to be used as the foundation for policies in widely diverse environments bears a striking similarity to Scott's concept of high-modern ideology (see also Ostrom 1989, 2002). The belief is so widespread that Holling, Gunderson and Ludwig (2002), after extensive review of many natural resource failures, refer to this as 'the trap of the expert'. They ask, 'Why does expert advice so often create crisis and contribute to political gridlock? Why, in many places, does science have a bad name?' (Holling et al. 2002: 7).

3. SOCIAL-ECOLOGICAL SYSTEMS AS COMPLEX ADAPTIVE SYSTEMS

During the same era in which the approach to develop natural resources outlined earlier in this chapter rose to dominance, another approach slowly evolved as more and more scholars and practitioners became disillusioned with the performance of policies based on the dominant views. Following in the steps of Simon (1989, 1996), Holling (1973), Axelrod (1984, 1997) and Holland (1995), this approach views a variety of social-ecological systems as complex adaptive systems.

In complex adaptive systems the components and the structure of interactions between the components are able to adapt themselves to internal and external disturbances. From this perspective, the simple models used by many resource managers are not all wrong. Rather, they are only partial models of much more complex interactions. In some cases, the partial models have been sufficiently useful as to form a good mental model for policy recommendations. In systems that are indeed more complex, there is a further need to understand processes of organisation and reorganisation including collapse and what is likely to happen after collapse. Does a system have one and only one equilibrium to which it returns after a major shock and temporary collapse? Or are there multiple equilibria with different characteristics? How easy is it for a system to flip from a desirable equilibrium to an undesirable one? These are crucial questions.

Order in complex systems is emergent as opposed to predetermined. The system's history is irreversible and future behaviour is path-dependent. The system's future is often unpredictable due to the non-linearity of many basic causal relationships. The variables that affect performance move both fast and slow. If information about slow-moving variables is not recorded for a long period of time, substantial surprises can occur when such a variable reaches a threshold. In social-ecological systems the social components are individuals and institutions. Individuals may change their relations with other individuals, may change their strategies and may change the rules they abide by. In fact, individual strategies and institutional rules interact and co-evolve, sometimes in unpredictable ways.

The complex adaptive systems perspective provides the view of individuals within a variety of situations structured by the biophysical world, institutional rules and the community in which they interact. Boundedly rational individuals trying to do as well as they can in uncertain situations continuously tinker with their strategies including trying to change the rules that affect particular situations. Within ongoing structures, individuals search out perceived advantageous strategies given the set of costs and benefits that exist and the strategies that others adopt. They may look for loopholes in the law – particularly if they think others are doing the same. They may check the level of enforcement by breaking rules occasionally. Those responsible for changing the rules of an institution also experiment with new rules and try to learn from others why some other institutional arrangements appear to work better than their own.

One of the concepts that can be used to evaluate the dynamics of complex adaptive systems is resilience (Gunderson and Holling 2002, Holling 1973). Resilience is defined as the amount of disruption needed to transform a system from one stability domain (characterised by a configuration of mutually reinforcing processes and structures) to another. The concept of resilience originates from ecology. Ecological resilience may be measured 'as the width of the desirable attractor' in a multi-equilibria system 'measured in units of the fast variable' (Carpenter et al. 2002). Resilience has been used to understand how to improve management of systems so as to reduce ecosystem vulnerability. Many examples have been documented where ecosystems have shifted from desired configurations (e.g. a productive rangeland or clean lake) to undesirable configurations (e.g. degraded rangelands or eutrophic lakes) (Scheffer et al. 2001). Human activities have reduced the resilience of these managed ecosystems over time, making them vulnerable to disturbances.

3.1 The Adaptive Cycle

The adaptive cycle is a heuristic model useful to understand the dynamics of complex adaptive systems (Gunderson and Holling 2002, Holling 1986). Figure 1 provides a stylised picture of the adaptive cycle. Although the adaptive cycle originated from ecology, it has also been applied to social-ecological systems (Gunderson et al. 1995, Gunderson and Holling 2002, Walker et al. 2002). Specific adaptive cycles will vary substantially in terms of temporal and spatial scale, the number of relatively stable domains that exist and the structure of fast-moving and slow-moving variables. The adaptive cycle involves the movement of a system through four phases: a period of rapid growth and exploitation (*r*); leading into a long phase of accumulation, monopolisation and conservation of structure, during which resilience tends to decline (*K*); a very rapid breakdown or release phase (*Ω*); and finally a relatively short phase of renewal and reorganisation (*α*). If, in this final phase the system is resilient and still sufficiently retains its previous components, it can reorganise to remain within the same configuration as before. This is also a time when novelty can enter – new species, new institutions, ideas, strategies, policies and industries – and the "new" emerging system, whether it is in the same or a different configuration, gains some degree of resilience.

The "forward" (*r* to *K*) and "backloop" (*Ω* to *α*) dynamics of the adaptive cycle correspond to managing for production and managing for innovation: both are important objectives. They can be likened, in the area of investment, to the part of the portfolio aimed at maximising income (*r* to *K*) and the part aimed at maximising flexibility to cope with and adapt to unexpected change in the market (*Ω* to *α*). Just as there are costs and benefits involved in diversifying an investment portfolio, so there are costs and benefits involved in building resilience. There are trade-offs and synergies between production and resilience. Achieving both objectives requires an understanding of when it is appropriate to try to increase production efficiency and when (and where) it is appropriate to work to ensure sustainability.

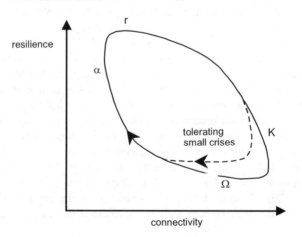

Figure 1. *The Adaptive Cycle*

The stages can be summarised as follows:

- *r to K*. There are strong controls and the system changes slowly. Regulatory policies and efforts to increase efficiency may be appropriate, although careful experimentation is sometimes critical. Application of techniques such as optimal control can be useful in this stage. However, resilience can be lost through gradual changes in underlying slow variables.

- *Ω to α*. The system changes rapidly. In this phase, no equilibria exist, there is turbulence and consequently novelty can enter. What is the appropriate approach to research and management? How can creative and potentially resilient new practices be discovered? The system is susceptible to loss of resources (soil erosion, species, human and financial capital) and measures to conserve capital are appropriate. It is also vulnerable to entering a potentially undesirable configuration. Guidance is needed. Influential ideas ("good" and "bad") can become entrenched and guide subsequent evolution of the system.

In the *r* to *K* phase, the connectivity increases, but resilience decreases and the system becomes vulnerable to disturbances. All systems face small to large disturbance from external or internal sources. Very few systems can be "controlled" and disturbance-free for long periods of time. In fact, efforts to control small disturbances may reduce a system's resilience to large disturbances. However, it is not always possible to judge whether a disturbance is small or large. Like immune systems are not perfectly able to attack all harmful invasions, and we might sometimes get ill, resiliency of systems is also determined by the ability to cope with the errors we make in judging the type of disturbance. A system trained in dealing with disturbances is more likely to anticipate when a disturbance is severe than systems that suppress disturbances.

A risk of decreasing resilience is that a crisis will render the system unable to recover into the same configuration; the system then flips into an alternative stability domain. For example, suppressing forest fires causes an accumulation of fuel on the forest floor and an accumulation of tree biomass. When a fire finally occurs, it will be hot and intense, affecting soil conditions and the capacity of the forest to recover from fire events. The system has flipped. Management can reduce the risk of flipping into an undesirable stability domain by tolerating small crises in order to prevent a big one. The draconian fire-suppression doctrine of the US Forest Service, the Smokey the Bear programme, was a clear and understandable mission to fight all fires in national forests. Tragically, it was a clear policy, but one that was clearly wrong (see Pyne 1982, 1996). A political example of losing resilience is an authoritarian regime which represses all contestation and debate at an early point and then finds itself embroiled in a full-scale civil war.

During the *Ω* to *α* phase institutional innovation can be significant. This is the phase when public entrepreneurs may be able to find new combinations of inputs so as to move the entire system to a more productive functioning and broaden the range of the attractor for positive system performance (or the resilience of the system) (Kuhnert 2001, Schneider et al. 1995).

3.2 Multiple Scales

A social-ecological system does not consist of just one kind of cycle at one scale. It functions as a nested, hierarchical structure, with processes clustered within subsystems at several scales (e.g. the farm, region and state). Different subsystems, at different scales, may be in different phases and may change at different rates (Gunderson and Holling 2002). The subsystems are semi-autonomous, but cross-scale interactions do occur. Particular attention needs to be paid to these cross-scale interactions.

Connections between the different levels of scale can be labelled as "trigger" and "remember". In the original ecological description of the adaptive cycle at different scales, the term "revolt" was used to convey the suggestion that fast and small events overwhelm slow and large ones, and the term "remember" was used to indicate that the experience accumulated at the larger level can be used to stimulate renewal at the smaller level (Gunderson and Holling 2002). However, we argue that in social-ecological systems the two connections can affect both the smaller and the larger scale systems. Therefore, instead of "revolt", we use the term "trigger" to denote a destabilising factor from one scale to another.

Figure 2 shows the adaptive cycle for three scales. Crisis at a smaller scale may trigger collapses on a larger scale. For example, the Asian financial crisis started in 1997 when some Thai companies could not pay the interest on their debts, leading to a cascading fall-down of financial institutions. The more than 200 unfinished skyscrapers in Bangkok are still visible monuments of the fast rate and large scale of the crisis. On the other hand, disturbances can also be triggered by developments at a larger scale, as we see in the globalised world of today. The *Human Development*

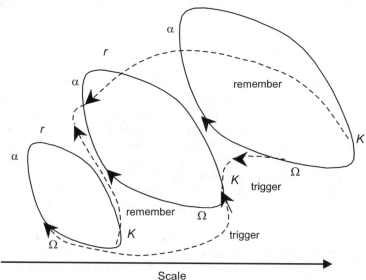

Figure 2. *The Adaptive Cycle at Three Scales*

Report (UNDP 2002), for example, stresses that while many national governments in developing countries have formally become democracies, they have not yet overcome major problems of political favouritism and corruption. Tragically, over 60 countries had lower mean per capita incomes in 2002 than they had in 1990 and more than 3.6 million lives have been lost in the same era due to civil wars. As Mark Malloch Brown of UNDP indicated in an interview with Barbara Crossette in the *New York Times* of 24 July 2002, 'The concern is that one multiparty election does not a democracy make.' He reflected that '[t]he international cheerleaders for democracy have underestimated what it takes to build a functioning, properly rooted democracy'.

Once a major disturbance occurs, memory from higher and lower scales may speed up and direct recovery. Memory is particularly important, since most complex adaptive systems are structured by at least three to five major interacting components that change at a fast, moderate or slow pace. Surprises and unpredictability are endemic in these systems. Responding to rapid change may become routinised, and may or may not improve the ability of a system to respond to moderately or slowly changing variables. Without effective memory at some level, however, responding to slow changes may involve experiencing a major disaster before an adequate set of responses is found anew. Thus, the importance of having relatively independent response capabilities and stored memories at multiple levels is one of the key lessons of recent research on multi-scale, human-ecological systems. Opening management systems to gain information and perspectives from multiple sources – in addition to scientific experts – is another important lesson (Gunderson and Holling 2002).

4. MULTI-SCALE ADAPTIVE PROCESSES IN THE HISTORY OF POLITICAL-ECONOMIC DEVELOPMENT

In an effort to understand how development may be enhanced by multi-level (or polycentric)[6] governance involving the capacity to initiate or veto action at multiple scales, we turn to two examples from Dutch history. The first is the role of waterboards in the historical development of the Netherlands. The second is the importance of smaller scale indigenous irrigation institutions in Bali and how "experts" working for both the Dutch and Indonesian governments during the past century misunderstood the operation of these institutions.

4.1 Dutch Water Management Through the Ages

A typical Dutch institution, namely the waterboards, serves to illustrate the evolution of successful multi-level governance.[7] Before 800 A.D. the inhabitants of the precursor of the Netherlands used non-structural measures to keep their feet dry. Such measures, like manmade hills or abandoning areas when there was danger of flooding, were the result of decisions made by individual households. Increased population pressure, technological know-how and finance led to the development of more structural water-control measures after 800 A.D. These measures, including dikes and sluices, required cooperation within a community to construct and maintain. Farmers whose lands bordered directly on the dikes agreed to commit themselves to the construction work and to maintenance activities afterwards.

At almost the same time, drainage activities were developed as well. These made lowland areas habitable by getting rid of superfluous water. Small dams and sluices were built and maintained, based on agreements similar to those for the flood-protection systems. The noticeable difference with the inputs for dike maintenance was that from the beginning all beneficiaries had to pay for the benefits received from the drainage activities.

Originally, the local communities in the countryside were in charge of all general collective interests and took responsibility for water management. Around 1100 A.D., however, a new adaptation occurred when water management tasks gradually began to be separated from the general public tasks. The reason was probably increased occurrence and severity of flooding as well as a growing interdependence and complexity of the hydraulic works that began to stretch out beyond the local scale.

At the end of the eleventh century and the beginning of the twelfth century the first public bodies for local and regional water management appeared on the scene and the phenomenon of the waterboards was born. The purpose of these boards was to ensure safety from flooding through the construction and maintenance of dikes and dry feet through drainage by means of the necessary hydraulic structures. The establishment of the waterboards was recognised by the higher, regional authorities who still held themselves responsible for water management but resigned from their administrative water management duties.

Each of the waterboards differed in design and the way it implemented physical structures as well as rules. They were also confronted with different problems. They were not always successful in preventing floods or draining areas effectively. During the Republic of the United Provinces of the Netherlands there were severe floods, and extensive peat-digging caused unintended artificial lakes and other management problems. Still, the waterboards survived. Dolfing (2000) argued that the main reason for the waterboards' long-term adaptation and survival is the institutional arrangements. The rules for the waterboards were designed based on the shared norms and values of the population. Although the boards were not always successful in maintaining safety and dry feet, preserving the institutional arrangements that they were familiar with and could adapt was seen as more important than switching to new and unfamiliar institutional arrangements. Perhaps the roots of the shared norms in contemporary Dutch society go back to those people who found ways to make the land inhabitable by developing institutions based on reciprocity (see Toonen 1996).

The history of the waterboards shows a continuous tinkering with the rules at different scales. Disturbances like floods and the results of peat digging triggered new rules and structures. The Dutch waterboards illustrate how local-level governments can evolve into a resilient collaboration of multi-level governance when national institutions recognise the importance of smaller governance units and work with them rather than destroying them.

Does the Dutch tradition of water management help the country's engineers and officials to manage water elsewhere? An illustrative example where national officials for some time did not recognise the value of local institutions is the recent history of rice production in Bali, a small island east of Java of the former Dutch colony Indonesia.

Bali's complex irrigation system has existed for more than a thousand years. Irrigators in Bali face a coordination problem. On the one hand, control of pests is most effective when all rice fields have the same planting schedule. On the other hand, the terraces are hydrologically interdependent, with long and fragile systems of weirs, tunnels, canals and aqueducts. To balance the need for coordinated fallow periods and use of water, a complex calendar system has developed which states what actions should be done on specific dates. These actions are related to offerings to temples: the small temples at the rice terrace level, the temples at the village and regional levels, up to the *Pura Ulun Swi*, the "Head of the Rice Terraces" temple of the high priest Jero Gde, the human representative of the Goddess of the Temple of Crater Lake. This crater lake feeds the groundwater system which is the main source of water for irrigation. These offerings were to reciprocate for the use of water that belonged to the gods.

Bali consisted of many kingdoms before its conquest by the Dutch around 1900. The Dutch saw these offerings in a different light, namely as a royal irrigation tax. The fact that during the nineteenth century there were a number of kingdoms in Bali was a sign to the Dutch that the institution of kingship had weakened over time, with one powerful kingdom disintegrating into a number of little kingdoms. The Dutch wanted to restore centralised government; in particular they wanted to use a revived royal irrigation tax to improve the irrigation system. The Dutch administrative reorganisation failed, partly due to lack of funding, but also, as a historical analysis conducted during the 1930s demonstrated, because there was no tradition of a centralised government from the past. After World War II, Indonesia became independent from the Netherlands, though the colonial bureaucratic systems were taken over by the new independent government.

During the late 1960s the Indonesian government made self-sufficiency in rice a major national development goal. In the same period the Green Revolution began in Asia, the spread of new rice-growing technologies that promised a dramatic increase in rice production. Bali was one of the first targets of the Green Revolution. In contrast to the earlier Dutch attempts to modernise rice production, this time the engineers were well-funded.

The function and power of the water temples were invisible to the planners involved in promoting the Green Revolution. They regarded agriculture as a purely technical process. Farmers were forced to switch to the "miracle" rice varieties, which would lead to three harvests a year instead of the two of the traditional varieties. Farmers were stimulated by government programmes that subsidised the use of fertilisers and pesticides. The farmers continued performing their rituals, but now they no longer coincided with the timing of rice farming activities. Soon after the introduction of the miracle rice, a plague of plant-hoppers caused huge damage to the rice crop. A new variety was introduced, but then a new pest plague hit the farmers. Furthermore, there were problems of water shortage.

During the 1980s an increasing number of farmers wanted to switch back to the old system, but the engineers interpreted this as religious conservatism and resistance to change. Lansing quoted a frustrated American irrigation engineer, 'These people don't need a high priest, they need a hydrologist!' (Lansing 1991: 115). It was Lansing who unravelled the function of the water temples and was able to convince the financers of the Green Revolution project on Bali that irrigation was best coordinated at the level of the water temples. Lansing built a computer model of the artificial ecosystem and showed that for different levels of coordination, from the farmer level up to central control, the temple level was the scale at which decisions could best be made to maximise rice production (see also Lansing and Kremer 1994).

As this story suggests, the complex irrigation systems and the role of the temples have evolved over a long history of local adaptations, at different levels of scale. The water temples played a significant role in coordination of the use of water, but also in providing technical advice and mediating conflicts between different *subaks,* or local cooperative groups of farmers, on water use. The offering to the different temples made the farmers aware of the interconnections between the water flows at different levels. Due to Lansing's insight and analysis, some of these systems have evolved still further and avoided the fate of many self-organised systems of this kind when experts declared them defunct and constructed new infrastructures without paying much attention to local property rights, ecology, culture and tradition.

5. CONCLUSION: COPING WITH COMPLEXITY

The last half-century has witnessed scholars from a variety of disciplines adopting a belief system that Scott (1998) calls "high modernism". High modernists try to suppress complexity through the design of unitary governments that rely on the advice of experts to optimise progress towards a preferred social goal. In the developing world, the advice of experts has led to the suppression of indigenous institutions that evolved over centuries, leaving post-colonial governments as virtual monopolists of official power. Without facing independent and strong organisations in the public or economic sphere (except for multinational corporations which frequently bribed their way to autonomous operation), political leaders did what monopolists usually do – maximised their own short-term interests. While chunks of national coffers were transferred to private bank accounts, the majority of their own citizens have been left to live on less than two US dollars per day (UNDP 2002). With regard to the protection of natural resources, the approach has led to creation of parks on paper rather than real-life functioning biodiversity reserves (Brandon 1995). It has also produced major losses of ocean fisheries, increasing vulnerability to drought, and pest control efforts that have paradoxically resulted in chronic pest outbreaks (Holling et al. 2002).

It is time to declare this belief system bankrupt!

Fortunately, we do not need to start over. While proponents have extolled the virtues of high modernism, researchers in multiple disciplines have worked on a variety of approaches to the study of living complex orders. This chapter discussed the fruitful approach of viewing ecological systems as complex, multi-tiered, adaptive systems that vary in their capacity to be resilient in light of diverse temporal and spatial disturbances. A complementary approach is the study of polycentric systems (Toonen 1988). Ostrom's definition of a polycentric order is one 'where many elements are capable of making mutual adjustments for ordering their relationships with one another within a general system of rules where each element acts with independence of other elements' (Ostrom 1999: 57).[8] Market systems, judicial systems, water management, irrigation systems and many natural resource systems that have evolved over long periods of time – like the irrigation systems of Bali – are polycentric or multi-level in structure.

Polycentric systems are in no way guaranteed to cope successfully with all of the problems of complex human and human-ecological systems. No system of governance is. Substantial research has, however, been conducted on the performance of governing units serving metropolitan areas within the United States. A consistent finding is that metropolitan areas characterised by large, medium and small public and private agencies with considerable autonomy but which also face incentives to seek out opportunities for complementary efforts tend to outperform metropolitan areas served by a few large-scale units (see McGinnis 1999a, 1999b, 2000 and literature cited therein). Frey (1994) and Pommerehne (1990) have undertaken extensive studies providing strong evidence of the performance of complex governance systems.

Ashby (1960) recognised long ago that to achieve any level of regulation in a system, one needs to design in as much variety in the response capabilities as exist in the relevant system. What we have learned is that the hot debates about opposites – small-scale versus large-scale, centralised versus decentralised, top-down versus bottom-up – lead nowhere. Resilient adaptive systems need attributes of all of the above. What we do need is careful empirical research that helps us to better understand how multi-level or polycentric governance systems work, how they adapt over time, what are the major threats to their continued resilience and how we can build even better resilient, learning, complex systems in the future.

NOTES

1. This chapter was prepared for delivery at the ISS Conference on 'Environmental Degradation, Institutions, and Conflict,' 8 and 9 October 2002. An earlier version was presented at the Annual Meeting of the American Political Science Association, Boston, Massachusetts, 29 August to 1 September 2002. The authors appreciate support from the National Science Foundation and the Resilience Alliance. We also appreciate the careful work of Sarah Kantner in preparing this manuscript and of Laura Wisen in helping run down elusive citations.

2. According to World Bank poverty data, the absolute number of individuals subsisting on less than US $1.00 per day has grown, but the percentage of the population living in poverty in developing countries has been slightly reduced since 1987 (World Bank 2000).

3. In examining the problem of aid dependence, Bräutigam (2000: 1) stressed, 'Large amounts of aid delivered over long periods, create incentives for governments and donors that have the potential to undermine good governance and the quality of state institutions.'

4. As reflected in a World Bank report, 'A 10 per cent bribe on the cost of a good public investment project depresses the project's rate of return only slightly. A bribe that saddles the country with a white elephant investment may result in economic costs far exceeding the corrupt payment, particularly if the policy environment causes a value-subtracting investment to appear nominally profitable' (World Bank 1997b: 57).

5. Ostrom (1989) stressed that this type of thinking was responsible for an "intellectual crisis" in the study of American public administration.

6. We are using the terms multi-level and polycentric as having similar meanings. Hooghe and Marks (2001) consider polycentric governance as one or two major forms of multi-level governance.

7. This brief history of the waterboards is mainly based on Dolfing (2000) and Kaijser (2002).

8. Another useful approach undertaken by legal anthropologists is that of legal pluralism. For a recent overview see Meinzen-Dick and Pradhan (2002).

REFERENCES

Andersen, M.S. (2001) *Economic Instruments and Clean Water: Why Institutions and Policy Design Matters*. Paris: OECD.
Ascher, W. (2001) 'Coping with Complexity and Organizational Interests in Natural Resource Management', *Ecosystems* 4: 742–757.
Ashby, W.R. (1960) *Design for a Brain. The Origin of Adaptive Behavior.* 2nd ed. New York: John Wiley.
Axelrod, R. (1984) *The Evolution of Cooperation.* New York: Basic Books.
Axelrod, R. (1997) *The Complexity of Cooperation.* Princeton: Princeton University Press.
Bates, R. (1998) 'Institutions as Investments', in S. Borner and M. Paldham (Eds) *The Political Dimensions of Economic Growth*. New York: Macmillan.
Bauer, P.T. (1984) 'Remembrance of Studies Past: Retracing First Steps', in Gerald M. Meier and Dudley Seers (Eds) *Pioneers in Development*, pp. 25–43. New York: Oxford University Press.
Berkes, F. (1999) *Sacred Ecology: Traditional Ecological Knowledge and Resource Management.* Philadelphia and London: Taylor & Francis.
Berkes, F. (2002) 'Cross-Scale Institutional Linkages: Perspectives from the Bottom Up', in E. Ostrom, T. Dietz, N. Dolsak, P.C. Stern, S. Stonich and E. Weber (Committee on the Human Dimensions of Global Change) (Eds) *The Drama of the Commons*, pp. 293–322. Washington, D.C.: National Research Council, National Academy Press.
Birdsall, N. and J. L. Londoño (1997) 'Asset inequality matters: An assessment of the World Bank's approach to poverty reduction', *American Economic Review* 87 (2): AEA Papers and Proceedings, pp. 32–37.
Boone, P. (1994) 'The impact of foreign aid on savings and growth'. Working Paper. London: London School of Economics.
Boone, P. (1996) 'Politics and the effectiveness of foreign aid', *European Economic Review* 40: 289–329.
Brandon, K. (1995) 'People, parks, forests or fields: A realistic view of tropical forest conservation', *Land Use Policy* 12 (2): 137-44.
Bräutigam, D. (2000) *Aid Dependence and Governance*. Stockholm: Almqvist & Wiksell International.
Bressers, H. (1988) 'A comparison of the effectiveness of incentives and directive: The case of the Dutch water quality policy', *Policy Studies Review* 7 (3): 500–518.
Bromley, D.W. (Ed.) (1992) *Making the Commons Work: Theory, Practice, and Policy.* San Francisco: ICS Press.
Burnside, C. and D. Dollar (1997) 'Aid spurs growth: In a sound policy environment', *Finance & Development* (December): 4–7.
Carpenter, S.R., W.A. Brock and D. Ludwig (2002) 'Collapse, learning and renewal', in L.H. Gunderson and C.S. Holling (Eds) *Panarchy. Understanding Transformation in Human and Natural Systems*, pp. 173–194. Washington, D.C.: Island Press.
Catterson, J. and C. Lindahl (1999) *The Sustainability Enigma: Aid Dependency and the Phasing Out of Projects: The Case of Swedish Aid to Tanzania*. Stockholm: Almqvist and Wiksell International.
Clark, C.W. (1976) *Mathematical Bioeconomics.* New York: Wiley.
Crossette, B. (2002) 'U.N. report says new democracies falter', *New York Times*, 24 July, p. A8.
Dales, J.H. (1968) *Pollution, Property, and Prices: An Essay in Policy Making and Economics*. Toronto: University of Toronto Press.
Dieckmann, U., R. Law and J. Metz (Eds) (2000) *The Geometry of Ecological Interactions: Simplifying Spatial Complexity.* Cambridge, UK: Cambridge University Press.

Dolfing, B. (2000) 'Waterbeheer Geregeld? Een historisch bestuurskundige analyse van de institutionele ontwikkeling van de hoogheemraadschappen van Delfland en Rijnland 1600-1800', Ph.D. Thesis, Leiden University.

Dollar, D. and W. Easterly (1999) 'The search for the key: Aid, investment and policies in Africa'. Policy Research Working Paper 2070. Washington, D.C.: World Bank.

Dollar, D. and J. Svensson (1998) 'What explains the success or failure of structural adjustment programs?' Policy Research Working Paper 1938. Washington, D.C.: World Bank Development Research Group.

Dolšak, N. (2000) 'Marketable Permits: Managing Local, Regional, and Global Commons', Ph.D. Diss., Indiana University, School of Public and Environmental Affairs and the Department of Political Science.

Edgren, G. (1995) 'Indo-Swedish development cooperation: Objective, issues and achievements', in V. Sahai (Ed.) *Sharing Challenges, The Indo-Swedish Development Cooperation Programme*, pp. 8-46. Stockholm: Ministry for Foreign Affairs.

Ehrenfield, D.S. (1972) *Conserving Life on Earth*. Oxford: Oxford University Press.

Elgström, O. (1992) *Foreign Aid Negotiations*. Aldershot, England: Avebury.

Ertman, T. (1997) *Birth of the Leviathan*. Cambridge: Cambridge University Press.

Finlayson, A.C. and B.J. McCay (1998) 'Crossing the Threshold of Ecosystem Resilience: The Commercial Extinction of Northern Cod', in F. Berkes and C. Folke (Eds) *Linking Social and Ecological Systems: Management Practices and Social Mechanisms for Building Resilience*. Cambridge University Press.

Frey, B. (1994) 'Direct democracy: Politico-economic lessons from Swiss experience', *American Economic Review* 84 (2): 338–348.

Gordon, H.S. (1954) 'The economic theory of a common-property resource: The fishery', *Journal of Political Economy* 62: 124–142.

Gunderson, L.H., C.S. Holling and S.S. Light (Eds) (1995) *Barriers and Brdiges to the Renewal of Ecosystems and Institutions*. New York: Columbia University Press.

Gunderson, L.H. and C.S. Holling (Eds) (2002) *Panarchy. Understanding Transformation in Human and Natural Systems*. Washington, D.C.: Island Press.

Hardin, G. (1968) 'The tragedy of the commons', *Science* 162: 1,243–1,248.

Holland, J.H. (1995) *Hidden Order: How Adaptation Builds Complexity*. Reading, MA: Addison Wesley.

Holling, C.S. (1973) 'Resilience and stability of ecological systems', *Annual Review of Ecology and Systematics* 4: 1–24.

Holling, C.S. (1986) 'The resilience of terrestrial ecosystems: Local surprise and global change', in W.C. Clark and R.E. Munn (Eds) *Sustainable Development of the Biosphere*, pp. 292–317. Cambridge, UK: Cambridge University Press.

Holling, C.S., L.H. Gunderson and D. Ludwig (2002) 'In Quest of a Theory of Adaptive Change', in L.H. Gunderson and C.S. Holling (Eds) *Panarchy. Understanding Transformation in Human and Natural Systems*, pp. 3–22. Washington, D.C.: Island Press.

Hooghe, L. and G. Marks (2001) *Multi-level Governance and European Integration*. Oxford: Rowman and Littlefield.

Kaijser, A. (2002) 'System Building from Below: Institutional Change in Dutch Water Control Systems', *Technology and Culture* 43 (3): 521–548.

Killick, T., R. Gunatilaka and A. Mar (1998) *Aid and the Political Economy of Policy Change*. London and New York: Routledge Press.

King, A. (1995) 'Avoiding ecological surprise: Lessons from long-standing communities', *Academy of Management Review* 20 (4): 961–985.

Kuhnert, S. (2001) 'An evolutionary theory of collective action: Schumpeterian entrepreneurship for the common good', *Constitutional Political Economy* 12: 13–29.

Lansing, J.S. (1991) *Priests and Programmers: Technologies of Power in the Engineered Landscape of Bali*. Princeton, NJ: Princeton University Press.

Lansing, J.S. and J.N. Kremer (1994) 'Emergent properties of Balinese water temple networks: Co-adaption on a rugged fitness landscape', in C.G. Langton (Ed.) *Artificial Life III: Studies in the Sciences of Complexity*, 201–23. Reading, MA: Addison-Wesley.

Martens, B., U. Mummert, P. Murrell and P. Seabright (2002) *The Institutional Economics of Foreign Aid*. Cambridge: Cambridge University Press.

258   ELINOR OSTROM & MARCO JANSSEN

Mauro, P. (1995) 'Corruption and growth', *Quarterly Journal of Economics* 110 (Aug.): 681–712.
Meinzen-Dick, R.S. and R. Pradhan (2002) 'Legal pluralism and dynamic property rights', CAPRi Working Paper No. 22. Washington, D.C.: International Food Policy Research Institute.
McGinnis, M. (1999a) *Polycentricity and Local Public Economies: Readings from the Workshop in Political Theory and Policy Analysis*. Ann Arbor: University of Michigan Press.
McGinnis, M. (1999b) *Polycentricity and Development: Readings from the Workshop in Political Theory and Policy Analysis*. Ann Arbor: University of Michigan Press.
McGinnis, M. (2000) *Polycentric Games and Institutions: Readings from the Workshop in Political Theory and Policy Analysis*. Ann Arbor: University of Michigan Press.
Mwangi, Samuel (2003) *Indigenous Knowledge, Policy and Institutional Issues for Collaboration Between Mountain Adjacent Communities and Management Agencies,* Nairobi: Kenya Resource Centre for Indigenous Knowledge
National Research Council (1998) *Sustaining Marine Fisheries*. Washington, D.C.: National Academy Press.
North, D.C. (1996) 'Epilogue: Economic performance through time', in L.J. Alston, T. Eggertsson and D.C. North (Eds) *Empirical Studies in Institutional Change*, pp. 342–355. Cambridge: Cambridge University Press.
Ophuls, W. (1973) 'Leviathan or oblivion', in H.E. Daly (Ed.) *Toward a Steady State Economy*, pp. 215–230. San Francisco, CA: Freeman.
Ostrom, E., C. Gibson, S. Shivakumar and K. Andersson (2002) *Aid, Incentives, and Sustainability: An Institutional Analysis of Development Cooperation. MAIN REPORT*. Sida Studies in Evaluation 02/01. Report for the Department for Evaluation and Internal Audit (UTV) and the Swedish International Development Cooperation Agency (Sida). Stockholm, Sweden: Sida.
Ostrom, V. (1989) *The Intellectual Crisis in American Public Administration*. 2nd ed. Tuscaloosa: University of Alabama Press.
Ostrom, V. (1999) 'Polycentricity', in M. McGinnis (Ed.) *Polycentricity and Local Public Economies: Readings from the Workshop in Political Theory and Policy Analysis*, pp. 52–74. Ann Arbor: University of Michigan Press.
Ostrom, V. (2002) 'Why were there so many constitutional failures in the twentieth century?' Paper presented at the 2002 meeting of the Public Choice Society held at the US Grant Hotel, San Diego, California (22–24 March).
Peluso, N.L. (1992) *Rich Forests, Poor People: Resource Control and Resistance in Java*. Berkeley: University of California Press.
Pommerehne, W.W. (1990) 'The Empirical Relevance of Comparative Institutional Analysis', *European Economic Review* 34 (2–3): 458–469.
Pyne, S.J. (1982) *Fire in America: A Cultural History of Wildland and Rural Fire*. Princeton: Princeton University Press.
Pyne, S.J. (1996) 'Nouvelle southwest', in *Proceedings from U. S. Forest Service Conference on Adaptive Ecosystem Restoration and Management: Restoration of Cordilleran Confer Landscapes of North America*, pp. 10–15. General Technical Report No. RM-GTR-278. Flagstaff, AZ: U. S. Forest Service, 6–8 June.
Repetto, R. and M. Gillis (Eds) (1988) *Public Policies and the Misuse of Forest Resources*. Cambridge University Press.
Scheffer, M., S. Carpenter, J. Foley, C. Folke, and B. Walker (2001) 'Catastrophic Regime Shifts in Ecosystems', *Nature* 413: 591–596.
Scheffer, M., F. Westley, W.A. Brock and M. Holmgren (2002) 'Dynamic Interaction of Societies and Ecosystems – Linking Theories from Ecology, Economy, and Sociology', in L.H. Gunderson and C.S. Holling (Eds) *Panarchy. Understanding Transformation in Human and Natural Systems*, pp. 195–239. Washington, D.C.: Island Press.
Schneider, M., P. Teske and M. Mintrom (1995) *Public Entrepreneurs: Agents for Change in American Governments*. Princeton, NJ: Princeton University Press.
Schuurman, J. (1988) *De Prijs van Water*. Arnhem: Gouda Quint BV.
Scott, J.C. (1997) 'Seeing Like a State. Legibility, Simplification, and the Creation of State Spaces', in *Yale ISPS* 1: 1–7.
Scott, J.C. (1998) *Seeing Like a State: How Certain Schemes to Improve the Human Condition have Failed*. New Haven, CT: Yale University Press.

Simon, H.A. (1989) *Models of Thought*. New Haven, CT: Yale University Press.

———— (1996) *The Sciences of the Artificial*. 3rd ed. Cambridge, MA: MIT Press.

Stillman, P.G. (1975) 'The tragedy of the commons: A re-analysis', *Alternatives* 4: 12–5.

Stone, R.D. (1992) *The Nature of Development: A Report from the Rural Tropics on the Question for Sustainable Economic Growth*. New York: Alfred A. Knopf.

Tietenberg, T. (2002) 'The tradable permits approach to protecting the commons: What have we learned?' in E. Ostrom, T. Dietz, N. Dolsak, P.C. Stern, S. Stonich and E. Weber (Committee on the Human Dimensions of Global Change) (Eds) *The Drama of the Commons*, pp. 197–232. Washington, D.C.: National Research Council, National Academy Press.

Toonen, T.A.J. (1988) 'Monocentrisme, polycentrisme en de economie van het openbaar bestuur', in A.F.A. Korsten and T.A.J. Toonen (Eds) *Bestuurs-kunde: Hoofdfiguren en kernthema's*. Leiden: H.E. Stenfert Kroese bv.

———— (1996) 'On the administrative condition of politics: Administrative transformation in the Netherlands', *West European Politics* 19 (3) (July): 609–632.

United Nations Development Program (2002) *Human Development Report 2002: Deepening Democracy in a Fragmented World*. New York: UNDP.

Van de Walle, N. and T.A. Johnston (1996) *Improving Aid to Africa*. Washington, D.C.: Overseas Development Council.

Walker B., S. Carpenter, J.M. Anderies, N. Able, G. Cumming, M. Janssen, L. Lebel, J. Norberg, G. Peterson and R. Pritchard (2002) 'Analyzing resilience in a social-ecological system: An evolving framework', *Conservation Ecology* 6 (1): 14. <http://www.consecol.org/vol6/iss1/art14>

White, H. (1992) 'The macro-economic impact of development aid: A critical survey', *Journal of Development Studies* 21 (2): 163–240.

———— (Ed.) (1998) *Aid and Macroeconomic Performance*. London: Macmillan Press.

———— (1999) *Dollars, Dialogue and Development: An Evaluation of Swedish Programme Aid*. Stockholm: Sida.

Willard, A.R. (2000) 'Book review', *Policy Sciences*. 33: 107–15.

Wilson, J., R. Costanza, B.S. Low and E. Ostrom (Eds) (2001) 'Scale misperceptions and the spatial dynamics of a social-ecological system', in R. Costanza, B. Low, E. Ostrom and J. Wilson (Eds) *Institutions, Ecosystems, and Sustainability*. New York: Lewis Publishers.

World Bank (1996) *Poverty Reduction and the World Bank: Progress and Challenges in the 1990s*. Washington, D.C.: World Bank.

———— (1997) *Helping Countries Combat Corruption: The Role of the World Bank*. Washington, D.C.: World Bank.

———— (2000) *World Development Report*. Washington, D.C.: World Bank.

Wunsch, J.S. and D. Olowu (Eds) (1995) *The Failure of the Centralized State: Institutions and Self-Governing in Africa*. 2nd ed. Oakland, CA: ICS Press.

CHAPTER 14

THE LIMITS OF INSTITUTIONS: ENVIRONMENTAL DEGRADATION AND KNOWLEDGE FRAMING

M. A. MOHAMED SALIH

1. INTRODUCTION

The study of the relationship between environment and politics is hardly new. The complexities of this thorny relationship are ever present in probably all human societies regardless of their level of socio-economic development. The questions that inspired me to revisit this relationship are (i) whether communities enter into conflict over a degraded environment and (ii) what role do institutions play in conflict perpetuation, mitigation or management. These questions are qualitatively different from the conventional thesis that environmental degradation causes conflict. The issues they entail are difficult and at the same time important, not least because they entice social and natural scientists to study a subject that cannot be approached without laying bare regimes of truth regarding real or imagined physical environmental degradation. Whether the framing of environmental knowledge is socially or physically driven defines the dominant policy orientations responsible for environmental degradation or environmental maintenance.

Unfortunately the debate linking environmental degradation and conflict has seldom been conclusive, as many resource-rich regions have not been immune to environmental conflicts similar to those taking place in environmentally degraded locales. But both natural and social scientists, curiously, some even weary of the implications of this relationship, contend that there must be some indirect, if not direct relationship between the two enduring problems of environmental degradation and conflict. More unsettling is the study of where conflict generates environmental degradation and where it creates conditions for environmental regeneration. The relationship between environment and politics provides a powerful basis for discourse questioning the certainty of science as well as the theoretical/ideological tenets of social science and the disciplinary biases on which natural and social scientists explain the causes and outcomes of human-environment interactions.

Max Spoor (ed.), Globalisation, Poverty and Conflict, 261 282.
© *2004 Kluwer Academic Publishers. Printed in the Netherlands.*

This chapter reviews the institutional premise on which knowledge linking environmental degradation and conflict has been framed. Instead of affirming the prominence of institutions in environmental conflict management, it argues that institutions themselves could be sources of conflict and that they should be treated as factors contributing to environmental conflicts. In other words, institutional politics and conflicts could impose limits on the institutions' abilities to offer workable frameworks for environmental conflict management. In this respect, institutions operate within socio-economic and political constraints that often limit their ability to realise their potential role in building social norms and conferring legitimacy and predictability in either peace or conflict situations.

In the developing countries institutional conflicts ensue particularly when western-style institutions are superimposed on communal institutions and local resource management arrangements. In such cases, institutions become a source of conflict rather than an instrument for improving efficiency and cooperation.

2. FRAMING ENVIRONMENTAL KNOWLEDGE

Social scientists' anecdotal premises have tended to accentuate lay-claims of a causal relationship between environmental degradation and conflict. In today's world, high intensity environmental conflicts are rife in regions endowed with strategic environmental resources such as oil (Angola, Chad, Iraq, Nigeria and Sudan), diamonds and gold (Sierra Leone and the Democratic Republic of Congo). Though this does not preclude the existence of environmental degradation, it does mean that, on the whole, little serious effort has been made to disentangle environmental degradation and environmental change. Environmental change results from the intensity, scale and quality of human interventions in the environment, through the expansion of agriculture and livestock production, irrigation, deforestation, industrialisation and urbanisation, to mention but a few. In other words, the use of environmental functions beyond their carrying capacity may in some circumstances transform environmental change into severe environmental problems or degradation.

The main issues associated with what constitutes environmental degradation and what does not are framed by institutional identities, arrangements and science-based policies. These inform the debate on the relationship between human intervention and environmental change. To that extent, whether the degree of environmental degradation is over-stated or under-stated, the relationship between knowledge and the interests behind its production and translation into policies are also institutionally framed and socially constructed.

The institutional framing of environmental knowledge asserting whether degradation exists or is a problem rather than a symptom is not neutral. After all, the institutional framing prescribes the environmental policies. Answers to questions of what level of intensity can be described as environmental degradation[1] leave the door open for social, economic and political interests to frame the type of explanation perceived conducive to particular interests. After all environmental change occurs with 'reference to the demands and requirements that different systems of production place on the environment' (Goldblatt 1996: 29). Ecosystems of production are not abstract entities but representations of real people with real differentiated and contested interests behind them. 'Access to and the distribution of environmental public "goods and services" be they cultivable land, fuelwood or clean water is uneven' (Adams 1992: 87).

Because of the contested nature of environmental change and its potential consequences, at least two competing domains of science have emerged: *conventional science* which according to some has betrayed science and *environmental science* linked to the environmental movement and "green" public. The institutional framework for scientific knowledge generation stipulates that science itself is guided by competing interests in the definition and nature of environmental degradation – also part of the relationship between knowledge and human interests (Mohamed Salih 1994). However, the institutional framework governing the generation and social interpretation of conservation science is as ideological and value-ridden as social science. Therefore reliance on the internal validity of environmental data proving or disproving environmental degradation should be approached with a great sense of scepticism. Because our knowledge of the natural world depends on natural scientists, Swanson (1984: 246) alerted us,

> large institutions of science and technology that are entrusted with generating knowledge about environmental degradation and how to regulate it often use their increasing knowledge of public policy regulations to subvert regulation.

If manipulative human interests bind the production of scientific knowledge, it is then a simple matter to establish the relationship between scientific knowledge, whether social or natural, and the power structures and interests that propagate it. The institutional arrangements that confer legitimacy on scientific knowledge and validate its possible transformation into policy directives is neither neutral nor without limits.

On the other hand, if environment and development are linked, the overall scene of society-nature-development becomes more complex because (i) this trio integrates development into the social and natural domains thus contributing a new equally contested dimension and (ii) the institutional framework integrating the three is riddled with institutional politics. Essentially, the association between environment, development and conflict could be viewed as self-explanatory regardless of whether environmental degradation is pertinent. The reason is that conflict is a political resource and an expression of competing interests in environmental resources, hence environmental politics. In the developing countries, environmental politics is essentially development politics because livelihoods depend on the environment, locally understood to denote nature and natural resources in their benign forms (Mohamed Salih 1999a, 1999b). Therefore the institutional arrangements needed in this context are more complex and much broader. They encompass the dialectical relationship between major stakeholders: society, epistemic community, state, market and external actors.

To be sure, environmental politics and the institutions informing it are expressions of an environmental consciousness brought about by a nature-social transformation leading to conflict and cooperation, regardless of whether the natural resource base is degraded. This enduring relationship implies that there would be more intense conflicts over a healthy environment than over degraded resources.

In light of the well-established linkages between environment and development, it could be argued that environmental conflicts cannot be simply swept under the carpet of environmental degradation. Development is not just another agent of environmental degradation or conflicts devoid of institutional interventions and a development urge to create new socio-political and economic arrangements. Development interventions are often accompanied by institutional arrangements different if not contradictory to those of the receiving society. Development politics is therefore immersed in an institutional and environmental politics that is as important an outcome of development as the conflicts that it may generate. It is near impossible to argue that environmental conflicts are the direct results of misconceived development policies that engender political coercion and exacerbate economic pressures without locating these within their institutional frameworks.

My position is that when destructive development ensues, the first casualty is local institutions on which communities depend for resource conflict management, because their role is often taken over by state-sponsored institutions and agents. Excessive reliance on externally prescribed policies and remedies (whether national or international) leave little if any space for communal institutions to play their role in managing their natural resources, including the commons. Communal institutions use identity politics as a resource in confronting conflicts emanating from impaired environmental functions. In Tainer's words, responding 'to external pressures determines much of the political and cultural behaviour of institutions' (Tainer 1998: 174–175).

Hence institutional competition and even conflict between state-sponsored and communal institutions is inevitable.

Paradoxically, the same institutions that have sponsored environmental degradation-inducing policies are those entrusted with negotiating problem-solving scenarios with communal institutions, of which they are scornful, if not doubting their capacity to produce negotiated settlements for environmental conflicts. Therefore, policymaking and knowledge-generating institutions (i.e. state-sponsored epistemic communities) and those whose interests are served through local knowledge and communal institutions come on a collision course. The need to maintain a defensive posture to protect its members, to preserve territory or merely to remain competitive then becomes more enduring.

Environmental conflicts influence identity politics and the identities of those engulfed in these conflicts. It is in this much deeper perception of resource conflicts and projection of their potential outcomes that we find the most intriguing processes linking political resources and environmental resources as a conflict metaphor. Ostrom (1990), however, warns, 'policies based on metaphor are harmful'.

In a sense, the prominence of institutions such as state, market and community and their competing interests in shaping, managing and controlling the attributes of the physical environment are "politically" motivated. Social and natural scientists provide the knowledge base that responds to the institutional demands and arrangements within which environmental knowledge is translated into the environmental policies used or abused by policymakers. Epistemic communities and institutions also deal with how the relationship between environment and politics is perceived, projected and acted upon, while responding to demands espoused by those involved in environmental policy as well as environmental management vis-à-vis local communities' collective actions and environmental struggles expressed as livelihood struggles.

Because the state controls the production and distribution of resources, resource conflicts are waged in the political realm, making the state the focus of social conflict. Such struggles or conflicts often invigorate institutional arrangements that from the outset appear as atavistic remnants. Customary and communal-based institutions become dialectical opposites of state-centred institutions and the policy instruments they prescribe. By demanding accommodation and acceptance by the state, community-based institutions may begin to negotiate with other stakeholders, including the state and corporate interests, even when they recognise that the outcomes they seek may be unattainable.

3. THE ENVIRONMENTAL DEGRADATION CAUSES CONFLICT THESIS

During the early 1990s with its emergent sustainable development ethos, throwing anything environmental at the state would yield political acceptability and secure a place in the agenda of "high politics". Environmental security, which formed an important contribution to the debate at the time, was highly profiled by some main international research think-tanks.[2] A number of publications emerged which added urgency to the linkages being established or thought to have been established between environmental degradation and insecurity.[3]

Although linking environmental degradation to national and international security issues is plausible, the idea has endured much critique[4] and continues to attract attention. As an example of how the trio of population growth, environmental degradation and conflict still loom large in policymaking circles, the United Nations Environment Programme (UNEP) entered the fray in 1999, publishing a report entitled *Environmental Conditions, Resources and Conflicts: An Introductory Overview and Data Collection*. The report, co-authored by Schwartz and Singh, was greeted with great enthusiasm in policy and academic circles. In his forward to the report, Klaus Toepfer, UNEP Executive Director, stated, 'This publication attempts to give a brief overview of the most recent discussions surrounding environmental conditions and resources, and the linkages to conflict'(Schwartz and Singh 1999: 4).

Surveys were conducted in 13 countries (El Salvador, Ethiopia, India, Israel, Kenya, Mauritania, Mexico, Nigeria, Pakistan, Peru, Philippines, Senegal and Sudan). Although the dynamics of the presumed environment-related conflicts differed from one country to another, a single basic model ran through the studies and was summarised in a flow chart. Three main conclusions of the report are important (Schwartz and Singh 1999):

- Increasingly, national and intra-national conflicts appear to be linked to deteriorating environmental conditions and resources. This linkage reflects deteriorating environmental conditions worldwide, but recognition of the linkage is due in large part to a post Cold War shift in discourse at the international level; a shift away from traditional "high politics" issues of war between major powers, to "low-level politics" issues such as poverty, population growth and environmental degradation.

- The nexus between the environment and conflict is not always straightforward. Most often, environmental factors are enmeshed in a complex web of social, economic and political factors that function together to engender conflict. Nevertheless, if it is by understanding the nature and causes of conflict that we will enable the fabrication of solutions, then both the theoretical and empirical evidence suggest that environmental conditions and resources should be a focal point for conflict resolution and conflict prevention at both international and national levels.

- The resolution and prevention of environmentally related conflicts can be facilitated by a typology (presented in the report) which provides policymakers and researchers with a concise guide to the literature within these increasingly complex fields. Familiarisation with the literature and issues should allow policymakers and researchers to engage in more informed and fruitful discourse of these issues, which in turn, should facilitate conflict resolution and prevention.

The report was not completely off the mark in arguing that the nexus between environment and conflict is not always straightforward and that environmental factors are enmeshed in a complex web of social, economic and political factors. However, to claim that national and international conflicts are increasingly linked to deteriorating environmental conditions and resources is absurd to say the least. The main problem with this statement is that conflicts are blamed on environmental degradation without identifying those power relations and structures that caused the degradation in the first place. The most disturbing implication of arguing along these lines is that the problem is identified as environmental degradation, which implies that solutions should be found in technical fixes for what is misconceived as an apparently physical environmental problem. In fact, such an analysis is useless for local communities and institutions because it reinforces the very "technologies of administration" which are the main contributors to environmental degradation rather than redeeming its victims.

It is one thing to approach the social, political and economic issues that have contributed to environmental degradation as social, economic and political problems. It is a completely different thing to look at environmental degradation as the cause. In actuality, it is the trigger or rather the symptom of more serious socio-economic and political anomalies. Obviously, environmental degradation is inseparable from its social, economic and political drivers.

In response to the UNEP report, we may examine the literature related to three specific cases from countries (Ethiopia, Sudan and India) where, according to the report, conflicts have emanated from environmental degradation. First is the Ogaden region of Ethiopia, which is a centre of conflict between Ethiopia and Somalia for reasons originating from their colonial past. The conflicts between the Isaq and Ogaden Somali clans over this rugged and inhospitable land are due to land alienation, particularly of the rich seasonal river basins. Neither human nor animal population pressure could offer a plausible explanation for the centuries old conflicts between these two rival clans. In most cases, it was alliances with the Ethiopian or the Somali government for clan or individual gains that proved lethal. There is also Somali nationalism and the Somali people's quest to fulfil their dream of creating a state as the home of all Somali, not only those who live in Somalia, but also those scattered in northern Kenya, Ethiopia and Djibouti.

Second, land, timber and recently oil have played an important role in the war in southern Sudan and the Nuba mountains. Southern Sudan is one of the least populated regions of the Sudan, with its population dispersed over large areas except for some concentrations of internally displaced people. There are some 4 million hectares of large-scale mechanised agricultural schemes in the Upper Nile, though most of these are not operational due to the war. The Nuba mountains is a straightforward case of land alienation. Northern Sudanese merchants called Jellaba have come to the area and, supported by the Sudan Agricultural Bank and Islamic banks, acquired more that 16 million hectares of the most fertile lands. True, the conflict in the Nuba mountains is an environmental one in that environmental resources, land in particular, are appropriated and reallocated to wealthy and powerful northern Sudanese business interests. It can hardly be explained away as a war against or engendered by environmental degradation.

Third, in the Uttaranchal region of India low intensity conflicts between the local communities and the state are a result of uneven development and land distribution. Seventy per cent of the land is under forests (and not degraded), with 10 per cent allocated to Indian defence forces after the 1976 war. The 1976 Forest Act advocated the idea of forests without people. The act goes against the traditional privileges of local communities, putting national parks and wildlife conservation over the interests of people.[5]

Environmental degradation might have occurred in the regions of Ethiopia, Sudan and India, as mentioned in the UNEP report – there being no evidence to the contrary. There is, however, no conclusive causal relationship between the prevalence of conflict and environmental degradation in these regions. The evidence available does suggest that all three cases have their peculiar histories which have shaped both the relationship between people and their environment and that between people and the external forces that intervened in the making of their history. It is through these peculiar histories, which are to a large extent marred with land alienation and incredible exploitation of human and natural resources, that environmental conflicts occur – again regardless of whether the environment is degraded.

Ten other case studies not presented in the UNEP report illustrate another pattern in regions where conflicts occur: namely, the fact that most conflicts are in resource-rich areas. These case studies, from Sierra Leone, Tanzania, the Democratic Republic of Congo, Zimbabwe, Madagascar, Thailand, Ecuador, Colombia and Brazil, as well as Ethiopia, Sudan and India, point to six common features:

- Resource conflicts tend to involve the control of regions endowed with a concentration of highly demanded and strategic resources. Some of these resources are "traditional", such as land, water, forests and fish, and others are "modern" such as oil, gold, diamonds and other highly valued minerals.

- Resource conflicts are commonly caused by resource alienation. Interests at work may emanate from the state or state agents, national or foreign corporations and wealthy and powerful classes or individuals.
- Resource conflicts are by and large between local entities attempting to protect historical or hereditary rights over environmental resources and external forces and developments.
- Resource conflicts are largely caused by external interventions implemented under the pretext of development, conservation, national security or the generation of "wealth for all". Contrary to their promise, these interventions have proven unable to attain most of their objectives and instead contributed to destitution and impoverishment of local communities.
- External interventions to claim environmental resources, whether for conservation or development, exhibit particular types of authoritarianism even in states that have pronounced themselves democratic. In a sense authoritarian development and resource alienation go hand in hand, and both marginalise local institutions and impair their capacity to solve local problems they have for centuries been able to cope with.
- Resource conflicts comprise what are commonly termed "livelihood struggles" because most local communities affected by development-deficit interventions depend on the natural environment to eke out their living.

These six features indicate that environmental change, particularly degradation, has occurred in all 13 cases discussed (see also Annex 1). In some cases environmental degradation has been so severe as to impair some environmental functions, while in other cases it is as yet within a tolerable range and has not created situations where livelihood conditions are totally impaired. Exceptions to this generalisation are the cases where oil spills and gas flares from oil wells contaminate land, such as in Ecuador and Nigeria among others. All the case studies tell us that environmental conflicts are caused by resource alienation and ecological poverty rather than by environmental degradation. Ecological poverty and livelihood poverty cannot easily be disentangled.

Another observation is that local communities' environmental and livelihood struggles take place in an interdependent world. Keohane and Ostrom (1995) and Goldman (1998) drew our attention to the intractable, complex constellation that shapes the global-local nexus. The heterogeneity and multitude of actors and the multi-layered institutional mechanisms (Keohane and Ostrom 1995) that bring the global and local levels together is clear case for the primacy of institutions in resource management and sustainability. Institutions are also concerned with conflict resolution as alluded to in Schwartz and Singh (1999).

More often than not, environmental degradation strikes not because local communities have relinquished their concern with environmental maintenance, but because in reality the policy environment does not lend itself either to sustainability or to sustainable demands on environmental resources. While policymakers are expected to deliver alternative administrative, economic, political and technical interventions and solutions, they are often constrained by weak institutions and ill-defined governance responsibilities.

Opschoor (2001: 31–32) pointed out three governance failures associated with institutional failure, namely correction failure, intervention failure and administrative failure. These inadequacies and failures, which are so common in both inter and intra-institutional arrangements, shed doubt on the capacity of state-sponsored governance and institutional arrangements alone to overcome the constraints associated with the uncertainty of policy outcomes and the non-sustainability of the livelihood conditions associated with it. Obviously, multi-stakeholder institutional failure is a recipe for conflict, often leading to the subversion of the public goals on which institutional behaviour is predicated.

In today's world, no quest for environmental sustainability, as a dialectical opposite to environmental degradation that causes conflict, is simple or predictable within the immediate confines of institutions, including self-governing institutions. Neither degradation nor sustainability is predictable in the present historical experience, as they draw upon embedded knowledge and tested experiences that span decades if not centuries of society-nature interactions. The tendency towards complex sustainability even in seemingly simple systems is not surprising. In the words of Clayton and Radcliffe (1996: 12), 'In general, complex systems generate outcomes that depend on numerous interactions. As a result, many complex systems are highly sensitive to the precise starting conditions and loading of factors.'

In other words, the physical environment is susceptible not only to immediate policy and technological interventions, but also to similar or other interventions that took place during several bygone decades. The very external factors that augment environmental degradation are the ones that maintain a certain degree of presence in local communities' current attempts to create alternative solutions to environmental decay.

The exercise of power outside a responsive institutional framework legally (or customarily) responsible for the disposition of legitimate authority often ushers in authoritarianism, resistance or outright rejection:

> [I]nstitutions hold society together, give it a sense of purpose and enable it to adapt. [This] applies both to the structures of power and positional relationships – as found in organisations with leader, chief, information flows, resources for mobilisation and the usual distortions of communication to suit their mandate (O' Riordan and Voisey 1998: 34).

The difference between the argument that environmental degradation causes conflict and the argument that conflict relates to resource-rich environments, is obviously relevant to my opening statement. That was that the institutional framework of knowledge production privileges the interests which that particular type of knowledge is intended to serve. In this regard, the 1999 UNEP report was not an isolated incidence of a lone academic query. It fits into a pedigree of powerful knowledge-generating regimes of truth that inform the manner in which environmental conflicts are constructed and deconstructed at will to serve certain interests. Not only environmental conflicts, but environmental degradation itself has developed into an ideological instrument used and abused to entrench the power structures themselves responsible for degradation.

4. INSTITUTIONAL INTERVENTIONS: DEVELOPMENT AGENTS OR SOURCES OF CONFLICT?

As both a symptom and the sum total of impoverishment, disempowerment and destitution, environmental degradation is used as the *raison d'etre* to absolve those responsible for livelihood destruction from their responsibility. Concomitant with this mind-set is an apparent attempt to mute the debate on environmental maintenance by treating it as an organisational matter amenable to the rationality of bureaucratic procedures. Yet it is in the realm of conflict resolution that institutional politics and institutional hierarchies exhibit their worst inefficiencies. Because power determines the flow of resources, it also gives weight to the institutional arrangements that establish what institutional arrangements should obtain.

Although institutions play a significant role in almost every aspect of human life, they do have limits. Treating them more critically and demystifying what they can and cannot do is therefore fruitful. The main reason for taking this line is that a large number of social scientists and policymakers present institutions as a panacea and a solution to all socio-economic and political problems, from failed economic policy reforms and delivery of justice to resource conflict management. As a result "institutional" approaches to development, social enquiry and problem-solving have gained considerable strength and respectability – to the extent that "institutionalism" has become an industry in its own right with an attentive audience and formidable power support structures. In the tradition of critical theorists, the uncritical belief in the "institution" must be criticised, for idealising it would entail betraying human capacity for transformative action (Jay 1996).

Yet institutions do matter, and they should be taken seriously. Moreover, institutions should be preoccupied with institutional politics; that is, institutional conflicts that would sacrifice the public good for institutional or even individual gains. The following sections reflect on why institutions should be criticised and not idealised, particularly if we hope them to be capable of fulfilling their promise as a regulatory instrument of human behaviour.

4.1 Understanding the Antecedence of Institutionalism

The antecedence of contemporary institutional approaches to the study of social problems is relatively new and can be traced only to nineteenth-century Europe and the United States, where the social sciences began to emerge from the embryo of moral philosophy. Already, however, the dominance of logical positivism had been inspired by the Industrial Revolution and the triumph of science also meant for some social scientists that the research methods applied to the natural world would also be applicable to the social world. Early institutionalists were rightly concerned with formal politics, rules, rights and procedures. They also analysed the evolutionary development of economic and political systems and the possibility of reconstructing institutional forms (Scott 2001: 6). Concern with logical positivism and formal political institutions was understandable given the newness of constitution-making in most western democracies and the emergence of political parties and interest associations heralded by the struggle for liberty vis-à-vis despotic rulers. Hence from a political science viewpoint, these political developments contributed to the leap of faith in formal institutions and their capacity to create norms of "stable" political engagement. As part of this enlightenment, modern institutions were accepted as a matter of fact and much faith was put in their capacity to deliver the promise of a progressing society.

Institutions were important instruments for cementing the values of the Industrial Revolution. Indeed, new administrative arrangements and predictable organisational structures were needed to safeguard the stability of early industrial society. In terms of political development, formal institutions informed the dominant systems of governance of the time, aided by the fact that it is more predictable to be ruled by established norms of behaviour than by absolutist rule. But this was also a period of great institutional transformation, from the ethos of the pre-industrial society to new forms of industrial organisation of production, labour, trade unions, intellectual clubs and systems of administration and government.

Students of public administration and political science as well as colonial officers and administrators conceded to institutional approaches and wished to transfer their almost magical institutional qualities to the colonies. The colonial period could best be described as the era of great institutional transfer to and transformation of large parts of the developing countries. Western-style institutions predominated over local institutions whereby earlier contacts between the local organisations were shrouded in difficulties and misunderstandings if not outright conflict. The debate on what institutional arrangements are best for development or resource management is still ranging more than four centuries on, with new, probably more complex, questions and problems than ever before.

With the departure of the colonial powers, institutional "transfer" was almost complete. The nascent institutions were left behind to be administered by the emergent political elite.[6] With the inauguration of the first decade of development during the 1960s and in the late twentieth century, interest in institution building for development began to shift from formal institutions to exploring the continuum between formal and informal, state-centred and society-centred. The conception of stability gave rise to the possibility to conceptualise institutions and link that conceptualisation to the theory of knowledge or epistemology. As early as the mid-1960s, Taylor (1956: 25–28) contended that institutions embody three distinct conceptions linked to the theory of knowledge. In the first conception, institutions constitute the basis of individuals' thinking, feeling and acting. In this situation the institution is sacred and the thought forms embodied in it have an eternal validity guaranteed by an ontology deriving its authority from some transcendental stadium (e.g. God). Strictly speaking the objectivity of this situation is best described as a norm of truth. The second conception views the institution as separate from the individual. The institution, in a pejorative connotation, becomes the object of critical reflection and analysis. Objectivity is no longer grounded in the institution but in the cognitive act of a rational subject. In the third conception there is a sociological reconstruction of the significance of the institution, but upon a platform of validity measured by secular criteria (e.g. science).

Obviously, every theory of knowledge employs subjective and objective methods of knowing and as the three crude classifications offered by Taylor (ibid.) suggest, "social behaviour" is considered an important aspect of what institutions attempt to develop into a norm, an attitude or an acceptable way of behaving and acting. These aspects of change were a matter of great sanctity, not only for colonial administrators and missionary societies, but also for the colonial technical staff working in departments of trade, agriculture, forestry and land administration.

With particular reference to environmental resource management institutions, obviously they cope with regimes of truth, cognitive acts of rational/irrational subjects as well as sociological constructs.[7] These constructs are important because people use them to measure their success and validate or negate their failures. In this respect the institutions entrusted with environmental management are not an exception and have played with vigour their role in informing the value theories of knowledge closely associated with the power structures peculiar to them.

The rigid dichotomy that once informed the ideological debate on what motivates human societies to develop in terms of the methodology of liberal individualism or social collectivism no longer holds much sway. March and Olson (1989: 3) lament:

> Institutional arrangements often produced diverse and sometimes contradictory assumptions about social behaviour and how it affects the functioning of institutions and their capacity to channel compliance with the norms they espouse.

Within this perspective, modernisation theory must have grown out of behaviourism, particularly its insistence on changing people's values as a prerequisite to achieving economic development. In institutional terms, replacing traditional institutions with modern ones was the essence of modernisation and as such was not devoid of conflict.

Most of the development debate between individualism and collectivism has now been relegated to the domain of the history of development theory, and it would hardly be helpful to rehash it here. In *Rediscovering Institutions*, March and Olson (1989) explored the main perspectives of the debate and, in the process, demystified the inherent logic on which the different institutional schools of thought were constructed.

In applying challenging perspectives such as reductionism, utilitarianism, instrumentalism and functionalism to state-centred natural resources management, obviously, the institutional approaches used in the developing countries have their antecedence in western schools of thought. The earliest experimentation with the above attributes of western institutionalism in the developing countries was during the colonial period. The nineteenth century was the century of a more resource-conscious colonial expansion and with it came the exploitation of the human and natural riches of the colonies. While this earlier form of globalisation was less pervasive in terms of human interaction, it did create worldwide economic opportunities in the metropolitan centres. Colonial expansion generated interest in governing the colonies with institutions of government similar to those of the colonial powers and, simultaneously, the study of the institutions of the colonies. It was a period in which anthropology was designated as the discipline concerned with the study of other cultures, primitive government or people without government (Mair 1962, Beattie 1966, Barclay 1980). Although these connotations are no longer used, their transformation into various forms of paternalism in the local-global nexus of governance arrangements is still present.

Ironically, even after the departure of the colonial powers, colonial institutional models, from government to natural resources management and administration, continued to haunt researchers and policymakers in the developing countries. Institutional development and comparative institutions, rather than institutional rediscovery and resurrection, was the norm. Some intellectuals and policymakers advocated modernisation, cautioning that the use of western-style institutional norms was on a collision course with "traditional" customary institutional rule and norms. Others dismissed the developing world's own endogenous institutions, considering them backward relics of the past and unsuited for the more sophisticated complex modern world. Those who agitated for the respect of endogenous norms were either victimised or given a cold shoulder by mainstream institutions and structures of the powers that "matter".

The reason for tracing institutionalism to its western antecedence is to illustrate that the inclination to impose western institutional rationality and optimism on the developing countries should be viewed with great care, even cynicism. More than romanticising institutional modelling as the ultimate solution to resource management and resource conflicts, it is more appropriate to start with a critical posture examining what institutions are actually capable of doing.

Ostrom (1990), aware of the negative consequences of imposing institutions, advocated freeing self-organised collective-action groups of individuals from externally imposed action (i.e. action organised by the state and firms). Even when using models, Ostrom (ibid.: 56) was modest in expressing, 'Obviously, I do not know if these appropriators reached optimal solutions to their problems. I strongly doubt it.'

Institutionalists old and new are sensitive to the fact that policies are fundamentally value-driven, process-oriented, instrumental, evolutionary, activist, fact-based and democratic (Petr 1984: 1, 4). These characteristics make institutions relevant to real problems facing real people. However, these essentialist claims in a sense privilege the institutionalists position, since their basic assumption is overly cloaked in optimistic humanism at its best. Nonetheless, Galbraith in his seminal work on the significance of institutions in modern life commented that we are victimised by a system of illusions (Hession 1972), and Petr (1984: 13) commented that institutionalists have optimistically asserted the ability of humanity's rational creative intelligence to surmount institution-based limitations.

The cautious optimism of Petr and the pessimism of Galbraith are two sides of the same coin. Their contemporary relevance to natural resource conflict and management is therefore often better understood as a new wine in old bottles. But, are institutions really a rational humanistic creation capable of solving the problems of the politics within, let alone the politics without?

4.2 Institutional Politics and Conflict

Having briefly introduced some institutional perspectives in general and their relevance to managing environmental conflicts, we may understand why some institutions are backed by respected natural and social scientists adamant that environmental degradation causes conflicts. This is a complex matter that requires diverse and large evidence and cannot possibly be fully comprehended in the space available to this contribution. What I will attempt here is to look into the institutional politics that inform the positions held by the "environmental degradation generates conflict" scenario.

To begin with, institutions (local, national, regional and even global) are called upon to intervene for environmental management and maintenance as well as to minimise the effects of production on resources. The resultant environmental degradation is the sum of inappropriate land management systems often imposed by the state. These include zoning regulations, extension of large-scale export crop cultivation, extension of irrigated agriculture, wildlife parks and reserves, settlement schemes, mining, oil prospecting and other types of production.

Because state institutions are not uniform and represent diverse interests, different line ministries directly or indirectly concerned with environmental management and maintenance often carry out these interventions. These institutions are not abstract entities but representations, as Petr (1984) argued, of real people with real interests or real people facing real problems. While the agents and agencies of a ministry of water and soil and of water conservation are all interested in water and soil, those of the ministry of agriculture are interested in land productivity, including the use of pesticides and fertilisers. Resources competition and "the rival interests" that institutions generate (or incline to deploy) to safeguard these interests, is at the core of troubles marring their efficacy. This, in my view, is the arena of institutional conflict. While the ministry of finance is concerned with timber logging to generate financial resources for "development", the ministry of forestry and wildlife is interested in biodiversity, soil and water conservation and the maintenance of the minimum forest cover subscribed to in international treaties and conventions.

Counter to the often assumed rationality of institutions is the reality that the state, the most formidable of all institutions, is neither uniform nor without serious problems in getting institutions to work together. Despite this obvious handicap, Boyer and Hollingsworth (1997: 451–454) remained convinced that institutions are rational agents capable of building mutually advantageous norms and values voluntarily. The emphasis here is on extra-economic institutions to build trust and forms of economic governance within institutions operating according to market principles and rationality in which agents will seize the opportunities presented. In the same vein, Ostrom (1994) argued that neither the market nor the state is the answer to the commons dilemma and the conflicts associated with it. The idea of self-governing institutions with relative autonomy from the state and not completely subsumed under market principles may deliver better results than depending on these two externally imposed institutions.

However, in modern institutional forms of governance power is at the core of institution building. Modern institutions are heavily dependent on traditional norms, the social constructs responsible for the establishment of successful economic institutions in a cultural community. Economic agents use both constraints and opportunities created by political action to "cook" governance modes. In this view economic (*or environmental*) institutions are inherently political and social constructions and polity, rather than being mere economic rules of the game, which channel agents' behaviours in certain directions (Boyer and Hollingsworth 1997: 454).

It is true that most institutions dealing with public environmental goods operate with less emphasis on market principles, or they become market-dependent by engendering competition among alternative strategic preferences for sustainability. The danger in such a system, according to Boyer and Hollingsworth (1997: 447), is that capitalism is a dynamic form of economy and the market – if not contained – will erode traditional institutional arrangements, *including traditional environmental management institutions*. Yet many traditional arrangements are necessary to provide the trust on which transactions among economic actors depend. On the other hand, if traditional institutional arrangements are too strong in imposing social obligations on individual actors, the potential dynamic of the market will be stifled.

At another level, if modern institutions are so muddled up in institutional politics and conflicts, how can we trust them in their dealings with communal institutions that employ institutional arrangements with norms and values dissimilar to theirs? Indeed, the institutions that interfere in resource management and in resource-use patterns maintain their existence through conflict rather than through coordinated rational coexistence, as some institutionalists would like us to believe. If modern institutional norms, values and instruments differ it is because they are embedded in and duly informed by a particular type of socio-economic and political conditions. In one sense, institutional rationality is born out of the circumstances that nurtured it.

However, it would be mistaken to consider resistance to external interventions an exclusively local community affair. Goldman (1998: 13–14) observed the following characteristics amongst the institutional elements of new European ecopolitical movements and struggles:

> They organise for socialised control over use values without being solely anti-commodity or pro-subsistence. They challenge dominant techno-sciences for being partial and situated knowledge. They work as internationalists while being located. They mobilise to gain control over ecological use-value for livelihood and social justice rather than for more efficient ways to reproduce conditions for capital expansion. They transcend the most politicising discursive dualism – local/global, traditional/modern, capitalist/non-capitalist relations, ethno-science/techno-science and nature/culture.

Because institutions are so pervasive, their contribution to resource conflict management, an equally pervasive subject, should be greeted with caution and care. Institutions, at best 'have a mind of their own' (Douglas 1987) and at worst are greedy (Coser 1974: 6) and exercise pressure to weaken or prevent alliances with other institutions, even if they are not competitors. Institutions may also be incapable of accommodating or accepting individual attempts to defy their norms or, for that matter to have "minds of their own". So strong is the grip of institutions on the lives of modern or traditional citizens that the tyranny of institutions is impeccable. As we have seen, institutions make claims, serve interests and regimes of truth vis-à-vis other institutions with which they compete for social, material or financial resources. In such circumstances, institutional culture and institutional politics are a hindrance to conflict management and typically draw on value-laden norms, themselves susceptible to generating rather that solving conflict.

In essence, what this section has tried to highlight is that institutions are inherently political and as such cannot be trusted. This is mainly because, according to Colebatch and Degeling (1986: 344),

> [o]rganising is inherently a political process; it entails and reproduces relationships of power and advantage. For this reason, analysis should focus on what it is that people are doing as they structure their relationship with others, the meaning and interests embodied in their actions, the social practices through which they are expressed and how some of the meanings and interests embodied in action have come to be institutionalised. It follows therefore that policy and organisation are not separate phenomena.

With knowledge placed in its proper frames and in light of such claims of the primacy of institutional values and norms, institutional development seems to be more about consolidating institutional gains even if conflict is considered a resource at the disposal of institutions. Environmental degradation as well as environmental regeneration and "wealth" could, depending on the circumstances, become causes or triggers of conflict.

5. CONCLUSION

The limits of institutions are self-imposed. Whether we love them or hate them, institutions provide the link between people and the socio-economic, political and environmental transformations that impact on them. Because of this pervasiveness, institutions are sometimes overloaded and are therefore expected to do more than they can cope with or were originally created for. Nowhere in human interactions has this been more evident than in the field of natural resources management. The persistence of faulty conceptions about what causes human misery (poverty, destitution, displacement, etc.) would always find institutional defenders whose interests lie at worst in mystifying and at best in deconstructing reality.

Knowledge-based solutions to environmental questions are also power-based solutions. Therefore institutions and the social forces and epistemic communities behind them cannot afford to undermine their own power base and interests. Institutions' rationality has its own dynamics and cannot possibly be attributed to the coherence or ability to stand above their own inadequacies, let alone those deficiencies imposed on them by the structures of power of which they are, in most cases, proactive cronies. By paying more attention to conflicting institutional interests and the power structures informing them, no matter how well-intentioned institutions are, they may become instruments for subverting public goals and hence proactive agents for hideous private and personal gains. Because of these inherent limits, institutions should be demystified rather than glorified.

ANNEX 1. RESOURCE CONDITIONS AND CAUSES OF CONFLICT

Source	Regional/Local Resource Conditions	Causes of Resource Conflicts
Wolde-Semait (1989) and Mohamed Salih (1999)	Ethiopia: Tigray region. Largely arid, poor soils, climatic unpredictability, recurrent drought.	Could be described as periphery-centre conflict (1970s to 1990s) over autonomy, democracy and land reform. Currently, low level conflict between different resource users.
Christiansson and Tobisson (1989)	Tanzania: Eastern Mara region. Northern highlands are fertile, dwindling timber firewood resources, grazing and crop production potential.	Competition between Serengeti game park, crop production, livestock and human populations. Soil erosion due to tree-cutting for the market and domestic fuelwood use.
Witte (1993)	Democratic Republic of Congo: Fertile, contains 2.5 per cent of the world's remaining tropical forests, rich in minerals. Land is abundant and still largely owned by subsistence farmers with complex yet environmentally sustainable traditional framing systems. Logging and signs of degradation around gold and diamond mines.	Severely divided state along ethnicity, class and regional lines. These have translated into internal and regional rivalry between the political and military elite over mineral and state resources. Warlordism is rife along with the intervention of neighbouring countries with interests in DRC riches.
Mohamed Salih (1997, 1999b)	Sudan: Southern Sudan and the Nuba mountains. Land is fertile and there is no landlessness despite the expansion of large-scale privately owned farms in the Nuba mountains and Upper Nile. South Sudan is generally fertile and most of the south-western regions are located in a moderate climate, with patches of rainforest.	Conflicts are generated by land alienation, construction of the Jongeli canal in the south to reclaim water, oil production in the south without redistribution and calls for self-determination. Religious conflict between the Muslim north and the Christian south is used as an ideological base to justify resource conflicts.
Moore (1996)	Zimbabwe: Eastern Highlands, Nyanga district. There are signs of land degradation in subsistence farming areas. Climate is moderate, with protected national parks and wildlife reserves.	Colonial land policies shaped a dual agricultural economy. Large-scale European commercial farmers and small-scale African peasants live on relatively densely populated poor soils. Land is used to bolster the waning support for President Mugabe and his party, the Zimbabwe National Front.
Jarosz (1996)	Madagascar: Fertile lands except where excessive cash-crop cultivation and deforestation of tropical forests prevails. Some densely populated areas are degraded due to population concentration and prohibition of shifting cultivation.	Forest reclamation for cash-crop cultivation (rice, coffee, sugarcane) and livestock raising for export. Transformation of land tenure with land policies that allowed for large timber harvesting concessions as well as the creation of game parks and wildlife reserves vis-à-vis the local populations.
Lohman (1993)	Thailand: Relatively fertile, with varied climate and topography. Some areas are densely populated as a result of forest "colonisation" and exclusion of local communities from forest (land) resources.	Forcible eviction from national forest reserves and displacement of local communities, land appropriation and concentrations of populations on lands, loss of soil fertility in places as a result of excessive logging and deforestation.
Rangan (1996)	India: Uttaranchal region. No signs of excessive land degradation. Most communities are scattered subsistence farmers, with centuries-old traditions of water and soil conservation. Seventy per cent of the land area is under forests, with 70 per cent of the people depending on the forest for their livelihoods.	Uneven development, lack of infrastructure. Scattered subsistence peasants and forest dwellers affected by the expansion of private cash-crop producers. Ten per cent of the land is allocated for defence purposes. The Forest Act is biased towards national parks, wildlife and forest reserves without people.

Source	Regional/Local Resource Conditions	Causes of Resource Conflicts
Kimerling (1996)	Ecuador: Oriente region. The land outside the petrol industries is fertile and cultivable. There are no reports of severe environmental degradation in the rest of the region. Environmental protection is lax due to state inaction.	There is clear environmental degradation in areas close to the petrol industries. Soil degradation has resulted from opening the Amazon region and increased logging and deforestation. However, there is no single major case of reported environmental degradation
Alexander (1996)	Colombia: Pacific coast region. Tropical rainforest, mangrove swamps, endowed with fertile land and rich biodiversity, rivers and tributaries. The areas surrounding mining sites and oil exploration fields are degraded. Government introduced basic sanitation and protection of biodiversity in 1992.	The Indian of the Chocos and the Afro-Colombian slaves are under pressure from developers. Most local communities, both indigenous and of African origin, depend on subsistence farming, fishing and hunting and gathering. However, there is also *Plan Pacifico* (hydroelectric power generation), gold mining, oil exploration and road construction as part of Pan-American Highway.
Monbiot (1993)	Brazil: Particular reference to the Amazon region which is generally fertile and biodiverse, except on overgrazed and degraded lands around mining pitches. Deforestation by powerful business/corporate interests.	Land concentration and land alienation (43 per cent of the land is owned by 0.8 per cent of the population while 53 per cent of the population owns 2.7 per cent of the land).
Rankin (1996)	Argentina: Chaco region. Semi-arid, thinly forested with running rivers during the rainy season. Crop production and fishing are the main sources of sustenance. Land degradation due to enclosure of large tracks of Indian land by settlers.	Well-armed settlers occupied the land of the Wichi Indian population who also suffer discrimination. In 1987, the governor of Salta province passed a law which forcibly recognised the settlers rights to the large land enclosures.

NOTES

1. Tellegen and Wolsink (1998:1–15) classify common environmental problems into three categories: (i) exhaustion or depletion of renewable and non-renewable resources; (ii) pollution, or the transformation of matter or substance into solid, fluid or gaseous waste which is then released into the environment, (iii) disruption of the environmental functions caused either by natural or human induced interventions.

2. Including among others the Woodrow Wilson Institute, the Swiss International Peace Foundation, the United States Agency for International Development, the American Association for the Advancement of Science and the Worldwatch Institute.

3. As examples, the following were influential in shaping the concept: Boulding (1992), Brown (1992), Fleischman (1995), Homer-Dixon (1991), Homer-Dixon, Thomas and Percival (1996), Ornas, Hjort and Salih (1989), Westing (1986), Kakonen (1994), Porter (1992), Barnett (2001) Woodrow Wilson Center (1995) and Prins Gwyn (1993).

4. See, for example, Dalby (1992), Deudney (1990, 1991), Deudney and Mathew (2000).

5. See Gadgil and Guha (1995).

6. Of particular relevance to institutionalism and contemporary development, is the work of Weber, who among other social scientists asserted that cultural rules, ranging from customary mores to codified constitutions or rule systems, define social structures and govern social behaviour, including economic structures. The primacy of institutions in human life could not be more pronounced. No scholar in my view has promoted institutionalism in its current multi-facetted form more than Max Weber.

7. Not surprisingly, the behaviourists entered the fray much later and argued that politics could be fully understood only when the role of the informal institutions in the distribution of power, attitudes and political behaviour are equally investigated. Naturally, behaviourists focused their attention on the

study of voting behaviour, party formation and public opinion. According to Peters (1999) behaviourism was reinforced and deepened by the rational choice and game theorists who brought about fundamental changes in the application of economic assumptions to political behaviour. Again, the process of ascertaining public opinion is not essentially different from that which resource management researchers use to attest to how individual or group value preferences affect resource productivity, sustainability and degradation.

REFERENCES

Adams, W. (1992) *Green Development: Environment and Sustainability in the Third World*. London and New York: Routledge.
Alexander, L. (1996) 'Colombia's Pacific plan: Indigenous and Afro-Colombian communities challenge the developers', in Helen Collinson (Ed.) *Green Guerrillas: Environmental Conflicts and Initiatives in Latin America and the Caribbean*. London: Latin American Bureau.
Alexander, S. (1993) *The Waorani: People of the Ecuadoran Rainforest*. New York: Dillon Press.
Barclay, H. (1982) *People Without Government*. London: Kahn and Averill.
Barnett, J. (2001) *The Meaning of Environmental Security: Environmental Politics and Policy in the New Security Era*. London: Zed Books.
Bauer, D. (1990) 'The dynamics of communal hereditary land tenure among the Tigray of Ethiopia', in Bonnie J. McCay and J. M. Acheson (Eds) *The Question of the Commons: The Culture and Ecology of Communal Resources*. Tucson: University of Arizona Press.
Beattie, J. (1966) *Other Cultures Aims, Methods, and Achievements in Social Anthropology*. London: Routledge and Kegan Paul Ltd.
Boulding, Elise (1992) *New Agendas for Peace Research: Conflict and Security Re-examined*. Boulder, Colorado: Lynne Rienner Publishers, in association with the International Peace Research Association (IPRA).
Boyer and Hollingsworth (1997) *Contemporary Capitalism: The Embeddedness of Institutions*. Cambridge: Cambridge University Press.
Brown, Lester (1992) 'Redefining security', Worldwatch Paper No. 14. Washington D.C.: Worldwatch Institute.
Chambers, R. (1988) 'Sustainable rural livelihoods: A key strategy for people, environment and development', in C. Cornoy and M. Litvinoff (Eds) *The Greening of Aid: Sustainable Livelihood in Practice*, pp. 1–17. London: Earthscan Publications.
Charles H. Hession (1972) *John Kenneth Galbraith and His Critics*. London/Reeks: Mentor Books.
Chessen, Betsey (1998) *Rainforest*. New York: Scholastic.
Chichilnisky, G., G. Head and A. Vercelli (Eds) *Sustainability: Dynamics and Uncertainty*. Dordecht: Kluwer Academic Publishers.
Christiansson, C. and E. Tobisson (1989) 'Environmental degradation as a consequence of socio-political conflict in the eastern Mara Region of Tanzania', in A. Hjort, Af Ornas and M.A. Mohamed Salih (Eds) *Ecology and Politics: Environmental Stress and Security in Africa*. Uppsala: Scandinavian Institute of African Studies.
Clayton, M.H. and N. Radcliffe (1996) *Sustainability: A System Approach*. London: Earthscan Publications.
Clochester, M. (1993) 'Guatemala: The clamour land for land the fate of the forests', in M. Clochester and L. Lohman (Eds) *The Struggle for Land and the Fate of the Forests*. London: Zed Books.
Colebatch, H. and P. Degeling (1986) 'Talking and doing in the work of administration', *Public Administration and Development* 6 (4): 339–356.
Collinson, Helen (Ed.) (1996) *Green Guerrillas: Environmental Conflicts and Initiatives in Latin America and the Caribbean*. London: Latin American Bureau.
Coser, L. A. (1974) *Greedy Institutions: Patterns of Undivided Commitment*. New York: The Free Press.
Douglas, Mary (1987) *How Institutions Think*. London: Routledge and Paul Kegan.
Earth Council (1997) *Rio+5 Regional Reports, Africa East*. San Jose, Costa Rica: The Earth Council Secretariat.

Fleischman, R. (1995) 'Environmental security: Concept and practice', *National Security Studies Quarterly* (2) 1: 2 (Spring).

Gadgil, Madhav and Ramachandra Guha (1995) *Ecology and Equity: The Use and Abuse of Nature in Contemporary India*. London: Routledge.

Goldblatt, D. (1996) *Social Theory and the Environment*. Cambridge: Polity Press.

Goldman, M. (1998) (Ed.) *Privatizing Nature: Political Struggles for the Global Commons*. London: Pluto Press.

Hession, C.H. (1972) *John Kenneth Galbraith and His Critics*. New York: New American Library.

Head, G. M. (1998) 'Interpreting sustainability', in G. Chichilnisky, G. Head and A. Vercelli (Eds) *Sustainability: Dynamics and Uncertainty*, pp. 3–22. Dordecht: Kluwer Academic Publishers.

Hjort, A. Af Ornas and M. A. Mohamed Salih (Eds*) Ecology and Politics: Environmental Stress and Security in Africa*. Uppsala: Scandinavian Institute of African Studies.

Hollingsworth, J. R. and Robert Boyer (1997) *Contemporary Capitalism: The Embeddedness of Institutions*. Cambridge: Cambridge University Press.

Homer-Dixon, Thomas F. (1991) 'On the threshold: Environmental change as causes of acute conflict', *International Security* 16 (Fall): 16 (2): 76–116.

Homer-Dixon, Thomas F. and Valerie Percival (1996) *Environmental Security and Conflict: Briefing Book*. Toronto: American Association for the Advancement of Science, University College and University of Toronto.

Jarosz, L. (1996) 'Defining deforestation in Madagascar', in Richard Peet and M. Watts (Eds) *Liberation Ecologies: Environment, Development, Social Movement*. London/New York: Routledge.

Jay, Martin (1996) *Dialectical Imagination: A History of the Frankfurt School and the Institute of Social Research, 1923–1950*. Second Edition. Berkley: University of California Press.

Jordan, A. and T. O'Riordan (1997) 'Social Institutions and Climate Change: Applying Cultural Theory to Practice', CSRGE Working Paper, 97-14. Norwich: University of East Anglia.

Kakonen, J. (Ed.) (1994) *Green Security or Militarized Environment*. Brookfield: Dartmouth Publishing Co.

Keohane, R. O. and Elinor Ostrom (Eds) (1995) *Local Commons and Global Interdependence: Heterogeneity and Interdependence in Two Domains*. London: Sage Publications.

Keohane, R. O. and Elinor Ostrom (1995) 'Introduction' in: R. O. Keohane and Elinor Strom (Eds) *Local Commons and Global Independence: Heterogeneity and Cooperation in Two Domains*. London: Sage Publications.

Kimerling, J. (1996) 'Oil, lawlessness and indigenous struggles in Ecuador's orientale', in Collinson Hellen (Ed.) *Green Guerrillas: Environmental Conflicts and Initiatives in Latin America and the Caribbean*. London: Latin American Bureau.

Lohman, L. (1993) 'Thailand: Land, power and forest', in M. Clochester and L. Lohman (Eds) *The Struggle for Land and the Fate of the Forests*. London: Zed Books.

Mair, Lucy (1962) *Primitive Government*. Harmondsworth: Pelican.

March, J. G. and Johan P. Olsen (1989) *Rediscovering Institutions: The Organisational Basis of Politics*. New York: The Free Press.

Martell, L. (1994) *Ecology and Society: An Introduction*. Cambridge: Polity Press.

Meadows, D. H., D. L. Meadows and J. Randers (1992) *Beyond the Limits: Global Collapse or A Sustainable Future*. London: Earthscan Publications.

Mohamed Salih, M. A. (1994) 'Social science and conflict analysis', in Arvi Hurskainen and M. A. Mohamed Salih (Eds) *The Role of Social Science in Conflict Analysis*. Helsinki: Helsinki University Press.

Mohamed Salih, M. A. (1997) 'Global ecologism and its critics', in C. Thomas and P. Wilkin (Eds) *Globalisation and the South*. Basingstoke/New York. Macmillan and St. Martins Press.

Mohamed Salih, M. A. (1999a) *Environmental Politics and Liberation in Contemporary Africa*. Dordrecht: Kluwer Academic Publishers.

Mohamed Salih, M. A. (1999b) 'Introduction: Environmental Planning, Policies and Politics in Eastern and Southern Africa', in M. A. Mohamed Salih and S. Tedla (Eds) *Environmental Planning, Policies and Politics in Eastern and Southern Africa*, pp. 1–17. Basingstoke and New York: Macmillan and St. Martins.

Monbiot, G. (1993) 'Brazil: Land ownership and the flight to Amazonia', in M. Clochester, and L. Lohman (Eds) *The Struggle for Land and the Fate of the Forests*. London: Zed Books.

Moore, D. S. (1996) 'Marxism, Culture, and Political Ecology', in Richard Peet and M. Watts (Eds) *Liberation Ecologies: Environment, Development, Social Movement*. London/New York: Routledge.

Myers, Norman (1989) 'Our endangered Earth: Environment and security', *Foreign Policy* 74 (Spring): 23–41.

Myers, Norman (1993) *The Ultimate Security: The Environmental Basis of Political Stability*. London: W.A. Norton and Company.

O'Riordan T. and H. Voisey (1998) 'The politics of Agenda 21', in T. O'Riordan and H. Voisey (Eds) *The Transition to Sustainability: The Politics of Agenda 21*, pp. 31–56. London: Earthscan Publications.

O'Riordan, T. (1993) 'Politics of sustainability', in R. K. Turner (Ed.), *Sustainable Environmenal Economics and Management: Principles and Practices*, pp. 37–69. Chichester: Wiley.

O'Riordan, T. and H. Voisey (Eds) (1998) *The Transition to Sustainability: The Politics of Agenda 21*. London: Earthscan Publications.

Odegi-Awuondo, Casper, Haggai W. Namai and Beneah M. Mutsotso (1994) *Masters of Survival*. Nairobi: Basic Books Ltd.

Opschoor, J.B. (2001) 'Towards security, stability and sustainability oriented strategies of development in Eastern Africa', in M. A. Mohamed Salih, T. Dietz and A. G. Mohamed Ahmed (Eds) *African Pastoralism: Conflict, Institutions and Government*, pp. 23–38. London: Pluto Press.

Ostrom, Elinor (1987) 'Institutional arrangements for resolving the commons dilemma', in Bonnie J. McCay and J. M. Acheson (Eds) *The Question of the Commons: The Culture and Ecology of Communal Resources*, pp. 250–265. Tucson: University of Arizona Press.

Ostrom, Elinor (1990) *Governing the Commons: The Evolution of Institutions for Collective Action*. New York: Cambridge University Press.

Ostrom, Elinor (1994) 'Neither market, nor state: Governance of common pool resources in the 21st century'. Paper presented at the International Food Policy Research Institute. Washington, D.C., 2 June.

Ostrom, Elinor (1995) 'Constituting social capital and collective action', in Richard Peet and M. Watts (Eds) *Liberation Ecologies: Environment, Development, Social Movement*. London/New York: Routledge.

Petr, J. L. (1984) 'Fundamentals of an institutional perspective on economic policy', in Marc R. Tool (Ed.) *An Institutionalist Guide to Economics and Public Policy*. Amonk/New York: M. E. Sharpe.

Port, P. L., Roberto Scazzieri and Andrew Skinner (Eds) (2001) *Knowledge, Social Institutions and the Division of Labour*. Cheltenham: Edward Elgar.

Porter, Gareth (1992) 'Economic and environmental security in the US national security policy'. Roundtable discussion on environment, economics and security in the post Cold War world. Washington, D.C.: Environment and Energy Study Institute.

Prins Gwyn (Ed.) (1993) *Threats Without Enemies: Facing Environmental Insecurity*. London: Earthscan.

Rangan, H. (1996) 'Development, environment and social protest in Garhwal Himalayas, India', in Richard Peet and M. Watts (Eds) *Liberation Ecologies: Environment, Development, Social Movement*. London/New York: Routledge.

Rankin, A. (1996) 'The land of our ancestor's bones: Wichi people's struggle in the Argentine chaco', in Helen Collinson (Ed.) *Green Guerrillas: Environmental Conflicts and Initiatives in Latin America and the Caribbean*. London: Latin American Bureau.

Redclift, M. (1992) 'Sustainable development and popular participation: A framework for analysis', in D. Ghai and J. M. Vivian (Eds) *Grassroots Environmental Action Plans*, pp. 23–49. London and New York: Routledge.

Richards, P. (1996) *Fighting for the Rainforest: War Youth and Resources in Sierra Leone* London: James Currey.

Schwartz', Daniel and Ashbindu Singh (1999) *Environmental Conditions, Resources, and Conflicts: An Introductory Overview and Data Collection*. Nairobi: United Nations Environment Programme.

Scott, W. R. (2001) *Institutions and Organisations*. Second Edition. London: Sage Publications.

Swanson, L. (1984) 'Shifting the burden of environmental protection', in Marc R. Tool (Ed.) *An Institutionalist Guide to Economics and Public Policy*. Amonk/New York: M. E. Sharpe.

Tainer, J. A. (1998) 'Competition, expansion and reaction: Foundations of contemporary conflict', in M. N. Dobkowiski and I. Wallimann (Eds) *The Coming Age of Scarcity: Preventing Mass Death and Genocide in the Twenty-first Century*. New York: Syracuse University Press.

Tellegen, E. and M. Wolsink (1998) *Society and its Environment: An Introduction.* Reading, UK: Gordon and Beach Science Publishers.

Taylor, S. (1956) *Concepts of Institutions and the Theory of Knowledge*. New York: Bookman Associates.

Toepfer, K. (1999) 'Foreword', in Daniel Schwartz and Ashibndu *Environmental Conditions, Resources, and Conflicts: An Introductory Overview and Data Collection.* Nairobi: United Nations Environmental Programme.

Tool, Marc R. (Ed.) (1984) *An Institutionalist Guide to Economics and Public Policy*. Amonk/New York: M. E. Sharpe.

WCED (1987) *Our Common Future*. World Conference on Environment and Development. London: Oxford University Press.

Westing, Arthur (1986) *Global Resources and International Conflicts: Environmental Factors in Strategic Policy and Action*. Oxford: Oxford University Press.

Wolde-Semait, B. (1989) 'Ecological stress and political conflicts in Africa: The case of Ethiopia', in A. Hjort, Af Ornas and M. A. Mohamed Salih (Eds) *Ecology and Politics: Environmental Stress and Security in Africa*. Uppsala: Scandinavian Institute of African Studies.

Woodrow Wilson Center (1995) 'Environmental change and security'. Project Report No. 1. Washington, D.C.: Woodrow Wilson Center.

CHAPTER 15

BEYOND STATE-COMMUNITY POLARISATIONS AND BOGUS "JOINT"NESS: CRAFTING INSTITUTIONAL SOLUTIONS FOR RESOURCE MANAGEMENT[1]

SHARACHCHANDRA LÉLÉ

The question is no longer whether decentralized collective action can be effective, but under what circumstances it is appropriate, and how positive synergy between the state, market and civil organisations can most efficiently and fairly supply public goods (Uphoff 1993).

1. INTRODUCTION

Decentralised collective action is no longer on trial in the current academic debate on natural resources management.[2] What still needs to be worked out are the conditions for its success and its optimal nesting in or relationship with other institutions. A vast body of research has addressed the question of under what (external) conditions decentralised collective action will succeed (for a recent review, see Agarwal 2001, specifically in the context of forests, see Ostrom 1998). But there seems much less debate, let alone agreement, on the question of relationships between local-level collective action institutions and other institutions at different levels. From the perspective of those interested in proving that community-level institutions of resource management can and do work, often the only role the state can (and should) play is "non-interference" or at most "legitimisation".[3] If the need for "nested enterprises" has been recognised, it is only in the case of common-pool resources that are part of a larger system (Ostrom 1990: 90). Even here, the precise meaning and form of "nestedness" are yet to be elaborated.[4] The focus has remained on the conditions of success for local-level collective action in which the state implicitly has a minimal role.

Others have recognised that the neat separation of state, community and private property regimes is only a theoretical one, and that in fact resource rights or tenure regimes are better characterised as variations in the manner in which different strands

283

Max Spoor (ed.), Globalisation, Poverty and Conflict, 283–303.
© 2004 Kluwer Academic Publishers. Printed in the Netherlands.

of the property rights "bundle" are distributed between the state, local communities, individuals and other actors (Ciriacy-Wantrup 1963, McKean 1998).[5] The concepts of "co-management" and "joint management"[6] that emerged in the 1990s occupy the vast (and fuzzy) middle ground between pure state control and pure community control. While there has been a virtual flood of research on the "performance" of these joint or co-management arrangements as implemented in various parts of the globe, there has been little discussion of the conceptual basis for joint management (see Lélé 1998b). Authors have identified variants of co-management that differ in the extent of devolution of state power to the community (Pomeroy and Berkes 1997). But the discussion appears confined to the distribution of power between the "state" and the "local community" without much interrogation of these concepts. Alternative forms of defining and organising the so-called local community are hardly ever mentioned.[7] The current structure of the state is taken as given, and other institutions appear to have no place in the conceptual framework.[8]

This chapter goes beyond the analytically loose (and practically often troublesome) concepts of joint management and co-management, to contribute to the debate on why and how multi-layered systems of resource governance should be designed. Section 2 defines "institutions" and points out how the "appropriateness" of institutions depends upon one's normative concerns. This is followed by a discussion of the broad normative concerns that should underpin institutional interventions in society, the actual variation in normative underpinnings of the current body of research on institutions for natural resources management and the need for a common, broad-based approach. Section 3 enumerates the ecosystem and social system characteristics that necessitate institutional intervention, given the broad normative concerns identified in section 2 and the kinds of institutions that might be required. This leads to the case for multi-layered structures for environmental governance. Section 4 outlines a process of designing these structures and identifies some design principles. The process is illustrated using the example of forests. The conclusion summarises the main points of this "institutionalist" exercise, and points out that institutional redesign is necessary, but not sufficient, to stem environmental degradation.

2. INSTITUTIONS FOR NATURAL RESOURCES MANAGEMENT: DEFINITIONS AND MORAL IMPERATIVES

If our goal is to craft appropriate institutions for environmental governance, then we must first define what we mean by "institutions" and then what "appropriate" means.[9] I begin by defining institutions in broad terms, indicating the scope of this definition and some useful distinctions within it. I then discuss the different values that shape people's concern about environmental or resource degradation, and show that this normative basis varies across different literatures, resulting in different institutional forms being privileged.

2.1 Defining Institutions

Institutions are generally defined as rules, regulations and conventions imposing constraints on human behaviour. They can be 'both enabling (in providing ways through which people negotiate their ways through the world) and constraining (in providing rules for action)' (Mehta et al. 1999: 13). From this broad definition, it follows that institutions include more than just self-organised collective-action groups operating at a local scale. Individual households at one extreme and the nation-state and international regulatory authorities such as the World Trade Organization (WTO) at the other are also institutions in that they enable and constrain individual human actions. This is also true of different forms of for-profit organisations (e.g. partnerships, cooperative societies and joint stock companies). Thus, there is no *a priori* reason to focus only on self-organised collective-action groups in this discussion.

This is not to say that all these institutions are identical in the role they play or their internal structure (and hence their performance). While a detailed taxonomy is beyond the scope of this chapter, it is useful to categorise institutions in a limited way. One obvious distinction is between *voluntary* associations of individuals (whether cooperative or corporate) and *involuntary* ones (e.g. nation-states).[10] Perhaps a more important distinction is between those that play a *productive role* and those that play a *regulatory role*. There is some correspondence between these two classifications: voluntary associations (cooperative or corporate) generally play a productive role, while the main role of the state is seen today as regulatory, *except* in the production of pure public goods.[11]

Furthermore, within each category, several institutional forms are possible. Voluntary institutions may be corporate, cooperative or non-profit trusts. Involuntary institutions also take many forms. For instance, after a phase of highly centralised and monolithic governments, it is now accepted that separation of productive and regulatory roles within the state and separation of the provision of common-pool resources from that of public goods might be necessary. Hence, regulation may be done either by state bureaucracies or by specialised regulatory bodies to which coercive power is delegated and which are accountable directly to the public (such as the utility commissions in the United States). Similarly, public goods may be produced by generalised state bureaucracies or specialised agencies or institutions jointly supported by state and civil society. Finally, one need not take the nation-state as a vertical monolith. Different levels of the state (national, provincial, local) with varying degrees of devolution are possible.

2.2 The Normative Underpinnings of Institutions

What then is a legitimate normative goal or moral purpose for which voluntary institutions are formed or for which nation-states may intervene in our lives?[12] While enumerating all possible purposes would be a daunting task, the umbrella terms *efficiency, sustainability* and *equity* seem to cover most of the concerns that drive public policy discussions and interventions.[13] In the context of natural resources management, these terms may be defined as follows:

- *Efficiency* is concerned with maximising current well-being derived from the natural world at minimum cost, whether measured in physical or monetary terms.
- *Sustainability* is concerned about the continuation of well-being into the future, either one's own or that of several generations to come. *Ecological* sustainability as a normative concern further assumes that there is some 'immutable biophysical basis to human well-being' that needs to be preserved (Lélé 1991).
- *Equity* is concerned with the *intra-generational distribution* of human well-being, across barriers such as class, ethnicity and gender,[14] including concerns about fairness of outcomes as well as processes. It is relevant in the context of sharing both the fruits of resource use and the externalities generated by resource extraction, processing and consumption.

I take the normative position that "environmentally sound development" must encompass these three categories of concerns.[15] Thus all institutional arrangements proposed for natural resources management must incorporate and be judged against all three concerns.

2.3 The Normative Underpinnings of Institutions for Natural Resources Management

Admittedly, many voluntary institutions (for resource management or otherwise) are formed with only efficiency gains in mind. One would expect, however, that debates on *public* policy regarding institutions for natural resources management would keep in mind all three concerns. But that is not always the case. In Hardin's famous essay itself, overgrazing is "bad" because it results in everybody getting less than they could out of the pasture. In other words, he, and the bulk of the common-pool resource literature that emerged in response to his essay, is primarily concerned about *under-production*[16] or *inefficiency* in resource use. The future *sustainability* of the resource seems to be an issue only insofar as this year's mismanagement affects next year's production, which is a rather short-term notion of sustainability. The question of *distribution* of benefits has also been given lower priority. While it is quite likely that the original response to Hardin was motivated by concern about the iniquitous effects of resource privatisation that Hardin was recommending, it is often simply assumed that collective-action institutions are equitable. But in Hardin's own example, graziers who own more sheep naturally stand to gain more in absolute terms, whatever the regime of pasture management. Even those who have demonstrated how collective-action institutions can and do avert Hardin's tragedy admit that these institutions are welfare-enhancing only in the sense of being "Pareto improving" (Menon 1999).[17]

Others, possibly forming a much larger section of the environmental community, are concerned about resource degradation because it compromises the ability of future generations to meet their own needs (*a la* the Brundtland Commission). For them, "degradation" means decline over time, leading to lower availability (in a quantitative or qualitative sense) of resources and environmental services for future generations. These sustainability scientists focus on changing current resource use practices to avoid excessive costs in the future (Pearce 1988, Costanza 1991). Certainly most conservationists, too, would like to forsake current resource use (at least by others!) in order to save some "pristine environments" for posterity. Thus, the institutional arrangements proposed by the sustainability- and conservation-oriented literature focus on striking a balance between sustainability and efficiency, or even give sustainability overriding priority, as in the case of protected areas.[18]

Equity concerns have been the ideological driving force behind state interventions in many sectors, but are perhaps least foregrounded in debates on common-pool resources (see Menon 1999 for a detailed exposition of this problem). This is unfortunate, because one of the fundamental concerns in natural resources management in developing countries should be equitable access to natural resources (Agarwal 1985). It must also be recognised that local resource use often leads to externalities in space and time. Underpinning the concept of "externality" is a normative concern for the *unfair* allocation of costs and benefits of the resource use process. Perhaps we are conditioned to think of environmental externalities only in the context of pollution or "brown" issues, and not in the context of common-pool resource management or "green" issues. But there is an urgent need to integrate these debates. The institutions that can address such concerns would usually be regulatory, quasi-state institutions rather than voluntary collective-action institutions (although some role for collective-action institutions cannot be ruled out). A variety of institutional forms needs to be brought onto the institutional "menu" for natural resources management.

In short, part of the current disagreement over what constitutes "appropriate" institutional arrangements for resource management may actually be a consequence of differences in normative concerns rather than in theoretical or empirical claims about the relative ability of these arrangements to meet similar social goals. It is therefore necessary to separate these two debates, to agree *ab initio* about the desired forms and combination of efficiency, sustainability and equity and then proceed to choose institutional arrangements from the array of possibilities. Of course, working out such a consensual definition of these norms is easier said than done. Nevertheless, in what follows, I assume that the reader shares with me a certain minimum concern for efficiency, sustainability and equity, including interest in equity and justice *for their own sake* and not just because they might further sustainability or efficiency.

3. THE NEED FOR INSTITUTIONS IN THE CASE OF NATURAL RESOURCES/ECOSYSTEM MANAGEMENT

Having made our normative position clear, we need to ask whether unregulated individualistic behaviour can ensure efficiency, sustainability and equity of resource use and environmental management. The answer is obviously in the negative. The common-pool resource literature has focused largely on characteristics of resources that make them common-pool and hence prone to "under-production", which is essentially an efficiency problem. But there are several other characteristics of ecosystem processes and also characteristics of social systems that require the coordination between and regulation of the actions of many actors.

3.1 Ecosystem Characteristics

3.1.1 (Non-)excludability and Subtractability

In thinking about ecosystem characteristics that require rule formation, the conventional focus has been on excludability and subtractability, resulting in the 2x2 classification of goods into privatisable, common-pool, toll and purely public goods. Many ecosystems – rangelands, fisheries, forests and groundwater – suffer from an inherent excludability problem (as in the case of groundwater) or have high exclusion costs (as in the case of rangelands). Given that their tangible products (water or grass) are subtractable, these ecosystems (or at least these products) can be classified as "common-pool goods".

Note, however, that some other "benefits" or valued features of these ecosystems seem to call for them to be classified as public goods or toll goods. For instance, the existence value of wildlife is non-subtractable, making wildlife more akin to a public good. On the other hand, to the extent that it is possible to exclude someone from enjoying the aesthetic value of wildlife (by controlling access to wildlife-rich habitats), these ecosystems are toll goods. Similarly, if biodiversity is valued for the genetic information it contains, then it is more of a toll good.

3.1.2 Scale

Excludability problems are a result of ecological connections between different parts of the resource or ecosystem. However, these connections do not extend indefinitely – each ecosystem process or component has its own "typical" boundaries. For instance, groundwater moves within an aquifer, fish or land animals within some typical range, and even air pollutants might be confined to some valley, whereas carbon dioxide entering the atmosphere essentially circulates all over the globe. As long as the scale or boundary is small enough for one actor to control access to the resource (i.e. for the actor to patrol the boundary), one would not need any institutional arrangements, at least to ensure efficient use. But many ecosystem processes do not operate at such convenient scales. Hence, it becomes necessary to set up collective-action institutions to coordinate human actions at that scale.

3.1.3 Regeneration Rate

If one is concerned about the sustainability of benefits obtained from natural resources, one has to worry about their regeneration rates. These range from none in the case of non-renewables such as minerals and slow regeneration as in the case of forests to relatively fast regeneration as for grasslands. When juxtaposed against time horizons typically employed by human beings in their decision-making, different resources require different approaches to ensure long-term sustainability.

3.1.4 Tradeoffs Between Uses and Users

Most discussions of common-pool resource management focus on situations where there is a single use of a resource: grasslands for grazing, forests for wood, or groundwater for irrigation. In fact, most resources have multiple uses within and across user communities. This interconnectedness between uses is increasing as human ability to manipulate resources expands. Grasslands in the Himalayas are important as tourist attractions; forests produce timber and non-timber products as well as provide watersheds; and irrigation water is diverted to meet urban needs. These situations differ from simple common-pool situations, because the issue is not one of coordinating actions to ensure efficient use but rather of equitable allocation across uses and users.[19] Note also that the different users or beneficiaries of a resource may be located at different distances from the resource, making simple, face-to-face collective action impossible anyway.

3.1.5 Unidirectional Externality Across Stakeholders

Perhaps the least recognised ecosystem characteristic in the debate on "fit" between ecosystems and institutions is that many ecosystem processes are unidirectional in nature, generating *asymmetrical externalities* between individuals or communities.[20] For instance, water flows in a river basin is unidirectional, hence upstream actions influence the well-being of actors living downstream but not vice-versa.[21] Similarly, while the atmosphere may appear to be a well-mixed ecosystem and hence akin to a conventional common-pool resource, in other cases air pollution may have strong directionality: only populations or countries downwind may be affected.

Unidirectionality of ecosystems does not create problems of efficiency, but affects equity. With downstream communities unable to "retaliate" through the biophysical process, there is no incentive for upstream communities to modify their use patterns. When combined with the fact that some externalities may operate over large scales (such as river basins) and hence introduce monitoring problems, this characteristic poses a major challenge to institutional design.

3.1.6 Complexity and Spatial Heterogeneity

It has been pointed out that ecosystems vary tremendously in their structure, species composition and hence functioning and response to human intervention. This introduces tradeoffs between the degree of centralisation in resource management and the sustainability or efficiency of management. The highly localised traditional ecological knowledge developed by communities over centuries of experimentation gives them a strong comparative advantage over centralised bureaucracies in resource management.[22] In certain situations, the repository of specialised knowledge may even be the individual or the household, rather than the community, making it the most suitable unit for day-to-day management decisions. In other cases, local (informal) knowledge may have to be judiciously supplemented with knowledge produced more rigorously by scientific institutions.

3.2 Social System Characteristics

3.2.1 Missing Public Goods

Even private property rights regimes must be enforced and protected. While individual protection of their own property might be possible to a limited extent, there are significant efficiency and equity gains if enforcement is seen as a public good to be provided by the state. The same holds true for conflict resolution mechanisms such as judicial systems. High-quality information about new technology, the condition of a resource (such as groundwater, see Ostrom 1995) or about the flows of pollutants or resources across a large spatial scale is another public good that will be under-produced unless some institutional arrangements are made for their provision.

3.2.2 Varying Concern for the Future

A substantial literature shows that discount rates used by individuals are higher than the rates they themselves expect the state to use in discounting socially productive investments (Dasgupta et al. 1972). In other words, if sustainability is a concern, individuals may expect society to take care of it rather than act individually to provide for it. In any case, it is certainly true that, in democratic societies, states are responsible for imposing social concern about long-term sustainability on individual actors who might simply not share this concern.

3.2.3 Initial Allocation of Resource Access

Even where it might be possible to manage resources at an individual level, some larger-than-individual institution needs to decide what the initial allocation of resources will be. Extending this argument, one can see that even if, say, forests are to be treated as common-pool resources at the scale of a village, the inter-village allocation of forests remains to be determined. Options include letting the villages work out the allocation on their own or asking some higher level authority to do it. The latter need not be the state; it could be a federation of village-level associations.

3.2.4 Other Market Failures

Communities that depend upon natural resources for income generation may be affected by market failures such as difficulties in obtaining credit (due to the lack of individual titles on the resource) and poor information on prices or thin markets (particularly in the case of perishables such as fish or thinly distributed resources such as timber and non-timber forest products). Societal interventions to correct such market failures can increase gains and reduce inequitable distribution of resource rents between harvesters and traders.

3.2.5 Pervasive Social Divisions

Societies are often fractured on lines such as class, ethnicity and gender. These divisions are of two kinds: "horizontal differences" where barriers to cooperation may exist but there is no hierarchy of power and "vertical differentiation" where in addition to social distance there is unequal distribution of power. The former may lead to efficiency losses, as groups may refuse to cooperate even in win-win situations. The latter may or may not result in lack of collective action, but it clearly results in less-than-equitable outcomes. Designing institutions that overcome these barriers and inequalities is a major challenge.

In sum, a variety of ecosystem and social system characteristics necessitate coordination *and* regulation of individual actions through rules (and institutions that make the rules) *if* efficiency, sustainability and equity in resource use are to be enhanced. This should not, however, be misunderstood to mean that there is no role for individual decision-making. Indeed, there are a variety of *other* socio-ecological characteristics, such as physical proximity to the resource, knowledge about the resource and maximum interest (direct dependence upon the resource) that favour giving maximum control to individuals or households living in close proximity to the resource.[23] Nonetheless, if we recognise that human beings are socially and ecologically interconnected, individual decision-making has to be circumscribed by ecological and social norms. In today's complex world, this cannot be achieved by informal, culturally internalised behavioural norms alone. It also requires formal institutional arrangements.

4. THE DESIGN OF MULTI-LEVEL INSTITUTIONS FOR NATURAL RESOURCE GOVERNANCE

Clearly, the different characteristics enumerated above demand different institutional arrangements. Common-pool resources bounded at the scale of a village can be addressed by village-level collective-action institutions (such as a village forest committee or water users' association), whereas unidirectional externalities in a river basin require a basin-scale regulatory institution (such as a watershed board or river basin authority) that would essentially protect the rights of the downstream stakeholders. Public goods such as knowledge and market information may be produced through collective action (such as fishers doing their own research) or by the state in a quasi-market situation where resource managers pay fees for the information. In all cases, institutions require at least some legitimisation by the state.

More important, these institutional requirements are not mutually exclusive but rather overlapping. Given the fact that virtually all resource use situations are characterised by multiple uses and users across local and regional scales, that most resource user communities are characterised by social divisions, and that market failures of credit and information are pervasive, purely self-organised collective-action institutions are unlikely to produce efficiency, sustainability and equity gains. Even the two-layered approach of most "co-management" or "joint management" arrangements would be insufficient. More than two institutions are clearly required (if for no other reason than to adjudicate over conflicts between the state and the local community!). There will also have to be some degree of "nesting", because functions such as conflict resolution or external policing support must be provided by institutions at one or more levels higher[24] than the level of the day-to-day management units.

The case for multi-layered institutions for natural resources is thus easily made. Assuming that one is concerned about efficiency, sustainability *and* equity to some extent, such a multi-layered governance structure would naturally have to contain productive and regulatory, voluntary and involuntary, specialist as well as generalist institutions. It could involve individual as well as community-level organisations, professional non-governmental organisations and different levels of the state. In designing such a multi-institutional structure, one has to specify not only the *internal design* of each institution, but also the *inter-institutional linkages* (i.e. the distribution of authority and responsibility across institutions).[25]

4.1 General Process and Some Design Principles

The challenge therefore is to develop a more systematic framework for structuring multi-layered governance under different situations. Generally speaking, this would involve five steps:

- identifying the ecosystem and social system characteristics peculiar to the particular resource that requires institutional intervention and the function to be performed by the intervention;
- identifying the possible institutional forms that might serve each of the functional requirements identified, and their efficacy in doing so;
- choosing the best fit between function and institutional form and specifying the inter-institutional linkages based on one's normative standpoint;
- fine tuning the arrangements to the specific socio-ecological context;
- designing the internal structures of each institution that is part of the governance structure.

Although much of the design process would necessarily be resource-specific, it is possible to identify some generic principles for multi-layered resource governance. A number of principles can serve as a starting point. The first is a *tight linkage between authority, responsibility and incentive.* In the context of institutions playing a productive role directly connected with the resource, this means that the day-to-day users (who have an incentive to be involved in resource management) must be given the responsibility of management and the authority to do so. Regulatory institutions or productive institutions not directly linked with the resource (such as those producing public goods) are seldom directly affected by the success of their efforts. Hence, for producers of public goods (such as monitoring, technical support or external policing), the linkage has to be a fiscal one, through partial support from user fees that are linked to performance. For regulators, separation from the state and direct electoral linkages with the multiple stakeholders is required to ensure accountability. In all cases, responsibility must be matched by authority.

The second principle is *jurisdictional parsimony.* In terms of scale, jurisdictions should be no larger than necessary. In other words, 'never globalise a problem when it can be dealt with locally' (Murphree 2000). Here, "can" is understood in both an ecological sense and a social sense. When common-pool problems (symmetrically) extend beyond the local scale (such as mobile wildlife populations in a forest), or in cases where public goods are to be provided at a higher scale (such as conflict resolution, information or policing), lower-level units should negotiate to determine which functions to delegate upwards and how to provide for them (ibid.). Higher level nation-states must not unilaterally usurp control simply because a particular resource transcends the jurisdiction of a local community, or because certain functions must be performed at a higher scale.[26] Regulation must be semi-participatory, quasi-stable, transparent and use a mix of standards and incentives. Although regulators must be accountable to those they represent and also involve those they regulate, they should be insulated from direct pressures from both quarters so as to remain relatively stable and fair. Mixing standards and incentives means that the regulators should attempt to set some basic standards that ensure minimal fairness, but then allow some room for negotiations in market-like situations.

A third design principle is *tight fiscal linkages*. Resource management institutions must be financially self-sufficient insofar as possible, working with funds raised largely from below rather than doles from above. When external funds are provided, they should be linked to the provision of specific public or off-site environmental benefits, and this linkage monitored in a neutral and transparent manner. It may or may not be possible to set up formal markets to enable such provisioning, but the principle of provisioning in proportion to performance must be maintained. *Maximum transparency* is an aspect of this principle. To build confidence in the governance structure, maximum transparency at all levels is paramount. Institutions can achieve this by being small in size to ensure face-to-face contact wherever possible. They must enshrine a right to information within their structure and make maximum use of information technology to realise this right.

I shall now illustrate the process of designing multi-layered resource governance systems by applying the process to the case of tropical forests.

4.2 The Ecological and Social Characteristics of Tropical Forests

Forests are a particularly interesting case because they embody many of the ecosystem and social system characteristics mentioned above. Forests generate multiple products, services and benefits: timber, firewood or fodder, soil erosion control, hydrological regulation, wildlife conservation and carbon sequestration. These benefits are not only distributed across different beneficiaries and scales, but their boundaries and directionality also vary (Lélé 1998b). For instance, plant products from forests, if harvested at low levels, have limited common-pool characteristics. Higher extraction can often be compensated by more intensive management efforts. But exclusion of people or livestock from one's use area may be difficult. Wild animal populations, when managed as game, are certainly common-pool resources at a multi-village scale. On the other hand, watershed services are positive externalities that accrue in a unidirectional manner to downstream populations (often living far away from the forest). The aesthetic value of wildlife can be said to be a toll good. The existence value of wilderness is virtually a purely public good. Because forests (particularly natural forests) are slow-growing resources, decisions taken today may take many years or decades to bear fruit. Tropical forests are also highly complex and spatially heterogeneous ecosystems that render modern, centralised knowledge systems highly inefficient, if not ineffective.

The social context of tropical forests is also one that requires institutional interventions. Although communities dwelling in and around forests are often highly dependent upon forests for their livelihoods and indirect benefits, they often lack well-defined or exclusive rights over forest products. Yet while local communities (or certain groups within them) often have the best knowledge about a particular forest ecosystem and its response to human use, this knowledge may be inadequate in the context of changing and intensifying use patterns. Protecting the boundary of a dense forest, even when done by local communities, is not an easy task, particularly when marauders are well-armed outsiders intent on poaching high-value, mobile and spatially heterogeneous resources such as ivory or sandalwood. The markets for many forest products are thin and non-competitive. Further, forest-dwelling communities are not always forest-*dependent*; thus, their interest in managing the resource may vary both within and across communities. This points to the fact that forest-dwelling communities are also not always homogeneous in terms of the distribution of social and economic power. Finally, current technologies and their costs bring significant economies of scale in collective protection, planting and harvest of forests, especially in densely populated regions of the tropics.

4.3 Match Between Functional Requirements and Institutional Forms

Having identified the functions that we want the institutional interventions to perform, we can identify alternative institutional forms for each function. For instance, day-to-day forest management may be carried out by individual households, user groups that are input-based or membership-based,[27] local governments,[28] or even corporate or semi-corporate[29] bodies. It could also be carried out by a specialised agency (such as a wildlife service) or a more all-in-one bureaucracy (such as the existing forest departments in many tropical countries). Initial allotment of access rights may be done by the user group, local government, a user group federation or a higher level government. Watershed communities can be ensured fair benefits by a generic state bureaucracy (as is the case in India today, at least on paper[30]) or by a separate watershed regulatory authority or similar institution.

Thus, one can create a matrix in which institutional function and form are the two dimensions, and try to map the "fit" between them. Table 1 provides such a mapping for the case of tropical forest management. Due to constraints of space, this discussion uses a subset of the possible institutional arrangements. (For instance, an NGO that performs important functions of public education, advocacy and demonstration would have some role in forest governance.)

Of course, the rankings in the table are only indicative. Apart from needing to know the specific context, we must acknowledge that our knowledge of these "fits" is limited. Moreover, since many functions embody more than one concern, such as equity and sustainability, and since there is always an interest in institutional efficiency (minimum transaction costs), we would ideally want to know the variation in "fitness" of an institutional form with respect to each of these concerns. This is not done due to constraints of space. In effect, I have given some implicit weights to the different concerns in coming up with the overall fitness rankings. Nevertheless, this provisional and partial mapping provides a more rigorous and comprehensive way of designing institutional arrangements than the current approach of tinkering with existing institutions or asking NGOs to perform unrealistic roles.

Table 1. *The Relative Appropriateness of Different Institutional Forms for Different Functions in the Context of Tropical Forest Management*

Function	Organisational Forms									
	Individual Household	Hamlet/ Village Committee	Coopera- tive or Self-help Group	Local Self- Govern- ment	Federa- tion of Village- level Bodies	Tribunal	Forest Protec- tion Squad	Watershed Regula- tory Board	Wildlife Regula- tory Board	Forest Con serv- ation Agency
Day-to-day resource management	+++	+++	+++	++						+
Intra-group allocation & enforcement		+++	++	++[6]			+			+
Intra-group dispute resolution		++	++	+++	+					+
External protection		++	++	+	+		+++			++
Inter-agency dispute resolution					+++	+++				
Enforcing "environ- mental" norms				+	++	++		+++[3]	+++	+++[4]
Enforcing "social" norms	+	++	+	+++[5]	+++[5]					
Marketing of products	++[1]	+	+++[2]	–[7]	+++					+
Overcoming credit problems	+	++	+++		+++			+	+	+

Notes: 1. When the product is a high-volume high-value one. 2. When the product is high-value but low-volume, or when it is a perishable. 3. When the upstream forests are proximate to and being used by villagers. 4. When the specialised forest management agency operates under the control of the state in areas that are remote from human habitation. 5. With support and pressure from higher levels of democratic governance, like the state government. 6. For inter-hamlet allocation of resources. 7. Mixing self-government roles with profit-making activities is likely to lead to corruption; hence the negative sign.

4.4 Tuning Institutional Arrangements to the Socio-Ecological Context

Our understanding of the finer effects of specific socio-ecological conditions on institutional performance in the wider sense is limited. Table 2 presents some propositions in this regard, more as speculation than as established fact. For convenience, these are posed as pairs of contrasting conditions without implying that real-world conditions can be so neatly pigeon-holed.

Now, using the design principles enumerated in section 4.1, it should be possible to design a multi-layered system of forest governance for a particular socio-ecological context. If my normative position, which involves a strong concern for equity and sustainability, and my understanding of conditions prevailing in the central Indian forest belt were adopted, the design might look like the following (see Table 3):

- Hamlet-level user groups would have secure rights of access, exclusion, management and sale of all products within sustainability norms wherever such

Table 2. *Fine Tuning Forest Governance System to Specific Socio-Ecological Contexts*

Socio-Ecological Context	Recommended Institutional Form
Dispersed settlements[1]	More individual rights,[1] representative democracy in user groups
Clustered settlements	Community rights, participatory democracy
Degraded lands	More investment required, more harvesting rights should be given to local communities
Dense forests	Timber and non-timber product marketing rights and organisation may be sufficient
Subsistence interest	User group may coincide with local village council
Commercial interest	User group (voluntary) should be separate from village council; latter should not be involved in economic transactions
Spatially and temporally heterogeneous resource	Community management of harvesting
Relatively homogeneous resource	Individual assignments of usufruct possible
High-value high-volume products	Individual marketing possible
High-value low-volume products	Cooperative marketing through village- and multi-village scale cooperatives

Note: 1. In the Western Ghats of India, where the rolling terrain leads to dispersed settlements or homesteads, we have documented the emergence and efficient, sustainable performance of individualised tenure regimes rather than community management regimes (Srinidhi and Lélé 2001).

groups come forward; remote areas would be managed by a forest conservation service.

- The initial allotment of forest patches would be carried out by district-level government in tandem with civil society organisations and local user groups.
- Marketing would be carried out by federations of user groups.
- Inter-village conflict resolution would be carried out by federations of user groups.
- Policing would be provided on demand by a specialised forest policing agency (separate from the forest conservation service) in return for some user charges that cover part of the agency's costs.
- Technical and monitoring support would be provided by independent research agencies, funded partly from user fees and partly by state and civil society.

Sustainability norms would be set partly by a watershed regulatory authority and partly by a wildlife regulatory authority. The former, at the river basin level, would also be responsible for regulating water use and would have representatives of all users in the river basin. The latter would be district level and include representatives from different walks of life, with links to higher level wildlife boards through fiscal transfers. Intra-user group equity would be ensured through norms laid down by district-level governments in consultation with user groups and be monitored by them jointly.

Table 3. *Socio-Ecological Characteristics of Forests, Underlying Concerns and Function to be Served by Corresponding Institutional Intervention*

Key Ecosystem Characteristic	Underlying Concern	Function of Institutional Intervention
Significant separability in plant product management	Efficiency of resource use	Individual or small group should have day-to-day control in plant product management
Spatial heterogeneity in resource distribution	Equity	A way of allotting initial access fairly and redistributing it from time to time
Fire management is non-excludable; NTFP availability highly variable in time	Efficiency of resource use	A way of coordinating fire protection and non-timber forest product harvest activities amongst the direct beneficiaries of such control
Unidirectional externalities of watershed benefits	Equity and sustainability	An efficient way of ensuring that the interests of downstream communities receive fair weightage in the decisions of upstream forest managers
Existence value of wildlife is a global public good	Equity and sustainability	A reasonably efficient way of ensuring that local, regional and global concerns for wildlife existence in the short and long-term get fair weightage in the decisions of forest managers
Slow regeneration rate	Sustainability	A way of formulating long-term sustainability norms that are fair
Local communities or households are generally highly dependent on forests, but not uniformly	Equity between uses/users, including within local community	Ensuring that local needs are met first before non-local demands (e.g. for timber) or that benefits from supplying non-local demands go to local, most needy groups
Economies of scale in local-level forest management	Efficiency	Ensuring that local communities have the opportunity to form collective-action groups to capture economies of scale
Markets are thin	Efficiency, equity	Ensuring that local forest product harvesters have opportunity to form marketing cooperatives
Protection is not easy	Efficiency, equity	Local groups must be involved in protection, outside protection support must also be available

Note: In the case of all institutional interventions, efficiency of the intervention itself in terms of transaction costs it creates will also be a secondary concern.

5. CONCLUSION: THE LIMITS OF INSTITUTIONS

In this chapter I have attempted to further the debate on multi-layered governance for natural resources management. I began by pointing out that, in the natural resources management context, we often seem to get stuck with very limited notions of "institutions" in relation to local-level collective-action institutions with an enabling state in the background. Even joint management is thought of as involving only these two entities. This leads to both continued polarisation (because the conceptual basis for the involvement of the other is unclear) and practical failure (because of faulty design). Some of this polarisation or confusion is also due to differences in underlying concerns. Whereas those who focus on efficiency in resource use would see local-level collective action as "sufficient" and the role of the state as minimal, those who are most concerned about "sustainability" of certain long-term benefits argue for state intervention as the only solution.

Our discussion sketched a consistent rationale for the presence and role of a variety of institutions that would be required for successful governance of natural resources and environmental problems. These include voluntary and involuntary institutions, productive and regulatory institutions, specialised and generalised institutions, single units and federations, state, quasi-state and civil society institutions, and institutions operating at various scales. Governance of natural resources involves identifying which institutions are required, what their linkages (or distribution of authority and responsibility across them) should be, and how they should be structured internally.

The case of forest resources was used to indicate how such a multi-layered governance structure could be crafted and what uncertainties remain. Of course, the structure that emerged is at great variance with that presently in use in most tropical countries. Arguments might therefore be made that it is better to promote some simple, practically feasible, changes in policy at the cost of conceptual rigour and detail. But the current experience with "joint" resource management arrangements in different sectors in many tropical countries suggests otherwise. First, the present level of oversimplification or vagueness leads only to co-optation (Lélé 2000, Sundar 2001). In fact, a clear case for multi-layered governance could win support from those who have taken extreme positions because they are currently reacting to a limited set of options. For instance, the conservationists may support the status quo of centralised bureaucratic management only because they see the need for significant non-local regulation. The communitarians may similarly oppose outside regulation because they see how oppressive the current regulatory apparatus (centralised state bureaucracies) is.

The proposed shift is really no more drastic than what has actually taken place in several other sectors in developing countries (e.g. the liberalisation and privatisation of the hitherto "nationalised" power and telecommunication sectors in India). "Unbundling" state control in other sectors has led to separation of policymaking, productive and regulatory roles, and private sector participation in productive activities. Of course, the direction of unbundling recommended here is different, being in favour of local communities and the poor rather than large corporations. Opposition to this re-crafting should therefore be seen as driven by normative concerns different from ours rather than by different perceptions on the performance of institutions.

The focus in this chapter was the re-crafting of institutions for natural resources management in the pursuit of environmentally sound development. Obviously, the underlying assumption is that modifying structures or rules that govern individual behaviour will substantially modify individual behaviour: that is, that institutions "matter" and they "work". It is important, however, to not get carried away by this faith in institutions to the point where we seem to be saying that *only* institutions matter!

The causes of environmentally unsound development are many. Institutional failure is one of them, but by no means the only one. Often, institutional failure is only a symptom of larger problems. At the very least, if *desirable* "rules of the game" are to emerge through a reasonably democratic and open process, then sufficient people must *desire* those rules – whether they relate to reducing fossil-fuel emissions or ensuring greater fiscal discipline. As Mahatma Gandhi said, 'there cannot be a system so good that the individual need not be good'. The best-crafted institutions can collapse if the values of the people within them are incompatible with institutional goals.

The environmental crisis is a product of a complex and inter-related set of factors (Lélé 1991). Technological change has given human beings the ability to dramatically and often unknowingly modify ecosystem processes. It has also increased the available array of consumption possibilities. This change is partly driven by a highly reductionist science and strongly fuelled by capitalist systems of production that thrive on and hence promote unbridled consumption. Unequal distribution of political and economic power within and across villages, cities, regions and nations enables the powerful to appropriate resources and externalise environmental consequences of unbridled consumption onto the relatively powerless. These powerful actors, who range from multinational corporations to entrenched national bureaucracies down to male heads of household, can resist attempts at regulation by the state or civil society. Traditional cultural sanctions against profligate consumption are breaking down and cultural mechanisms of redistribution are disappearing as the new culture of unlimited technological possibilism creates the morally bankrupt ideal of a "free" individualist consumer.

Determining the key elements of such a holistic approach to reducing environmental degradation would be very difficult and, in any case, is far beyond the scope of this chapter. It seems clear, however, that society will need to regulate not only the physical aspects of resource use, but also the rate of accumulation and movement of the capital that is used in converting natural resources into consumable products. Fair distribution of resource access and environmental impacts will require asking awkward questions about why resources are being used: survival, decent living or mindless luxury. In other words, the process of re-crafting institutions must be embedded in a larger movement for structural, epistemological and cultural changes. This movement will have to address local inequities as much as global ones, demand personal changes as much as institutional ones and seek spiritual value as much as economic efficiency.

NOTES

1. This paper draws upon two earlier efforts to deconstruct and reconstruct the concept of joint management in the context of forests (Lélé 1998b, Lélé 1999). I am grateful to Ajit Menon for comments. The Ford Foundation provided financial support for this research.

2. There is less consensus about this in the practitioner community. At least in the Indian forestry sector, it appears that the majority of forest officers continue to believe in the need for strong hands-on state control. Similarly, a sizable proportion of activists continue to believe in the need for complete community control (e.g. Rahul 1997). The concept of "joint forest management" emerged after 1990 and has been officially adopted by 23 states, but the gap between official rhetoric and reality is enormous (Saxena et al. 1997, Sundar 2001). "Joint management" is thus a pragmatic compromise, or a

way of state and donor co-optation of the idea of decentralised resource management, not a conceptually sound innovation that can stand the test of time.

3. This is not surprising. The common-pool reource research community was virtually born out of the need to demonstrate that Hardin's tragedy of the commons does not necessarily arise, that local communities can and do devise institutions for averting this tragedy. With the focus being on demonstrating the existence and successful operation of autonomous community-level resource management institutions, and given the antagonism of most governments towards such institutions, for a long time the only external "design condition" that was identified was the minimal requirement that 'the [local] users' right to devise their own institutions is not challenged [by the state]' (Ostrom 1990: 90).

4. In a later paper, Ostrom (1995) offered one or two general principles for what kind of support higher level institutions should provide – for example, public goods such as high-quality information on resource condition and cheap conflict resolution mechanisms – and should not provide – such as pumping large quantities of external funds into local institutions.

5. Srinidhi and Lélé (2001) developed this framework further and used it to characterise forest tenure regimes in the Western Ghats of India.

6. The two terms appear to have emerged almost simultaneously in two different parts of the globe and in different sectors: "co-management" in North America in the context of fisheries (Pinkerton 1989, Berkes et al. 1991) and "joint management" in South Asia in the context of forests (see Khare 1992, Poffenberger 1996). They are not exactly identical in practice: co-management 'involves the recognition and legitimization of traditional local-level management systems' (Pomeroy and Berkes 1997), whereas joint forest management or participatory irrigation management are based upon the creation of new institutions for joint control, often suppressing pre-existing forms of more autonomous, self-organised local-level systems (Sarin 2001, Patnaik 2002).

7. Townsend and Pooley (1995) were probably the first to systematically compare self-organised institutions, communal governance, cooperatives and their variants, including corporate (share-based) governance. But they too take the 'state' as a given and ignore the role of other institutions.

8. In practitioner circles, at least in the case of forests, debates on the role of *Panchayati Raj* (local self-governance) institutions, NGOs, academics and donor agencies abound (see, e.g. Anonymous 1999). But they all take the current (highly flawed) concept of joint management as a given (see Lélé 2001a for a succinct critique of this concept).

9. This aspect tends to receive inadequate attention. For instance, Pritchard Jr. and colleagues (1998) talk of the "fit" between ecosystems and institutions, but they do not provide any external measure of identifying a good fit. Similarly, Agrawal (2001), in reviewing the literature on common property institutions and sustainable governance, defined institutional success simply in terms of the survival of the institution, rather than in terms of its ability to meet any socially desired objectives (cf. Menon and Lélé 2003).

10. Even in democratic societies, while citizens may have the ability to shape the manner and extent to which the state intervenes in their lives, the situation is an involuntary one in the sense that no one can opt out of being part of some nation-state and its governance structures. Of course, there may be other involuntary institutions that are culturally imposed that are also difficult to "flout" or "opt out of". So the exact dividing line between voluntary and involuntary institutions is somewhat hazy.

11. Note also that "voluntary/productive" corresponds to the "enabling" role and "involuntary/regulatory" to the "constraining" role of institutions mentioned above. Voluntary associations can play a productive role by exploiting win-win situations. Regulatory institutions attempt to strike a balance between winners and losers and hence are unlikely to be voluntary. Voluntary associations generally cannot play a regulatory role, as they lack the coercive power and mandate. Occasionally they may evolve or appropriate some of this power (as when professional bodies have monopoly control over the license to practice that profession).

12. I use the word "legitimate" to distance the discussion from voluntary or involuntary associations formed for, say, theft, such as mafias or dictatorships, respectively.

13. See Lélé (1993) for a detailed exposition of these umbrella concepts.

14. It could even be said to subsume the deep ecology position, since depriving other organisms the right to live is essentially an unfair or unjust proposition.

15. The concept of "environmental soundness" encompasses both ecological sustainability and equity (see Lélé 1994, Lélé 1998a).

16. Common-pool resources, by being non-excludable and also subtractable, are particularly liable to under-production and hence need policy intervention (McKean 1998).

17. "Heterogeneity" has recently emerged as a concern, but only insofar as it may affect the possibility of collective action. Here too, recent developments in common-pool resource research that demonstrate the possibility of initiating and sustaining collective action in "asymmetric" situations have 'put a damper on equity' (Menon 1999).

18. This includes the literature on joint or collaborative protected area management (e.g. Pimbert and Pretty 1995), where it is taken for granted that certain areas should be preserved for posterity, and the focus is on how best to achieve this.

19. What Lee (1993) calls a "sectoral" externality.

20. I emphasise asymmetry to distinguish the situation from that of common-pool resources, where symmetric externalities are inherent. For example, the extraction of groundwater by one person creates an externality that affects other users of that aquifer. However, this is a symmetric externality, because others can impose similar externalities on the first person. It has been suggested to me that 'symmetric common-pool resources' and 'asymmetric common-pool resources' might be a better taxonomy than limiting the notion of common-pool resources to the case of symmetric externalities alone. However, as Farrell and Keating pointed out, common-pool is a useful concept only in symmetrical situations (Farrell and Keating 2000).

21. Some researchers have discussed the asymmetries created within a community resource management system by unidirectional flows, as in the case of canal irrigation systems (Ostrom and Gardner 1993). But the focus there has been on identifying conditions under which win-win situations can still emerge and lead to Pareto-improving collaboration between downstream and upstream communities. There is no discussion of the inherent *unfairness* of outcomes and institutional arrangements to avoid them.

22. Note, however, that this situation is not a static one. New technologies such as remote sensing and geographical information systems may enable monitoring-at-a-distance, with its consequences contingent upon the social context in which they are developed and applied (Lélé 2001b).

23. Which is different from indiscriminate privatisation that could include individuals or corporations distant from the resource.

24. That is, fully contain more than one local-level institution.

25. What Townsend and Poole (1995) called 'internal' and 'external' governance structures.

26. Agarwal (1998) made a similar suggestion to strengthen community forestry institutions in the face of state efforts to increase its own control.

27. That is, where returns are proportional to contributions or returns are distributed equally across members regardless of contribution.

28. Local governments are different from user groups because membership is involuntary, the institution is backed by the coercive power of the state, and the institution performs multiple developmental as well as allocative functions.

29. Joint stock companies whose shares can be purchased only by members of a certain community and/or only up to a certain number.

30. In India at least, although the forest departments are charged with implementing the national forest policy, which gives highest priority to maintaining watershed and other environmental benefits of forests, the forest departments know very little about the watershed effects of different land use practices, carry out no monitoring of watershed effects, and have no clear guidelines for ensuring these alleged benefits of forests.

REFERENCES

Agarwal, A. (1985) 'Politics of environment-II', in A. Agarwal and S. Narain (Eds), *The State of India's Environment 1984–85: The Second Citizen's Report*, pp.362-80. New Delhi: Centre for Science and Environment.

Agarwal, A. (1998) 'State formation in community spaces'. Paper presented at Crossing Boundaries: The Seventh Common Property Conference, International Association for the Study of Common Property (IASCP), University of British Columbia, Vancouver, Canada (10–14 June).

Agarwal, A. (2001) 'Common property institutions and sustainable governance of resources', *World Development* 29 (10): 1649–1672.

Agarwal, A. (2001) 'Participatory exclusions, community forestry, and gender: An analysis for South Asia and a conceptual framework', *World Development* 29 (10): 1623–1648.

Anonymous (Ed.) (1999) Proceedings: National Workshop on Joint Forest Management. Ahmedabad: Aga Khan Foundation (India), Gujarat Forest Department, Society for Promotion of Wasteland Development and VIKSAT.

Berkes, F., P. George and R. J. Preston (1991) 'Co-management: The evolution of theory and practice of the joint administration of living resources', *Alternatives* 18 (2): 12–18.

Ciriacy-Wantrup, S. V. (1963) *Resource Conservation: Economics and Policies*. Second Edition. Berkeley: University of California, Division of Agricultural Sciences.

Costanza, R. (Ed.) (1991) *Ecological Economics: The Science and Management of Sustainability*. New York: Columbia University Press.

Dasgupta, P., A. Sen and S. Marglin (1972) *Guidelines for Project Evaluation*. Vienna and New York: United Nations Industrial Development Organization.

Farrell, A. and T. J. Keating (2000) 'The globalization of smoke: Co-evolution in science and governance of a commons problem'. Paper presented at Constituting the Commons: Crafting Sustainable Commons in the Millennium, International Association for the Study of Common Property (IASCP), Bloomington, Indiana (31 May 31 to 4 June).

Khare, A. (Ed.) (1992) *Joint Forest Management: Concept and Opportunities*. New Delhi: Society for Promotion of Wastelands Development.

Lee, K. N. (1993) 'Greed, Scale Mismatch, and Learning', *Ecological Applications* 3 (4): 560–564.

Lélé, S. (1991) 'Sustainable development: A critical review', *World Development* 19 (6): 607–621.

Lélé, S. (1993) 'Sustainability: A plural, multi-dimensional approach'. Working Paper. Oakland: Pacific Institute for Studies in Development, Environment, & Security.

Lélé, S. (1994) 'Sustainability, environmentalism, and science', *Pacific Institute Newsletter* 3 (1): 1-2, 5.

Lélé, S. (1996) 'Environmental governance', Seminar 438: 17-23.

Lélé, S. (1998a) 'Resilience, sustainability, and environmentalism', *Environment and Development Economics* 3 (2): 251-255.

Lélé, S. (1998b) 'Why, who, and how of jointness in joint forest management: Theoretical considerations and empirical insights from the Western Ghats of Karnataka'. Paper presented at the International Workshop on Shared Resource Management in South Asia, Institute of Rural Management Anand (IRMA), Anand, India (17-19 February).

Lélé, S. (1999) 'Institutional issues in (J)FM(& R)', in anonymous (Ed.) *Proceedings of the National Workshop on Joint Forest Management*, pp.19-29. Ahmedabad: Aga Khan Foundation (India), Gujarat Forest Department, Society for Promotion of Wasteland Development and VIKSAT.

Lélé, S. (2000) 'Godsend, sleight of hand, or just muddling through: Joint water and forest management in India'. *ODI Natural Resource Perspectives* No. 53. London: Overseas Development Institute.

Lélé, S. (2001a) 'People's participation in forest management in Karnataka: Need for fundamental policy changes'. Paper presented at the State-Level Convention on Participatory Forest Management, Jana Aranya Vedike, Bangalore (10–11 December).

Lélé, S. (2001b) '"Pixelising the commons" and "commonising the pixel": Boon or bane?', *Common Property Resource Digest* (58): 1–3.

McKean, M. (1998) 'Common property: What is it, what is it good for, and what makes it work?', in C. Gibson, M. McKean and E. Ostrom (Eds) *Forest Resources and Institutions*, pp.23–48. Rome: Forest Trees and People Programme, Food and Agriculture Organization.

Mehta, L., M. Leach, P. Newell, I. Scoones, K. Sivaramakrishnan and S.-A. Way (1999) 'Exploring understandings of institutions and uncertainty: New directions in natural resource management'.

IDS Discussion Paper No. 32. Brighton, U.K.: Institute of Development Studies, University of Sussex.

Menon, A. (1999) 'Common property studies and the limits to equity: Some conceptual concerns and possibilities', *Review of Development and Change* 4 (1): 51–70.

Menon, A. and S. Lélé (2003) 'Critiquing the commons: Missing the woods for the trees?', *Common Property Resource Digest* (64):1–4.

Murphree, M. W. (2000) 'Boundaries and borders: The question of scale in the theory and practice of common property management'. Paper presented at Constituting the Commons: Crafting Sustainable Commons in the Millennium, International Association for the Study of Common Property (IASCP), Bloomington, Indiana (31 May to 4 June).

Ostrom, E. (1990) *Governing the Commons: The Evolution of Institutions for Collective Action.* New York: Cambridge University Press.

Ostrom, E. (1995) 'Designing complexity to govern complexity', in S. Hanna and M. Munasinghe (Eds) *Property Rights and the Environment: Social and Ecological Issues*, pp.33–45. Stockholm: Beijer International Institute of Ecological Economics; Washington, D.C.: World Bank.

Ostrom, E. (1998) 'Self-governance and forest resources'. Paper presented at the Conference on Local Institutions for Forest Management: How Can Research Make a Difference, Center for International Forestry Research (CIFOR), Bogor, Indonesia (19–21 November).

Ostrom, E. and R. Gardner (1993) 'Coping with asymmetries in the commons: Self-governing irrigation systems can work', *Journal of Economic Literature* 7 (4): 93–112.

Patnaik, S. (2002) 'Forest management: The "other" view', *Community Forestry* 1 (1 & 2): 8–12.

Pearce, D. (1988) 'Economics, equity and sustainable development', *Futures* 20 (6): 598–605.

Pimbert, M. P. and J. N. Pretty (1995) 'Parks, people and professionals: Putting "participation" into protected area management'. *Discussion Paper* No. 57. Geneva: United Nations Research Institute for Social Development.

Pinkerton, E. (1989) *Co-operative Management of Local Fisheries.* Vancouver: University of British Columbia Press.

Poffenberger, M. (1996) 'The struggle for forest control in the Jungle Mahals of West Bengal 1750–1990', in M. Poffenberger and B. McGean (Eds) *Village Voices, Forest Choices: Joint Forest Management in India*, pp.132–162. Delhi: Oxford University Press.

Pomeroy, R. S. and F. Berkes (1997) 'Two to tango: The role of government in fisheries co-management', *Marine Policy* 21 (5): 465–480.

Pritchard Jr., L., J. Colding, F. Berkes, U. Svedin and C. Folke (1998) 'The problem of fit between ecosystems and institutions'. *IHDP Working Paper* No. 2. Bonn: International Human Dimensions Programme on Global Environmental Change.

Rahul (1997) 'Masquerading as the masses', *Economic and Political Weekly* (February 15): 341–342.

Sarin, M. (2001) 'From right holders to "beneficiaries"? Community forest management, Van Panchayats and village forest joint management in Uttarakhand'. Center for International Forestry Research (CIFOR), Bogor, Indonesia (In Mimeo).

Saxena, N. C., M. Sarin, R. V. Singh and T. Shah (1997) 'Independent study of implementation experience in Kanara Circle'. Review Committee Report. Bangalore: Karnataka Forest Department.

Srinidhi, A. S. and S. Lélé (2001) 'Forest tenure regimes in the Karnataka Western Ghats: A compendium'. Working Paper No. 90. Bangalore: Institute for Social and Economic Change.

Sundar, N. (2001) *Branching Out: Joint Forest Management in India.* New Delhi: Oxford University Press.

Townsend, R. E. and S. G. Pooley (1995) 'Distributed governance in fisheries', in S. Hanna and M. Munasinghe (Eds) *Property Rights and the Environment: Social and Ecological Issues*, pp.47–58. Stockholm: Beijer International Institute of Ecological Economics; Washington, D.C: World Bank.

Uphoff, N. (1993) 'Grassroots organizations and NGOs in rural development: Opportunities with diminishing states and expanding markets', *World Development* 21 (4): 607–622.

IN CONCLUSION

CHAPTER 16

KNOWLEDGE SHARING IN SUPPORT OF HUMAN DEVELOPMENT[1]

J. B. (HANS) OPSCHOOR

1. BEYOND DEVELOPMENT?

Development and change have been the main foci of the Institute of Social Studies (ISS) since its establishment in 1952. During its tenth lustrum year, ISS explored what issues are relevant to development and what approaches can best be adopted to address them. In-depth debates were held at the institute about the notion of development, development strategies, migration and work, new avenues towards gender equity, and human rights. The ISS also held intensive discussions on issues related to poverty and poverty alleviation in the context of globalisation and conflict. The entire programme of activities[2] was designed to fit under the heading *Beyond Development? Meeting New Challenges*. Has enough effort been made towards development over these fifty years?

This conclusion chapter echoes some of the concerns that were reflected in the papers presented in this volume and more generally at the fiftieth anniversary conference of the ISS. It adds to these some observations on the future of development-oriented international tertiary education.

1.1 Development and Globalisation

Since Ovid we have known (or at least, we are told in his *Metamorphosis*), 'All things change;...there is nothing...which is permanent. Everything flows onward.' But that is not to say that this onward flow of society is, or should be, unidirectional. In fact, that is exactly what many people criticise in the notion of development as it has been depicted in the development literature since 1950: the association with modernity, economic progress along the path already taken by the industrialised countries.

Ironically, much of this critique comes at a point in time when the forces of globalisation along neoliberal lines are imposing uniformity, at least at the macro level

Max Spoor (ed.), Globalisation, Poverty and Conflict, 307–322.
© *2004 Kluwer Academic Publishers. Printed in the Netherlands.*

and at the level of institutions. Counterpoints to this in the real-life fugue of change are the phenomena of cultural diversity and alternatives in policy strategies that the ISS has explored and critically supported all along. In terms of an evolutionary analysis, it is the *mutations* that are interesting and which may hold the promise of societal sustainability and equitable human development. In biological reality these are nearly always spontaneous and unpredictable; in societal evolution this is not necessarily so. Yet, there has been critique of modernist and constructivist notions of development. I think we are labouring under the expectation of what one evolutionary economist (Joseph Schumpeter) once called *new combinations* of approaches that we have seen emerge in the past.

Some would argue that the road to economic progress is now clearly neoliberalism with good governance and that globalisation will deliver them. But is that road so unequivocal? And will globalisation deliver? Has it?

Let me try to be more specific. In its development plan of 25 years ago, the ISS explicitly stated that its focus was to be on 'the poorest 40 per cent'. Has the situation of this 40 per cent improved? Perhaps (though it is difficult to tell), in terms of absolute average income, but not for all poor, and the total numbers of poor have increased. Moreover, while incomes have risen, livelihoods have actually become more problematic for many. The picture is even less bright when we look at poverty in relative terms, an issue to which I shall return below.

While the effort to fight poverty and inequality continues, the world is waking up (although not as rapidly as many would like) to unresolved challenges in a number of other fields: the recognition, allocation and implementation of human rights; gender imbalances; access to livelihoods and resources on which livelihoods are based; labour standards and work opportunities; and participation in the processes of social transformation and governance, to mention only a few. So how can we talk about "beyond development"?

A key feature of globalisation is that it entails expanding activity, both in terms of the intensity of activity and its spatial diffusion, in a finite world. We observe a *globalisation of interdependence*. The dynamics of societal change are such that we will increasingly have to live with feedbacks, in terms of changing environmental conditions such as global warming and in terms of social and economic boomerangs. Most societal dynamics are the result of autonomous economic forces (*à la* Jürgen Habermas). In such a world, "development" – in the sense of deliberate, intended change – will increasingly turn out to mean anticipating such feedbacks and preventing or mitigating the adverse ones, adapting to the resultant changes, and optimising the beneficial ones.

In a recent attempt to summarise trends that seem to be apparent now and related to processes of development and change in the decades to come, UNESCO produced the following short list:

- the revolution in technology (ICT, bioengineering);
- increasing relative poverty and exclusion;
- new threats to peace, security and human rights;
- demographic changes;
- endangered world environment;
- inequality in access to information in an emerging information society;

- governance and participation in an era of globalisation;
- persistent gender inequality;
- new cross-cultural encounters;
- ethical challenges in science and technology (Binde 2001).

Thus, there is change and even development, but development that brings with it risks and adverse social impacts as well as opportunities. There must therefore be an ongoing effort to ensure that these changes will, on balance, lead to more opportunity, more participation and a more equitable distribution of results and access to resources; in other words, to a more sustainable and humanitarian development. No matter what name we choose to give it – "development", "social change", or "transformation", there will be work for social scientists.

1.2 Globalisation, Marginalisation and Conflict

Globalisation, technological progress and poverty are inter-related. Poor countries have difficulties in leapfrogging to technological change, as this demands a large pool of skilled labour, which most developing countries do not have. There are also financial constraints. With insufficient exports, foreign direct investment (FDI) and aid, and with low local investment, it is extremely difficult to close the technological gap.

Although globalisation is said to be accelerating the integration of the world, it also marginalises, leaving out many countries. During the conference we saw that some 11 developing countries account for 66 per cent of total exports from developing countries, as well as receiving the lion's share of FDI inflows. In fact, China, Mexico and Brazil account for half of all developing-country FDI inflows. Some conclude that 'globalisation has left out the poor countries, and the poor only feel its negative effects'.

Actually, the 1990s have shown a process of stagnation or deterioration (UNDP 2003), although some important developing countries, such as China, have fared quite well. The Millennium Development Goals for 2015 were formulated in order to promote concrete and measurable steps in development. Some thinkers (e.g. Pronk) are cautious about the feasibility of these goals. Others (e.g. Ritzen) put forward quite an optimistic and, compared to other contributions, a somewhat dissenting view on the important role of the private sector in delivering health and educational services and access to such services by the poor.

There is also an intensive debate raging on the issue of equality. The issue is important, not only in its own right, but also because it plays a role in the debate about development strategies, notably about the quality of neoliberal strategies in the context of globalisation. If, as a result of such strategies as have been recommended from Washington (World Bank, IMF) and Geneva (WTO), we were to see inequality decrease and absolute levels of income rise, then these strategies would have firm arguments in their favour. The realities in terms of average incomes were touched on above: there appears to be progress, in terms of *income* poverty (as the UNDP calls it). But what about inequality?

Between countries, inequality is still growing, although the picture looks better when expressed in purchasing power-weighted terms instead of based on exchange rate-based comparisons of income. *Within* many countries, inequality may be worsening, but some larger countries (in terms of population size) have seen improvements (e.g. Indonesia). What matters, then, is what inequality measures have to say in terms of distribution of wealth between people, rather than countries. With more than three-quarters of people in the world classified as poor, inequalities are greater than ever before. In terms of exchange rate-based measurements of incomes, population-weighted approaches show rising inequality, but the picture is less clear if based on attempts to capture real purchasing power. Refined analyses, however, do provide more transparency – and unfortunately show rising inequality rather than convergence (Milanovic 2002, Salah-I-Martin 2002).

Divergence and unequal distribution of wealth can lead to conflict. Redistribution mechanisms can therefore be employed – as they were in Europe – to avoid conflict. Given the relationships between conflict, inequality and poverty, the key to avoiding conflict is good institutions to manage it. Even a small reduction in the intensity of conflict can yield beneficial economic results (e.g. Murshed).

Growing inequality and poverty are not just issues related to so-called "Third World" countries. Poverty re-emerged in the 1990s in the countries in transition of the former Eastern bloc (from 14 to 147 million poor throughout the region) and inequality there has rapidly increased (the Gini coefficient rose from 0.25 to between 0.40 and 0.50). Although economists tend to see the transformation in terms of a transition from a planned to a market economy, there has been radical change at many levels, socio-political as well as economic. These transformations are marked by macroeconomic contraction, with catastrophic inflation rates and a marked withdrawal of the state. The growth observed in these countries benefits only a small elite. This has resulted in 'growth without equity' (e.g. Spoor). Sound policies must be pursued, aimed at growth with equity, or civil conflict could well be the result.

Finally, trade liberalisation does not lead to "convergence", and the international financial system is quite unequal, as analysed critically (e.g. Griffith-Jones and Raffer) as the Achilles' heel of globalisation in the past decade. Though changes have been made, they are as yet insufficient, given the scale of the crisis the world is facing. In addition, we must be critical of the asymmetrical relations in international markets and the dynamic effects these have when markets are globalising. From a development perspective a new or renewed financial architecture is needed to provide stable and sufficient flows of funds to support development and to prevent crises.

1.3 Governance, Civil Society and Poverty

An important question that emerges is whether the poor themselves can take action to remedy their dilemma. What is the role of civil society organisations (e.g. Reuben)? The concept of self-help, also discussed at the conference, is not without validity. But it must not be allowed to become an excuse for governments to wash their hands of the problem and leave the poor to their own devices. Identifying which groups among the poor should be targeted for poverty alleviation is important.

As Mkandawire pointed out in this volume, the challenge is to establish a social order that is developmental, democratic and inclusive. The concept of inclusive (versus exclusive) development, was also emphasised at the conference by Dutch Minister for Development Cooperation, Agnes van Ardenne. Unfortunately, the new democracies are doing no better than dictatorships in this regard. Where empirical evidence suggests a positive link between levels of income and democracy, this may be due to rent-seeking behaviour by selected groups. Models to stabilise economies are not development-oriented in the wider sense, nor do they engage in poverty alleviation. Democracy turns into "truncated democracy", which does not go beyond economic reforms. Pro-growth states have taken equity off the agenda. However, equity and growth must both be part of economic performance. If equity and growth are to be combined, redistribution mechanisms are needed. Therefore, changes in national and international governance are a precondition for development and poverty alleviation.

Globalisation is engendering greater interdependence. Yet conversely, it is also fragmenting society – between the rich and poor and between the empowered and powerless or marginalised and excluded. Livelihoods, even entire lifestyles, are affected by it, sometimes radically transformed. These changes, as has been said, are not always or necessarily undesirable, but often they are, or they are ambivalent. There is even ambivalence at a higher level. While in certain circumstances it may seem bad to be affected by globalisation, not to be included in it at all might be regarded as worse.

There are great disparities and asymmetries in the degree to which groups of people or countries participate in globalisation. There are both economic and social aspects to this, and the ISS has developed various activities in the fields of labour and employment and regional development, such as reflected in the contribution of Helmsing. There are also institutional and governance-related aspects which now demand attention. Of course, the disparities and asymmetries also give rise to increased international migration and mobility.

For the institutional setting and politico-economic system to become more conducive to development and the safeguarding of human dignity and rights, democracy must widen and deepen (UNDP 2002). Moreover, given the causal and functional links between these issues and concerns over equity, win-win situations may be expected. When the institutional framework improves, income poverty eradication may become easier and the vulnerable may become better able to access a productive resource base. Democracy and improved governance may help prevent people from exposure to economic and resource catastrophes or their consequences, as we know thanks to Amartya Sen, Nobel Prize winner and Honorary Fellow of the ISS. They may also, under certain conditions, expand social and economic opportunities.

In fact, there is a crisis of governance in many countries and regions. Conference participants suggested several ways of addressing this. Some argued that policymaking should be brought down to lower levels, allowing citizens to participate. However, in many cases this must be combined with synchronised activities at the central level, alongside those at the local and community levels. Ostrom even calls for multi-level organisation (especially in the field of natural resources management) as a way of avoiding institutional monopoly.[3] Multi-level governance is necessary because different scales of management problems require different scales of solution. 'Small is beautiful for some things, but not for everything', she said. The new "flex organisations", such as analysed by Wedel, in which the boundaries between the state and the private sector are blurring, show a decline of accountability.

The governance crisis is a problem of rules. In many countries (e.g. in Africa), and internationally, institutions need to be rebuilt, and we must get the rules of governance right if politically sustainable development is to be achieved (e.g. Mihyo). Effective government depends on a dynamic civil society and provides the environment in which it can function. The key is to fulfil the expectations of those represented and for government to be accountable to those it represents. As demonstrated by the case of Zimbabwe, achievement of economic democracy is a necessary condition; without that, political democracy and good governance will not work.

1.4 Environment, Institutions and Conflict

If incomes were rising and the income gap between rich and poor were closing, would the development problem be solved? Other issues are being put on the table. One them is the *sustainability* of development strategies and the patterns of consumption and production they imply.

Sustainability was added to the international agenda in 1987 by the Brundtland Report to the General Assembly of the United Nations. It was then picked up during the United Nations Conference on Environment and Development in Rio de Janeiro in 1992. The recently held World Summit on Sustainable Development in Johannesburg (2002) was intended to review progress made since 1992 and to develop a plan of implementation of the intentions and principles agreed at Rio. I am afraid, as a concerned academic and active participant at both of these mega-conferences, that the issue of sustainability – or rather *un*sustainability – has not been resolved. Steps were taken towards such resolution, not least thanks to the truly Herculean efforts of Jan Pronk (former Dutch Minister for Development Cooperation and most recently of Environment), but far fewer than hoped for and certainly not enough. There is, thus, a need for alternative strategies and policies, instruments, know-how and knowledge, and options for action at all levels, even when we only consider the aspect of sustainability (e.g. Lélé). As was appropriately reconfirmed in Johannesburg, even that one aspect has to be considered in at least three dimensions (referred to as the "pillars" of sustainable development): economic, social and ecological (see also Pronk).

The connectiveness between the social sphere and the biosphere (and the diversity underlying these) is highly dynamic and complicates analysis. The easy way out is to conduct static analyses. But we have to at least keep dynamics in mind in order to make a difference. At our conference much was said about the link between natural resources and institutions (see Ostrom and Janssen). I quoted some observations on that link when I discussed governance. There are, of course, limitations to what can be achieved through institutions. There are power relations inherent in institutions which can prevent them from functioning as they should. As one speaker (Salih) concluded, 'We must demystify institutions and not romanticise them.'

Social scientists concerned with "development" have to continue to work – analytically, educationally and in terms of policy advice – on the links between socio-economic configurations and regulatory regimes on the one hand, and transparency and the furthering of governance, human rights and development on the other. The ISS has to continue its activities aimed at empowering and invigorating accountable civil society forces in their capabilities to monitor both government and private sector performance regarding human rights and development and in developing alternative forms of political participation. These are not just juicy bones for academic top dogs to gnaw. They are issues with high relevance in the debate about strategies for development and change. They are issues that will have to stay on our agenda, and so they will.

2. KNOWLEDGE AND EDUCATION IN DEVELOPMENT

2.1 Education: A Public Good?

'Tertiary education… is a critical pillar to human development world-wide', wrote Mamphela Ramphele, Director of Human Development at the World Bank and Honorary Doctor of the ISS since its 45th anniversary. The report from which this quotation was taken (World Bank 2002) states,

> higher education, through its role in empowering domestic constituencies, building institutions and nurturing favourable regulatory frameworks and governance structures, is vital to a country's efforts to increase social capital and to promote social cohesion, which is proving to be an important determinant of economic growth and development.

Indeed, higher education, as Ramphele also suggested, confers important public goods that are essential to development and poverty reduction. According to the World Bank, two trends are especially pertinent to higher education:

- developing and transition countries are being increasingly marginalised in the world economy due to inadequate domestic knowledge-generation capacities;
- education is becoming borderless (with new providers and new modes of delivery) and is coming under the influence of market forces.

The World Bank describes education as a global public good but effectively recommends its provision in a market setting. It recommends the rapid removal of entry barriers to national markets for higher education. Though one may sympathise with the intentions behind this recommendation, I cannot say that I find the World Bank analysis in support of the removal of entry barriers sufficiently sensitive to some of the adverse consequences of such a strategy. In fact it may in many cases lead to significant social costs in terms of negative effects on emerging national providers of higher education, as, for instance, emphasised by South Africa's Minister of Education Kadar Asmal. Indeed, there *is* a case to be made for protection of infant industries as well as of emerging knowledge providers.

Returning to the World Bank's basic premise, we may ask whether higher education is a public good.[4] It is certainly not a public good in the economic sense of the term, if we consider criteria such as excludability of potential consumers. Indeed, it is increasingly clear that whatever public benefits higher education may bring, the private benefits are staggeringly high (Van der Wende 2002: 45). Yet, education may be a good with a high degree of associated positive externalities. But why should states have to care – and even provide – for the providers? Potential benefits of liberalisation include cheaper and therefore potentially more general access to higher education, especially in countries where governments lack the means to provide quality education themselves.

In the context of the negotiations on the liberalisation of trade in services (GATS), proposals have been made to enhance access to education providers, especially of higher, adult education and vocational training. Countries that consider themselves (or their national knowledge providers) to be competitive internationally (such as the United States, Australia and New Zealand) claim that this would entail commercial opportunities. The arguments here, however, are not entirely persuasive in terms of private profitability.

In the European arena, we see the space in which educational institutions operate opening up. Continental universities are now introducing the bachelor's and master's system – a system that the ISS has always worked with. We see new systems of accreditation coming. We are bound to see more competition between providers of higher education, and the current changes in education-oriented development policies will most certainly encourage that. However, the social costs of privatisation and liberalisation include losses of diversity. For instance, attention to "small" cultures and education in "small" languages will likely disappear. Existing (often good institutes) may lose their subsidies and incomes. Moreover, control over the quality and content of higher education will become more difficult. There is further a risk of "intellectual imperialism". Price increases might prevent potential students with low incomes from entering the market. The quality of teachers may erode if institutions yield to pressure to minimise costs; there may also be a loss of criticality within the educational system. Such social costs must be weighed against any real or envisaged benefits.

Regarding whether education is a *global* public good, most certainly the need for it is a global one. Therefore, there is a need to design systems that will deliver globally (and to all who warrant access). The least one can conclude is that there will be a manifest role for states and governments in the provision of higher education, nationally and internationally. Below, I explore this question in more detail.

2.2 The Future of International, Development-Oriented Higher Education

Global society has entered, or is entering, a knowledge-based development phase. The world economy is changing as knowledge becomes a key source of wealth. As knowledge becomes more important, so too does higher education.

One may distinguish four basic features of education. Looked at somewhat idealistically, apart from being a vehicle for allowing people to learn to *know* and to *do* things, education is – or should be – a vehicle for learning to *be* and to *live together*

(UNESCO 1996). In the more prosaic terms of the World Bank/UNESCO Task Force (2000) higher education should aspire to several aims:
- furthering economic development by inducing income growth, training enlightened leaders, expanding choice and increasing relevant skills;
- allowing countries to generate new knowledge and to engage in scholarly and scientific commerce with other nations;
- promoting an open, meritocratic and civil society based on inclusive public values that embody norms of social interaction, such as open debate, an emphasis on quality, autonomy and self-reliance and the rejection of discrimination.

Over the coming decades the level of activity in higher education will continue to rise globally. This is certain to be the case in the developing countries. The driving forces behind this include population dynamics and the emergence of knowledge-based development and globalisation. The latter two forces will cause a number of difficult qualitative changes in higher education. Moreover, there is a growing awareness that the non-economic functions that higher education can perform, as outlined above, are increasingly and urgently needed. Higher education is a growth sector but at the same time a sector in transition. Qualitatively, there are deep concerns over the impacts of large numbers of entrants and dwindling financial resources on the supply side. Quantitatively, the systems that provide higher education in developing countries will be unable to meet rising and changing demands. Moreover, the education on offer has not always kept up-to-date with shifting frontiers; knowledge systems have not always been upgraded satisfactorily and sustainably; and facilities have run down. These are deep problems juxtaposed against a need to ensure quality, relevance and access.

Disconcertingly, the gap between rich and poor countries, in terms of access to knowledge and potentials to generate it, is widening. This is illustrated by well-established indicators:
- The enrolment rate in higher education in industrial countries is five to six times greater than in all developing countries (Task Force 2000).
- Per capita, industrial countries have ten times more research and development scientists than developing countries and are spending four times the percentage spent by developing countries on innovation (World Bank 1998).

These gaps show up in stark contrasts in access to high-quality facilities, equipment and supplies, availability of well-trained teachers, links with the international scientific community and access to the global stock of knowledge.

Without more and better higher education, developing countries will find it increasingly difficult to benefit from the global knowledge-based economy: countries that are only weakly connected to the global knowledge system will find themselves increasingly at a disadvantage (Task Force 2000). Unless major programmes are put in place to address these asymmetries, the gap will continue to widen. There have been urgent calls (and there may be an emerging will) to reduce the growing gap in knowledge infrastructure and in access to knowledge between the rich and the poor countries, both by the higher education sector and by donors.

2.3 Globalisation and Higher Education

One aspect of globalisation in its present form (i.e. dominated by neoliberal strategies and paradigms) is that it forces countries that are in need of international financial support to accept conditionalities with respect to the macroeconomic and micro-economic conditions under which they operate. This often leads to reductions in government expenditure with associated pressure on the budgets of the spending ministries, including that of education. Yet at the same time, the emergence of global markets, trade liberalisation, privatisation and rapid technological change affect the demands manifested on labour markets. Production and management systems must be flexible and adaptable, and the same goes for human skills.

Higher education systems should aim to meet changes in demands for the human resources that they deliver. This suggests that lifelong learning and more varied educational packages in terms of intensity of delivery (e.g. evening classes) and place of delivery (e.g. distance learning) will have to be further developed and put on offer.

Several responses are already visible on the ground. For instance, in southern Africa the number of campuses is expanding (per country and often even per university). There is also steady growth in efforts to expand distance-learning facilities and their effective outreach. Universities are extending their teaching programmes beyond first degrees, to increasingly include master's programmes (and beyond). Graduate schools are emerging, and there are attempts at coordination to avoid duplication of costly efforts. There are exogenous initiatives to create centres and networks of excellence. Networks now link many southern faculties and departments with institutions in the North – links that are considered essential lifelines to the places where knowledge expands most rapidly. There is also the entry of outside providers, whatever their motives.

It is clear, however, that these efforts to improve higher education systems quantitatively and qualitatively in many developing countries and regions are still insufficient to meet the aggregated and changing needs of growing populations. Thus, given the needs for training and capacity, international efforts in higher education continue to be needed to boost the total supply.

The current international trend is towards intensification of lifelong learning or "permanent education". This is both a consequence of globalisation and rapid technological change and a response to concerns about sustained employability in a context of rapidly changing labour markets. Indeed, much of the demand for higher education on international markets comes from employed professionals from developing countries seeking to upgrade their skills and knowledge. But the need for refresher courses and new orientations through education is set to grow rapidly. The training needs of these mid-career professionals will increasingly influence future course contents and modes of delivery.

The effects of globalisation point to the need for international higher education systems to radically change in many ways. There will be profound changes in modes of delivery. Distance learning through correspondence and, in future, via the Internet will to a large degree replace classical forms of delivery. Far more modularised programmes (with perhaps shorter, cumulative blocks) will emerge, with students studying at more than one location in the context of a particular degree, using credit accumulation and credit transfer systems.

2.4 Development-Oriented Tertiary Education in the Context of Globalisation

Development-oriented international higher education is essentially involved in activities at two levels:

- human resource development for students from developing and emerging countries and countries in transition (the training of individuals, often professionals already employed by the kinds of institutions mentioned above);
- development (including maintenance and upgrading) of institutional capacities for engagement in cutting-edge human resource development and knowledge acquisition.

The need for both will remain, although demand for the latter will grow in relative terms. In student-oriented training, new modalities involving more partnership-based approaches with multi-locational delivery and more distance learning will emerge. There is also a growing demand for *specialised* master's degrees and, generally, for PhD training. The expected decline in funds available for these purposes must therefore be viewed with concern: it may simply come at too early a time. In international higher education, there is an observable trend towards multi-locational and multimedia delivery, with courses presented partly in the North, partly in the South and partly through distance learning. But, in terms of points of delivery, prioritisation and budget allocation, the centre of gravity is shifting away from the OECD region. This appropriately paves the way for a movement from northern supply-driven international higher education programmes to demands/needs-driven cooperation. Nevertheless, not all providers of tertiary education in the OECD region seem ready for that move.

Various forms of institutional capacity development have been identified. There is an apparent wish to establish centres of excellence at the postgraduate level at national higher education institutions, which would often operate regionally. In addition there is need for cooperation between countries that cannot set up complete higher education systems at the national level in setting up or encouraging regional centres. This wish is translating itself into a trend. However, newly established postgraduate programmes often cannot be maintained, especially in the poorer regions. Financial conditions are often to be blamed for this, but there is also a lack of effective management and, sadly, a lack of interest among potential employers in the "products" of such regional institutions. There is a need for more regional cooperation, most certainly in the less well-endowed areas, in order to achieve and sustain prestige more easily and cost-effectively than could be done at a national level. But it will take a long time to set that up in sufficient quantity and at sufficient levels.

Well-developed systems for research and knowledge generation continue to grow in importance. These allow developing countries to contribute to the global stock of knowledge and engage in scholarly and scientific commerce with other nations – an indispensable form of internationalisation in our globalising knowledge society. Any academic institution in the North with special expertise may be attractive as a partner, but especially those that have shown a sincere interest in capacity development in the South based on needs expressed there. Specialised institutes for international higher education can help bridge the knowledge gap and enrich dialogues between peoples. In such partnerships, arrangements can and should benefit both sides and are likely to generate spin-offs that can be extrapolated to institutions elsewhere.

There is a consistent and strong plea from higher education institutions in non-industrialised countries for long-lasting contacts with institutions in the OECD region. This long-term perspective not only ensures adequate time to develop activities, it also enables backstopping, quality control and maintenance in the implementation phases and beyond. North-South cooperation between institutes should, however, be based on genuine partnership. It should help to effectively strengthen institutional capacities in the South, promote resource sharing and the exchange of knowledge and reduce brain drain. International higher education has to be ready for longer term involvement in twinning (or other network) arrangements in order to strengthen and build up institutions in the South for knowledge gathering, processing and disseminating. North-South partnerships in higher education can be justified if the institutes involved are centres of excellence in particular fields of study and if the northern partners serve as facilitators or conduits for the developing-country institutes, enhancing their access to and interaction with global knowledge centres.

2.5 International Education in the Future

Addressing the future of international higher education, United Nations University Rector Hans van Ginkel (2000) observed that there are important niches for top-level, highly specialised relatively small institutions with good brand names, as long as excellent quality is guaranteed. The question is whether the international education system as it now operates has the right specialisations to remain in demand and whether it possesses the necessary comparative – and competitive – advantages.

Given the scarcities of financial and human resources and the prevailing priorities within higher education in the North and the South, niches will indeed remain for centres in the North: top quality and high-relevance training and research in relatively specialised and capital-intensive fields. Specialised institutes could work on the basis of substantive expertise or cutting-edge methodological advantages (e.g. in remote sensing techniques or modelling). Institutes in the North have found niches that are not yet accessible to higher education institutions in the South. From a development perspective, they should be ready to transfer the cores of these niches through institutional capacity development and, in the meantime, provide opportunities for human resource development at the level of individual students. Fellowship programmes and capacity development programmes thus seem relevant and should be maintained, if not expanded in real financial terms.

Small-scale institutions dedicated in terms of both development orientation and substantive specialisation may have the flexibility required to meet the various challenges ahead more easily than institutions for which *international* tertiary education is only part of their activities and their concerns and whose "core business" is delivery in the wealthier segments of the education markets. Institutions specialised in international education could become the vanguard through which a broader set of national and international actors in tertiary education (such as universities) become more effectively involved in development-oriented higher education. In particular, such specialised institutes could contribute to the "relevance" of any offer on the market for international partnerships in higher education, where both suppliers in combination can ensure its "quality".

Knowledge gaps show up in stark contrasts in a variety of characteristics of knowledge systems in the North and South. It has been suggested (Task Force 2000, UNESCO 1998) that these contrasts pose a sufficient argument for international cooperation between the North and South in higher education. Indeed, international higher education should view reducing these gaps as a major challenge. If true partnerships are achieved, academic cooperation with a development orientation should by definition not crowd out southern institutions but rather involve them more intensively, even if capacity levels remain asymmetrical.

In this light, the case seems clear for specialised institutes for *development-oriented studies* (broadly defined, *transformation* is the increasingly used term). Currently operational development-oriented institutes for international education have experience – and even a positive reputation – in understanding divergence in perceptions and the need to contextualise analyses of social and human phenomena. They have developed capabilities in handling these aspects. They have also learned to operate in a multidisciplinary or even a transdisciplinary fashion, thus addressing development issues far more effectively. Although thinking about the nature and content of development and change in the various regions of the world is an activity best left largely to the academic institutions based there, specialised OECD-based institutions may be well-suited to provide the comparative setting and the knowledge base for scholars from southern institutions to test and compare new knowledge and know-how.

Such specialised institutes for international education have an advantage over other higher education providers in that their faculty is typically much more international and experienced in development processes. Their staff easily link considerations of academic quality with those of relevance or pertinence – something appreciated much less in northern universities than in their southern counterparts. These specialised institutes may also be more ready to make the shift towards delivery of education in the South with southern counterparts. They may also have a better understanding of, and sensitivity to, the regional specificity required in developing joint activities in teaching and research.

3. CONCLUSION: RECIPROCITY AND PARTNERSHIP

This chapter began by looking at the pertinence of the notion of development, including issues emerging in relation to societal transformation in general. It also set a relevant agenda for social scientists and social science-based higher education. The second section analysed the impacts of globalisation and other driving forces on the demand for international education broadly speaking, and development-oriented international education in particular. That led to several ideas on the future role of such education for OECD-based internationally operating institutions such as the ISS.

Some forty years ago, the ISS was already engaged in research on "reciprocity", a notion that it regarded as an inherent quality of "society" and which may well be a crucial one to draw upon if we are to survive the challenges of globalisation and global interdependence. Reciprocity is a quality of and a value in human relations. According to philosophers as far apart as Jürgen Habermas and Martin Buber, reciprocal relations are the only durable, satisfying human relations. Egbert de Vries, the first full-time rector of the ISS, regarded reciprocity as defining the position of humans in society (De Vries 1966: 9, 19): man (or rather, humans, but I quote here) 'is not in the first place man, but fellow-man', with reciprocity as a dynamic force in the formation and development of society. In societies, humans have to coordinate their actions (cf., Habermas) in conditions where goals may be conflicting. Authority or dependency (power), award (exchange) and responsibility (care) are mechanisms or institutions to cope with that need. Reciprocity is especially connected with the latter of these. According to the dictionaries, reciprocity is behaviour between two people or groups in which each gives or concedes to the other or treats the other on the same basis as the other treats them. In De Vries' use of the term it has the overtones of symmetry or equivalence between the partners in the relationship and of a "joint venture". De Vries suggested that in a dynamic society the most vital relationships are born out of choice-based association, in the quest to realise shared goals. In order for an association to be lasting, there has to be more than the motivation of power or profit – there has to be reciprocity. He regards that as an alternative – if not an opposite – to commercialisation-driven human relations, in which the social content of relationships gets lost.

In modern societies, commercialising as they are, and in societies otherwise in transition, one main problem is to find new bases for reciprocal relations. In the early 1960s, the ISS research team surveyed and studied a number of organisations and activities to conclude, 'On an organised basis, systematically applied and at relatively large scale, we found little reciprocity. But in all ongoing, lasting and mutually beneficial and satisfying relationships, we found a reciprocal element' (De Vries 1966: 17).

De Vries found the road to reciprocity fraught with great practical problems caused by a number of factors, such as factual inequalities between partners and cultural distance distorting perceptions of positions and goals. That is true. Yet, reciprocity is the spirit in which the ISS wishes to develop itself and its activities in the years ahead. In the years behind us, we developed a range of activities undertaken from and offered at our home base in The Hague. These have included joint ventures in capacity development in developing and transitional countries. We have learned that *partnerships* may develop out of such joint ventures and we have tried to encourage that. We regard these partnerships as much more symmetrical than is implied by the notion of "capacity development". We hope to be able to continue along that road, solidifying and perhaps extending these partnerships to help bridge the gap between the available capacity and the need for higher education, human capital development and the elaboration of knowledge systems. We hope to be able to do so in a demand-based or needs-oriented way and in the spirit of reciprocity.

Many of ISS's current partners were represented at its International Conference on Globalisation, Poverty and Conflict of which this volume is the outcome. In fact, throughout our conference we took time out from the regular programme to discuss these notions. We see our joint ventures as a *South-South-North-East* network of partnerships and we want to explore and exploit their potential in development-oriented higher education. In terms of the trends in international higher education presented in this chapter, our network might provide not only an effective infrastructure for moving staff and students back and forth, but also an opportunity for the ISS to show its readiness to respond to the need to shift the focus of delivery and decision-making on higher education to outside The Hague and outside the OECD region. In fact, one could envisage sandwich-type master's programmes and short courses rotating over different campuses. This network would offer great potential for exploring and applying new modes of delivery of knowledge and know-how, such as distance learning. The network may generate substance for courses of much greater relevance in terms of contextual specificity and cross-cultural comparison than any of the partners could imagine producing on their own.

In conclusion I would like to quote British writer Robert Lacey. Reflecting on another fiftieth anniversary, he observed, 'If you go back to the religious origins of jubilees, their deliberate purpose was to renew the faith.' Transposing that to our academic scene, we should perhaps replace "faith" by "commitment".

At the close of the celebration of ISS's tenth anniversary in 1962, then-rector De Vries informed his audience that, at that point in time, the institute looked 'at the world with expectation and hope'. That was forty years ago. Now we are forty years older and, I hope, somewhat wiser. We are certainly, I am afraid, somewhat sadder. It is sobering to observe that, on the occasion of this anniversary, we felt it necessary to intensively discuss the question of why poverty endures. In what mood will we proceed? If not with expectation and hope, it should certainly be with *commitment*, from the needs-based grounds presented above, to continue to work towards the development and sharing of knowledge and know-how in the field of social change and sustainable human development.

NOTES

1. From the opening statement prepared for the fiftieth *Dies Natalis* of the Institute of Social Studies, The Hague, 9 October 2002, following the completion of the conference *Globalisation, Poverty and Conflict*. The occasion was sadly overshadowed by the death of His Royal Highness Prince Claus of the Netherlands on 6 October. Prince Claus had been an ISS Honorary Fellow since 1988. The *Dies Natalis* celebration started with an In Memoriam which commemorated the Prince's contribution to thinking on development and development policy.

2. During 2003 ISS organised a series of five academic debates on issues relevant in this context: globalisation, the Washington Consensus, migration, new gender perspectives, and rights-based approaches to development. It also hosted the International Conference on Globalisation, Poverty and Conflict of which this volume is the outcome. Moreover, it made a first attempt to explore the interface between culture and development.

3. Elinor Ostrom and Jan Pronk received an honorary doctorate during the celebration of the fiftieth anniversary of the ISS.

4. The following paragraphs draw heavily on Van der Wende (2002).

REFERENCES

Bindé, J. (Ed.) (2001) *Keys to the 21st Century*. Paris, UNESCO; New York Berghahn Books.
Ginkel, J. van (2000) 'Opening address academic year 2000 at ITC', ITC, Enschede, 28 September.
IBRD (1994) *Higher Education: The Lessons of Experience*, Washington, D.C.: World Bank.
Milanovic, B. (2002, draft) 'The Ricardian vice: Why Salah-I-Martin's calculations of world income inequality cannot be right' and paper at the ISS Conference on Globalisation and Poverty, 7–9 October.
Salah-I-Martin (2002) 'The disturbing 'rise' in global income inequality' and 'The world distribution of income'; both available on the Internet. <www.columbia.edu/~xs23/home.html>
Task Force on Higher Education and Society (2000) *Higher Education in Developing Countries: Peril and Promise*. Washington, D.C.: World Bank/UNESCO.
UNDP (2002) *Deepening Democracy in a Fragmented World*, Oxford: Oxford University Press.
UNESCO (International Commission on Education in the 21st Century) (1996) *Learning: The Treasure Within*, Paris: UNESCO Publications.
UNESCO (1998) *World Conference on Higher Education*, Paris: UNESCO Publications.
Vries, Prof. E. de (1961) *Man in Rapid Social Change*. London: SCM Press Ltd.
Vries, Prof. E. de (1966) 'Valedictory address', ISS, The Hague, 25 March.
Wende, M.C. van der (2002) 'Hoger onderwijs globaliter: Naar nieuwe kaders voor onderzoek en beleid'. Inaugural lecture UT, Enschede, 16 May.
World Bank (1998) *Knowledge for Development: World Development Report 1998/99*. Oxford: Oxford University Press.
World Bank (2002) *Constructing Knowledge Societies: New Challenges for Tertiary Education*, Washington, D.C.: World Bank.

CONTRIBUTORS

Agnes van Ardenne, Minister for Development Cooperation, The Hague, The Netherlands

Stephany Griffith-Jones, Professor, Institute of Development Studies, University of Sussex, United Kingdom

A. H. J. (Bert) Helmsing, Professor, Institute of Social Studies, The Hague, The Netherlands

Marco A. Janssen, Associate Research Scientist, Center for the Study of Institutions, Population and Environmental Change, Indiana University, Bloomington, United States of America

Sharachchandra Lélé, Center for Interdisciplinary Studies in Environment and Development (CISED), Bangalore, India

Paschal B. Mihyo, Professor, Director of Research & Programmes, Association of African Universities, Accra, Ghana

Thandika Mkandawire, Director General, United Nations Research Institute for Social Development, Geneva, Switzerland

S. Mansoob Murshed, Associate Professor, Institute of Social Studies, and Professor, Faculty of Economics, University of Utrecht, The Netherlands

J. B. (Hans) Opschoor, Professor and Rector, Institute of Social Studies, The Hague, The Netherlands

Elinor Ostrom, Professor, Workshop in Political Theory and Policy Analysis, Center for the Study of Institutions, Population and Environmental Change, Indiana University, Bloomington, United States of America

Jan Pronk, Professor, Institute of Social Studies, The Hague, The Netherlands

Kunibert Raffer, Associate Professor, Department of Economics, University of Vienna, Austria

William Reuben, Coordinator, NGOs and Civil Society, Social Development Department, World Bank, Washington, D.C., United States of America

Jozef M. Ritzen, Professor and President, University of Maastricht, The Netherlands

M. A. Mohamed Salih, Professor, Institute of Social Studies, The Hague, The Netherlands

Max Spoor, Associate Professor and Coordinator of the Centre for the Study of Transition and Development (CESTRAD), Institute of Social Studies, The Hague, The Netherlands, and Extraordinary Professor, Centre of International Relations and International Cooperation (CIDOB), Barcelona, Spain

Janine R. Wedel, Associate Professor, School of Public Policy, George Mason University, Pittsburgh, Pennsylvania, United States of America

INDEX